Extreme Tissue Engineering

Extreme Tissue Engineering

Concepts and Strategies for Tissue Fabrication

Robert A. Brown
University College London, UK

WILEY-BLACKWELL

A John Wiley & Sons, Ltd., Publication

This edition first published 2013, © 2013 by John Wiley & Sons, Ltd

Registered office: John Wiley & Sons Ltd, The Atrium, Southern Gate, Chichester, West Sussex, PO19 8SQ, UK

Editorial offices: 9600 Garsington Road, Oxford, OX4 2DQ, UK
The Atrium, Southern Gate, Chichester, West Sussex, PO19 8SQ, UK
111 River Street, Hoboken, NJ 07030-5774, USA

For details of our global editorial offices, for customer services and for information about how to apply for permission to reuse the copyright material in this book please see our website at www.wiley.com/wiley-blackwell

Library of Congress Cataloguing-in-Publication Data

Brown, Robert, 1952-
Extreme tissue engineering / Robert A. Brown.
    p. ; cm.
  Includes bibliographical references and index.
  ISBN 978-0-470-97447-6 (cloth) – ISBN 978-0-470-97446-9 (pbk.)
  I. Title.
  [DNLM: 1. Tissue Engineering. 2. Cell Culture Techniques. 3. Regenerative
  Medicine–methods. 4. Tissue Scaffolds. QT 37]
  612′.028–dc23

                                                        2012009774

A catalogue record for this book is available from the British Library.

Wiley also publishes its books in a variety of electronic formats. Some content that appears in print may not be available in electronic books.

Set in 10/12.5pt Minion by Laserwords Private Limited, Chennai, India.
Printed and bound in Singapore by Markono Print Media Pte Ltd.

First Impression 2013

Contents

Companion website

This book is accompanied by a companion website:

www.wiley.com/go/brown/tissue/engineering

The website includes:

Figures and Tables from the book for downloading

Preface: Extreme Tissue Engineering – a User's Guide

The most important first task of any User's Guide is to dispel possible misconceptions of what the 'user' might reasonably expect. In this case, there can be a small ambiguity in the title for those who are unfamiliar with 'tissue engineering'. So, if you are skimming through this book, expecting to learn how you could engineer extreme (or *weird*) tissues, it is recommended that you pop it back on the shelf and move to the science fiction movie section. On the other hand, if you want to learn how emerging concepts and strategies might be blended to revolutionize how we engineer (very familiar) tissues, then read on.

This is an unusual textbook in a field rich in books which come from the many sectors of activity which make up 'tissue engineering'. It is different, not least, because it tries to integrate these diverse viewpoints, rather than giving just one perspective. This tends to set it apart because of things that it is *not*. It does *not* come from the direction of explaining *specific* technologies or particularly useful cell types. Neither does it aim to describe *specific* control mechanisms or *target* tissue applications. The logic here is that there are far too many applications, body sites and permutations to treat them all in just a few hundred pages (and there are already many thousands of written pages out there anyway). It is also *not* a multi-author book, with single, specialist, research-level chapters.

It *does* seem to be unique in the field, as a single-author, basic textbook of advanced tissue engineering concepts, across disciplines. It *does* aim to enable newcomers (or the puzzled-but-interested) to understand much better what that mass of facts and data out there might really mean. First, though, it aims to provide a unified introduction, whether your original training is in cell biology, engineering, biomaterials, surgery or other contributing specialities.

It is illuminating to ask why, in two decades, there have been so few (if any) basic-concept texts, especially as the underlying idea *seems* to be so simple. Understanding this (discussed at length in Chapter 1) begins to explain what makes tissue engineering so *special* and, indeed, *illusive*. The answer is grounded in its position at the touch-and-merge point of so many well established disciplines.

It is arguable that tissue engineering barely exists unless its activity includes some form of integration or merger of ideas from two or more of its more established component subjects. These 'components' include cell biology, materials science, surgery or sensor biophysics, protein chemistry and bioreactor technology, to name a few. If this is true, then we cannot avoid the logic that the work we do in this subject overlap zone called tissue engineering, *must* be *integrated* or *merged*. We cannot claim that our new hybrid subject is generating real synergies if we work away at cell biology in January, materials science in February and surgical science in March, etc.

Worse still, it has proved all too easy to toil deeply within one of the single-component disciplines to solve questions which, in the end, have little logical value when applied outside that discipline. Imagine, for example, how far sustainable cities would be advanced if the Norwegian civil engineers developed new insulation for homes in Oslo with the aim of linking this with findings from the Turkish seismic geologists on tectonic plate movements around Istanbul. **If we claim to generate synergies between disciplines, deep integration is an essential**.

By analogy, those of you who have experienced the London public transport system in a hot summer rush hour will understand more clearly the critical importance of the 2–3 mm that separate 'close' from 'touching'. Arguably, the social, legal and emotional effects of being tightly squeezed into physical contact on the 6:00 pm Victoria Line tube in Central London are similar to those where academic disciplines merge at their edges. **Both events have more to do with anxiety and imagination than subject-matter or cold logic**.

This is why a complete chapter is dedicated to exploring what tissue engineering actually *is*, and where it originates. After all, if you *must* grapple with the Victoria Line tube, it is good to know about your fellow travellers.

The core trouble is that the tissue engineering 'concept' is, in effect, based on combinations of knowledge packets which are drawn from the simpler parts of its component disciplines. As a result, it is terminally tricky in tissue engineering to explain the *same*, very *basic* topics to individuals who have had a wide range of specialist training, for the simple reason that, at any one time, one or more sets of your readers will almost certainly become seriously bored. For example, where the text explains the basic concepts of one contributing discipline (e.g. cell culture, aimed at engineers, or stress-strain measurement to biologists) it becomes laughably simple – and terminally boring – for the expert group. However, leaving out any of these basic parts immediately compromises our aim of a single, *integrated* set of concepts (and we have already glimpsed the importance of integration).

Consequently, it is almost *guaranteed* that some readers will be bored (while others are learning) – and this is a seriously *undesirable* publication plan. Suddenly, it is easy to understand how we reach our present position of having a plethora of focused, specialist texts. Yet there is clearly a broad need for just such a 'doomed' textbook, explaining and integrating the basic concepts. Is it too high a hurdle, then, to explain the concepts and strategies of tissue engineering in an integrated, joined-up manner? Certainly, it would be helpful to students trying to understand the strategies and logics behind more specialist applications, such as engineering cartilage, nerves or blood vessels, **whatever sector they were trained in**.

But there is a way around this difficulty. The style of *Extreme Tissue Engineering* is designed to entertain and excite *all* reader groups, whether they are being (re-)introduced to their own discipline or to a new one. After all, if the claims of tissue engineering enthusiasts are even half true, there should be plenty of excitement to draw on at the various discipline interfaces. One way of generating engagement, and incidentally of helping with the learning process, is to use amusing, unusual analogies and extreme (even ridiculous) examples.

Extreme Tissue Engineering adopts this approach. It aims to *integrate* concepts from each of the component fields, often pulling together pairs of traditionally distinct subjects. At the same time, it actively approaches topics from new angles, drawing its logic threads from colourful starting points and illustrating this with recurring analogies. Wherever possible, these analogies bring to life abstract concepts by drawing on the everyday human world and its artefacts, or on familiar animals and plants. All of this allows us to understand tissue engineering from new perspectives (hopefully tracking where it is going) and why it must become *extreme* to get there. Indeed, the very process of producing a coherent explanation for tissue engineering logic inevitably highlights its paradoxes and identifies questionable assumptions.

On some occasions, these illustrative analogies help us to see the inherent flaws in current strategies. In others, they point us towards possible solutions. In all cases, their aim is to stimulate your own ideas on the problem and to cement the issues in your memory. First and foremost, it should be fun, refreshing and easy to remember.

But this all leads to a rather distinctive, even unfamiliar style. It really should generate controversy in areas where concepts and approaches are deep-rooted. For this reason, it is important that discussions around the logic and content are separated from reaction to its style, which is just a necessary tool for engagement. Where it makes our field easier to understand and explain, especially to newcomers, it may make a significant long-term contribution. If and when it successfully challenges or redirects worn and suspect strategies, it will have performed

---

Text Box 1 Author's personal note

I recently spoke at the Cheltenham National Science Festival to an enthusiastic lay audience. They were keen to hear about how tissue engineering and regenerative medicine *could* help in the health of ordinary people. Once our small panel had finished its story of enthusiasm and promise, we took questions. As many of us have found before, these questions were poignant and hit at the nub of chronic health problems, which we still tend to skirt around or back away from. Gradually, it became clear that some members of the audience were themselves threatened or were caring for treasured spouses, parents or birth-damaged children.

These are the real issues we must have in mind as we set loose our personal optimisms on a desperate world. We *must* be sure we can deliver before we speak – and not just under the 'scientific-eventually' caveat. Equally, it clearly becomes our duty, once we speak of these aims, to deliver to the very best *intellectual* level we can. In the face of false hopes offered to the mother of a damaged child, it is not enough only to point to long hours in the lab and a healthy grant income. Only extremely clear, joined-up thinking will do.

---

a more immediately useful role. Clearly, both objectives must be good for a subject as new and uncertain as tissue engineering. This is especially true where our list of success stories is so modest – and so impatiently awaited (see Text Box 1).

In short, this book is designed to leave you with an in-depth understanding of the overriding questions and problems of tissue engineering. As a bonus, you may also discover a selection of the possible solutions and routes to reach our tissue goals. It should transform how you see the rest of tissue engineering. In particular, it should make it easier for newcomers to understand and interpret the rich collection of specialist textbooks already produced by the many tribes of tissue engineering.

There is a liberal use of text boxes and footnotes throughout. These are included as 'asides' and caveats, designed to colour and enrich the logic without deflecting the reader from its main track (for example, see Text Box 1 above). Where these are successful, they will make it easier to follow the thread and to remember its key points.

In places, there are simple questions aiming to draw your thinking to new places after you have put the book down. These are designed not for repetition of message, nor to save your professors work; rather, they give you a chance to carry on with or extend the concept on the train going home, or when you are out enjoying the park. There are many sectional and chapter summaries which should allow you to recap on the main points at regular intervals and understand better where they are leading us.

Through the length and breadth of the book, you will find examples and analogies. Some are designed to inform, some to bring an idea or concept to life – even where it is not your favourite topic. Yet others are just embarrassing, even silly, as such images are perhaps the most effective way to lodge ideas and facts in the mind. An example of this mnemonic effect can be found in a highly successful UK/European advert series for car insurance. Clearly, insurance is one of the more challenging products when advertisers are required to generate 'customer excitement' or 'brand identity'.

For example, those of you who have experienced one recent campaign will now be deeply imprinted with a completely abnormal image-association based on the word '**meerkat**'.[1] You will almost certainly recall an image of fluffy Russian-accented puppets angry at a car insurance company for stealing their website name. Clearly, before this rather silly series of ideas were imprinted, the word would have brought to mind a more realistic image of jerky, mongoose-like mammals, sitting tall in the African veldt (if it is not shown in your area, you can get the gist from the campaign's website: www.comparethemeerkat.com).

Tissue engineering is too young to be brittle, and too much in need of successful translations to be

---

[1]You have just experienced an example of the very illustration under discussion. Hopefully, this circle-within-a-circle helps you to appreciate how potent these illustrations can be in leaving recallable ideas in our minds.

strategically fixed. The term 'extreme tissue engineering', then, has been coined here to reflect the target of generating a distinctive and challenging new approach. Its focus and reason for extremeness is the inescapable need for cross-discipline integration. This is an integration based on balance and equal voice, rather than a spurious democracy linked to the perceived 'size' of a contributing discipline. The number of workers in a discipline, or grants awarded to it, correlates depressingly badly with its success in solving the big problems of society. 'Integrations' which resemble the merger of shrimp and basking shark are not helpful to our cause. The trouble is, it is perfectly possible that the 'shrimp' (i.e. the minor discipline) might have *the* key answer to that log-jam problem which is holding us all back. Hence, integration on the basis of equality of voice is indispensible in our essentially tribal subject.

So, where you get the feeling that something you are reading sounds oddball, lateral or off-beam, remember that this is what we are hunting for. Where the analogies are puzzling, please persist, as they are designed to draw you along a logic pathway. Where you become downright embarrassed, enjoy the feeling, as these are the concepts and arguments that you will remember. Where you see some repetition, register it as necessary; these tend to be the points where the tribes and topics of *Extreme Tissue Engineering* truly touch and merge, so they are important for integration.

Here is one last thought to crystallize these points. Some years ago, a prestigious group (Lysaght & Hazelhurst, 2004) prepared a 'state of the nation' review of tissue engineering. This came on the heels of a series of major setbacks in the translation of tissue-engineered products to the clinic and the market. Interestingly, this review paraphrased Winston Churchill's famous speech to conclude that tissue engineering had reached '*the end of the beginning*' (without a question mark).

There is a danger, though, in taking such an optimistic (arguably complacent) position when there are so many other *alternative critiques* of progress in our subject which are possible. We are now several more years further on, and neither the review nor the additional years have substantially changed our tissue engineering paradigm. Given that this paradigm is now more than 20 years old, we might hope to have seen several evolutionary stages or even a couple of minor concept revolutions. Consequently, it is a core assumption of this book that a number of potentially revolutionary concepts *must* be discussed. To return to the Churchillian review; if we cannot *now* identify new strategies, it might be that we are not so much at '*the end of the beginning*' of tissue engineering as '*the beginning of its end*'!

Reference

Lysaght, M. J. & Hazelhurst, A. L. (2004). Tissue Engineering: The End of the Beginning. *Tissue Engineering* **10**, 309–320.

Further reading

Pretor-Pinney, G. (2006). *Cloudspotter's Guide*. Hodder & Stoughton, London. [Beautiful example of sneaking in dry-learning under cover of off-beam entertainment, analogies and stories.]

Bottom-up and top-down in 3D rock shapes.
Two forms of Portland Quarry Rock Art. Left is a partly exposed fossil ammonite (150 million years old). Right is a sculptured 'hat and hands' (*Tout Quarry Sculptors, Dorset*). The ammonite was always there, buried in rock which is being chipped away (arrows). It is a discovered beauty. The hat never existed until the artist imagined it and chipped it *into* the rock. The sculptor fabricated the hat *bottom-up*, from basics. Chipping the ammonite out is a *top-down* process, revealing what was there to be found (see 'Veselius').

1 Which Tissue Engineering Tribe Are You From?

1.1 Why do we need to engineer tissues at all?

As we are frequently reminded, tissue engineering and regenerative medicine (collectively called TERM) are new disciplines. Tissue engineering is widely considered to have its origins at the point of collaboration between (bio)materials scientists, cell biologists, surgeons and physical scientists/engineers (or any combination of these) towards generating therapeutic/tissue technologies.

This, we hope, is moving towards the distant dream of true therapeutic tissue *regeneration*. Regeneration is the key word here, and we shall be getting under the skin of its real implications later in this book, along with its near neighbours: repair, replacement, scarring and amphibian-limbs.

The target of initiatives in the two fields of tissue engineering and regenerative medicine was originally to produce successor treatments for both prosthetic (synthetic) implants and living tissue grafts or cadaveric transplants. Implantable prosthetic devices have had, and continue to have,

Extreme Tissue Engineering: Concepts and Strategies for Tissue Fabrication, First Edition. Robert A. Brown.
© 2013 John Wiley & Sons, Ltd. Published 2013 by John Wiley & Sons, Ltd.

an immensely successful history in many clinical and reconstructive surgical disciplines. Despite their many advantages, however, they still suffer the key limitation of never being more than a temporary substitute. They never work better than the day they are implanted; they are always foreign, artificial devices which the body tolerates – for a while – until they wear out or clog up.

Living tissue grafts and transplants – from heart and liver to skin, cornea and tendon – have all the advantages of natural systems which are missing in prosthetics, but these advantages also come with serious costs. Autografts, taken from one part of a patient's body and used to reconstruct another part, are not rejected and cannot infect the patient. They are used across the spectrum of plastic and reconstructive surgery, from rebuilding seriously injured or burned patients through to cosmetic body reshaping, but these approaches are also flawed. Relying on a single – usually injured – individual as the sole source of tissue is always a problem, as the available tissue pool tends to be unsuitable, insufficient or of poor quality. Worse still, the idea of adding *intentional* 'donor-site injuries' onto already severely injured patients (e.g. children, old people, burns victims) is clearly less than attractive.

Transplants or tissue allografts which get around this by being taken from donor individuals can be therapeutically excellent, as in the cases of kidney, heart or liver transplantation. However, donors are typically relatives, unknown or deceased persons, the tissues *will* be rejected without drugs[2] and they carry the risk of life-threatening infections. Needless to say, all of these are also in chronically short supply donor tissues.

The key shared feature of all these existing techniques is that, no matter how hard we work to improve them, they will always retain these same basic drawbacks. In fact, we now are finding 'worst scenario' examples, in which the *more successful* the procedure is, the worse their problems become. For example, as kidney and heart transplants became

successful and immune-suppression becomes better managed, the waiting lists for donors became inexorably longer. As we live longer and age better, suitable donors become ever more scarce – and this only gets *worse*.

Another example of a success-driven time-bomb can be found in the story of the prosthetic hip replacement. This is such a successful and long-lived operation that more and more patients across an increasing age range have been demanding it. As a result, the cumulative number of people (a) with steel and polymer hips and (b) living longer active lives has been spiralling up for many years. This would be fine, except for the base problem that no matter how well these prosthetics are made, they will *always eventually wear out* and fail. Consequently, there is now a parallel spiral in the number of patients needing much more complex, but much less successful, 'revision surgery' to remove and replace the worn implants. This represents a *major* healthcare-generated cost and problem which governments would prefer not to feed any longer than necessary.

1.1.1 Will the real *tissue engineering and regenerative medicine please stand up?*

How should we define tissue engineering and regenerative medicine?

It is customary, at a starting point such as this, to put forward a definition which captures the goals of the discipline or which lays out a theme that will recur through the book. Many short definitions have been proposed to sum up the targets and technological approaches involved in tissue engineering or regenerative medicine, and some examples of these are given at the end of this chapter as a guide to current concepts (see Annexes 1 & 2). However, this is not a simple or routine task. The next sections in this chapter will discuss why it is non-trivial, not least because an understanding of the paradoxes also provides essential insight into the nature of tissue engineering. So keep faith – definitions *will* emerge.

This section starts with an analysis of why it is perhaps unrealistic to expect a single, crisp

---

[2]Rejection is almost certain without the lifelong use of immunosuppressive drugs, which themselves can carry severe health side effects.

'definition', in its traditional sense, which we can really trust. The key factor here is that, while tissue engineering and regenerative medicine are two *new* subjects, they encompass several other *well-established* disciplines, all of which, by definition, are moving in their own independent directions. The big question, then, becomes: who can we trust to provide a sufficiently balanced perspective? In other words, however careful one author or another may be, (s)he will also be from one of the component *tribes* of tissue engineering and will tend to see the new discipline of tissue engineering as a derivative of their own speciality. Yet the idea that tissue engineering is just a branch of biomaterials, surgery, bioengineering or cell biology is probably the least acceptable of all options.

The theme of this book is to peek under the concealing conventions and to glimpse around the bend into the less well visited parts of the tissue engineering territory. We may as well start right at the beginning, then, by asking, "Why do we have *so much* trouble with definitions in *TERM*?"

1.1.2 *Other people's definitions*

There have been many formal attempts to define tissue engineering. Perhaps the fairest approach would be to go with the originator of the term itself. Fairness, though, is not necessarily a close acquaintance of 'useful'. The difficulty is that the most widely accepted (defining) feature of tissue engineering is that it is cross- or inter-disciplinary. This means that each discipline will have its own viewpoint on the subject. In particular, each will tend to consider, quite reasonably, their own discipline to be *the* critical and core essence of tissue engineering. This will include ideas on where the subject is going to and where it came from. Definitions with different starting points, perspectives and viewpoints tend to have patchy histories.

Defining concepts from different standpoints/disciplines can be highly problematic, but the practice is far from unique. Aside from the scientific world, most of the current 300 million US citizens utilize volume measures based on 'the gallon'. Most of their northern and Southern neighbours are

obliged to convert these measures to litres, one US gallon being defined as 3.785 litres.

However, the gallon was originally a British measure. It seems though, that some spillage may have occurred on those early transatlantic voyages to America, as UK gallons are 4.546 litres – 1.2 times bigger! Now, that is a serious perspective-dependent shift for a definition. But, despite some disappointments among British visitors to US beer-houses, it did work reasonably well over 2.5 centuries of transatlantic trade (see Box 1.1 for a more accurate historical analysis). The system seems to have been made to work by the simple expedient of nomenclature-sub-division, which resulted in the persistence of the 'US gallon' and the increasingly rare 'Imperial gallon' into 21st century life. This 'name sub-division' may be what is happening with tissue engineering and regenerative medicine, though hopefully it will not take 250 years to bring clarity!

While this may sound like gentle avoidance of the hard question, it is not. The key point here is to understand *why* definitions in this field only ever get us into the foothills of the mountain range. Foothills, of course, are fine, so long as we do not mistake a gentle information hump for journey's end and a peak in the Sierras (more of mountain analogies later). Another way of getting a realistic initial taste is to be obviously reductionist about our definitions. Thus, one foothill-walking approach is to stick literally to the words we have in the subject titles.

What, then, is literally the meaning of *'tissue engineering'*? Perhaps, in reductionist terms, we should be happy with the idea that this describes activities aimed at the engineering of living tissues – but there is a small ambiguity here within the term 'engineering'. As a verb, it could be used to signify either fabrication/construction of new structures from basic elements, *or* modification/alteration of pre-existing structures. In more conventional terms, this might be seen as the difference between designing, testing and fabricating a completely new model of, say, Land Rover, as opposed to engineering an existing petrol-fuelled Land Rover model in order to allow it to run on liquid gas (LPG). The special challenge of our definition, however, comes when

---

Text Box 1.1 Gallons and gallons

Just to illustrate more fully the confusion that can follow when definitions are not really definitions but viewpoints, let us look a little deeper into the many guises of 'the gallon'. Bear in mind that this is supposed to be a unit of measure whose main claim to utility is its constancy and predictability between people (merchant, sailor, scientist, clinician). It is essential to know precisely what is being offered or demanded in such a measure, and there will clearly be tears if the definition shifts, depending on what substance the volume refers to and where or when it is used.

    The real story of 'gallons' is, in fact, more surprising and informative than we implied earlier. At the time of the American War of Independence, both sides recognized no less than *three* forms of gallon, used for different substances. These were the *corn gallon*, for measurement of dry materials (i.e. the dry gallon, of around 4.41 litres), the *wine gallon* (also quaintly, though unhelpfully, known as Queen Anne's gallon – approx. 3.8 litres) and the *ale gallon*, which was around 4.62 litres (perhaps reflecting its greater water content).

    Not prone to tinker with a perfectly functional system, the thrifty Americans basically stuck with both the dry and the liquid gallons. Consequently, the present-day US (along with a number of Central and South American Republics) measure petrol and cola in US gallons, which approximate to the old British 'wine gallon' (happily, few of us need to barter in US corn, so do not have to wrestle with dry to liquid gallon conversions). Meanwhile, in 1824, the less conservative British Parliament succumbed to a wave of decisiveness and drained off all except the ale gallon (i.e. *not* the version used in the USA – relationships were a little prickly at the time). This was renamed the 'Imperial gallon' and, true to that name, it was used liberally over the British Empire, including Canada and a number of Caribbean islands. Not surprisingly this has caused the Canadians some difficulties and, in the 20th century, after briefly flirting with their own 'gallon' redefinition, they sensibly opted to switch to litres.

    After more than 250 years, though, these jelly definitions may finally be resolving. With the UK now pumping and drinking metric volumes, only a few Caribbean states retain the dilemma of the flexi-gallon and the rest of us talk *either* litres or gallons.

---

we try to team this activity with that most biological of terms, '*tissues*', and all that it implies about hierarchical, biological structure, sub-structure and molecular interplay*.

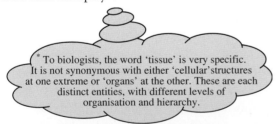

    * To biologists, the word 'tissue' is very specific. It is not synonymous with either 'cellular' structures at one extreme or 'organs' at the other. These are each distinct entities, with different levels of organisation and hierarchy.

1.1.3 Defining our tissue engineering: fixing where we are on the scale-hierarchy

It clearly will not be possible to achieve fabricated structures at one level of scale without first selecting where we want to be on the scale-hierarchy spectrum. This runs from cells up to organs. In other words, what *scale* do we need to focus on to engineer our tissue of choice?

In biology (arguably more than anywhere else), hierarchical levels of structures and systems are the source of much of the famous 'bio-complexity' and are notoriously difficult to view in isolation. For example:

1. Molecular and atomic level forces are critical to the specificity of binding between larger bio-molecules (i.e. at the sub-nano to nano scale). These are essential to the shape – and so the function – of proteins and genomic DNA, providing the exquisitely complex molecular recognition patterns which drive aggregations from:
   (i) nucleotide base-pairing to the DNA double helix and gene folding; or,
   (ii) cell surface receptor proteins (integrin-subunits) to physical connection of the internal cell skeleton (cytoskeleton) with its surrounding 3D extracellular matrix;
   (iii) antibody-antigen recognition in the immune system.

2. However, examples (i) and (ii) also merge beyond this into the next layers of the scale-hierarchy (i.e. meso- and micro-scale). For example:

   (i) Nucleotide base-pair binding operates throughout the structure of genes, then chromosomes and up to the complete nuclear structure. This is most obvious during cell division (mitosis), when all of the nuclear gene content is perfectly duplicated and then pulled apart into two identical halves, one for each daughter cell. Nucleus and cell division processes most definitely operate at the micron scale, yet are still governed by atomic level (nano) forces.

   (ii) The same hierarchical continuity is present in the protein-protein recognition between receptor and substrate molecule surfaces (i.e. nano-scale), but hundreds of thousands of the same interactions will allow a multi-micron diameter cell to move millimetres through its tissue matrix.

3. Finally, we can follow the example of the organisation of the extracellular matrix. The most important tensional load-bearing protein here is collagen, a protein consisting of three chains held in a triple helix spiral by millions of the same nano-scale bonds. However, to generate functional mechanical properties in our connective tissues (skin, bone, tendon, eye), this humble molecule is packed together in countless repetitions of 3D spirals and the same exquisite bond-recognition patterns (Figure 1.1). The longest linear tissue dimension of the largest living creature we know is probably the skin of a blue whale (the largest animal ever). From lip to fluke tip, this can reportedly approach 40–50 m (0.05 km). So, in this case, we potentially have a functional structure at the sub-km scale, made up of repeating nanometre-scale structures, all assembled in interdependent hierarchies. For collagen, then, these functionally inseparable hierarchies (i.e. they are all *physically joined*) nominally span 11 orders of size-magnitude, from ≈0.5 to 50,000,000,000 nm (100 billion:1).

This is important, as it means that when biological members of the tissue engineering community come

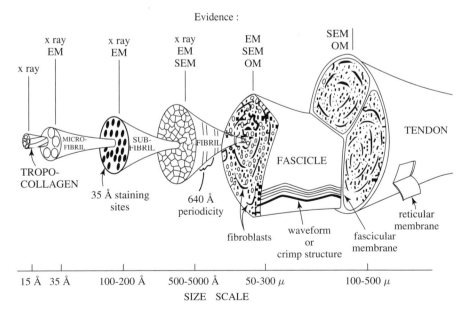

Figure 1.1 Collagen size-scale figure/drawing from tendon. Reproduced from J. Kastelic, A. Galeski, and E. Baer 'The Multicomposite Structure of Tendon', Connective Tissue Research, 1978, Vol. 6, pp. 11–23.

to discuss (i) cellular or (ii) organ engineering, they mean two different but completely interdependent things. We have created a bit of an artificial conundrum, because the design of vertebrate biology, based on natural evolution, means that *only* the whole, intact organism is capable of sustained, independent existence. So, in nature, the individual is the *de facto* functional unit. The conflict comes because, in the lab, we can now keep and manipulate isolated organs, tissues cells and even sub-cellular organelles and protein-systems.

In order to maintain their sanity and to make rational progress, bio-scientists over the centuries have described or invented numerous hierarchical levels or classifications. In the past, these have been largely based on microscopic structures of cells and tissues, and more recently on cell expression of proteins and gene-based classifications.

Examples of these include the identification on structural grounds of cells in different tissue layers of major arteries, in the skin and in the nerves. Differentiation hierarchies of stem and progenitor cells in the marrow (i.e. hematopoietic cells, generating blood-borne immune cells) are well understood. But the functional understanding of haematopoietic stem cells which underpins bone marrow transplantation was not enough to prepare biology for the shock of adult (stem) cell plasticity or reprogramming. Similarly, the treatment of haemophilia with factor VIII was worked out based on an understanding of the coagulation cascade (another protein hierarchy), but few other protein-replacement treatments have been as simple or successful. The problem is, we are only just finding out which bio-hierarchies are and are not *functionally* valuable divisions, where they are oversimplified or exaggerated.

So, what scale (or hierarchical level) of new body parts should we focus on making? Do we aim really high on the complexity spectrum for (say) a whole, beating heart, a factory-fabricated (ready to inflate) lung or a 4 kg mass of hot, living liver? The alternative, further down the hierarchical line, is to fabricate smaller and simpler spare parts, such as muscle strips, hollow tubes for nerve guides and blood vessels, tough sheets or rods to rebuild and refurbish

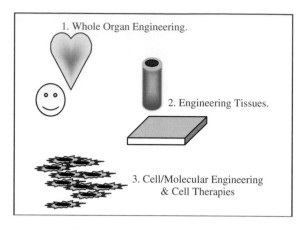

Figure 1.2 Three 'levels' of tissue and cell engineering. We can call them size scales, hierarchies or levels of complexity – but they are different.

more complex structures (Figure 1.2). This would be analogous to patching up and refitting your car's wing mirror, after an accident, with small parts such as a mirror, plastic casing and wires, rather than acquiring a complete new factory-sealed mirror unit, or even a complete new door and mirror.

Down at the really minimal end, we might aim to deliver as little as a small bolus of special cells (with suitable control factors) by injection to the injury site. The aim in this case is that the cells would be pre-programmed with all the information and vigour needed to regenerate a new body part completely. Examples might include: injecting articular cartilage cells – chondrocytes – into joint defects; corneal stem cell and keratinocyte therapies; or injection of olfactory ensheathing cells (OECs) into spinal lesions. This can be reduced and summarized (Figure 1.2) into (1) organ engineering, (2) tissue engineering and (3) cellular engineering and cell therapies.

The various merits and choices represented by this huge simplification will come up again and again as we progress. Since there tend to be few hard answers available to this question at present, we tend to work, pragmatically, across many such levels, from sub-cell and cellular to tissue and organ engineering. Yet, for convenience, we stick with the same language and classifications to describe what we are doing, and here we generate misunderstanding. By analogy, perhaps, we are not yet specifying whether we are using US or Imperial gallons.

To illustrate this, when we design and fabricate engines for transport, we look for input from mechanical, chemical and electrical engineering specialities to bring together bearing surfaces, hydrocarbon mixtures and ignition control systems. In addition, though, it is essential to understand the detailed function of that engine – specifically, will it be used in the automotive, shipping, rail, aircraft or toy industries? So, with engineered tissues designed to correct failing cardiac function (note, the target here is a *function*), should we put all efforts to fabricate an entire heart, as it is logically a 'unit' muscle with very specific application and function? Or, alternatively, should we try to adapt or multiply the functions of many lesser small muscles, as engineers might adapt an array of automotive diesel engines to power a small ship or adapt an aviation turbine for use in a train.

Members of the 'biology tribe' of tissue engineering might tend towards the perfection of the whole heart. In contrast, the 'engineering tribe' could argue the case for flexibility. After all, developing a small, generic muscle (able to be combined in all sorts of multiples) has the potential of solving many more problems than 'just' cardiac dysfunction (e.g. finger and arm movement, eye turning, sphincter (valve) control in the digestive/urogenital systems, or even breathing).

So what is so special and difficult (the engineers may be fidgeting here) about engineering '*a tissue*'? Just because it contains living, self-replicating structures (cells), why can it not simply be assembled, like anything else?[3] Indeed, this is a basic question we shall return to repeatedly. The equally simple answer (from the biological side) comes in two parts. They are special because:

- they *are* living and dynamic (and we should not make light of what that implies);
- when replacing a bio-component at *any* (hierarchical) level, we do not control the removal and re-fitting processes as we do in engineering.

1.2 Bio-integration as a fundamental component of engineering tissues

The second of the two points made at the end of the last section is so fundamental to us that it is easy to miss its significance. We ourselves, are, after all, living beings, and we take that for granted. We cannot unscrew or unplug a discrete 'unit' of a biological organism, for example one layer of the hierarchy. The surgeon *cuts out* what was once part of a structural-functional continuum when he removes one hierarchical part of the patient's body unit – hopefully, the defective part of a tendon, skin-patch or vein. This is clear, because, once the piece is removed, the patient's wound margins bleed, give pain (nerves are cut) and often physically retract. In the reverse direction, surgeons generally cannot 'clip in' a new bio-spare part. They must offer it into the host site in such a way that it might 'grow' into the existing biological structures and hierarchies. This in-growth and reconnection is an immensely complex, poorly understood and variable process, collectively termed 'integration'.

Integration comprises *vascular*, *neural* and *mechanical* (marginal) attachment into the rest of the body system. Superficially, it resembles the reattachment of the oil/fuel/water lines and the electrical cables and then bolting down a replacement engine into your car after a major refit. But these engineering steps are only equivalent to the surgeon's use of suture threads, screws, wires and glue to 'fit' the bio-implant into position. In automotive engineering fitting/implantation is the end of the process. *The car drives away*. For the patient, it is only the start, as bio-integration can *only* occur at the cell level with participation of the surrounding (wound margin) tissue surfaces. After all, it is the intimacy of this integration-linkage that

---

[3]In UK bio-science, this is called a 'Mrs. Lincoln question'. At the exit to Ford's theatre after that tragic performance of April 1865, a journalist caught Mrs Abraham Lincoln with the question, "Aside from that, Mrs L., how was the play?" This is a question where the caveat is so big that it becomes irrelevant, even embarrassing.

made it necessary for the surgeon to cut (rather than unclip) the tissue 'unit' from its hierarchical position in the patient.

In other words, the tissue in question never *was* a 'unit' as we normally consider the term in engineering. Consequentially, the concept of fabricating a replacement 'unit' is somewhat flawed from the outset. In particular, we cannot yet escape a heavy reliance on the natural tissue repair processes for integration. The integration process (hopefully) restores our engineered body part into its place in the hierarchy of the real functional unit, i.e. the patient.

Since we do not fully understand how some of these processes work (and especially how they work *together*), direct engineering or assembly of tissues starts to look daunting. This hurdle becomes clearer (and more scary) to the engineers and physical scientists as they begin to ask their questions about, even basically, *what exactly is it that we are being asked to fabricate*. Characteristically, the answers start to come back with what sound like enormous caveats, variations and unquantifiable flexibilities. It is here that the fresh-entry engineer learns the *real* meaning of 'cross-disciplinary working'. Despite the huge leaps in understanding of biological mechanisms in recent times, biological mechanisms rarely come with the precision, reproducibility and limits of tolerance that engineers and physical scientists take for granted.

1.2.1 Bio-scientists and physical scientists/engineers: understanding diversity in TERM

We are painting a very real intellectual chasm between the biological and the engineering tribes of tissue engineering. In essence, this hinges on the need for engineers to define almost all points that they touch (making fine control possible) and, as far as possible, avoiding those points where precision is not possible. The modern biologist, in contrast, has evolved to cope with the opposite, particularly in multi-cell systems. Biomedical scientists of all shades would make absolutely no progress at all if they avoided the indefinable.

Biotechnology to date has been very effective in simple cell systems, where the potential for systems interactions can be limited and so conditions can be controlled to some extent. However, in complex, multi-component or 3D systems (e.g. beyond fermentation-like processing), the potential to increase the permutations and system complexities increases exponentially. Such systems in nature seem to be controlled by innate cell-to-environment 3D feedback regulation, which is presently understood only loosely. The source of exponential complexity in tissue engineering is implicit and built into the need to put cells, often of different types, into three-dimensional, hierarchical structures such as layers or zones. 3D structure with multiple cell types is at the core of 'tissue' function, and so is a largely unavoidable source of control complexity.

Despite the prickly, scary nature of these points, it is a helpful analysis to make as it leaves us with clear concepts of the size of the problem. While it may, at first, seem daunting, it is essential for any rational strategy that we map out the key drivers and blocks.

For example, it would seem like basic good practice for us to work out rationally (and before we start) whether our clinical problem is best tackled using a cell, tissue or organ engineering approach (or a composite). From this, we can hope to identify:

(i) which clinical/non-clinical applications really can be solved, simply and incrementally by techniques we currently have;

(ii) which targets are just too far ahead at present, as these will demand that we first answer intermediate or even basic questions nearer to our real position;

(iii) which technologies and approaches really are 'too simple' to help with the larger problems, or have already been used to deliver just about all the useful applications they can.

To biological scientists, there is a tendency to interpret every compromise and simplification of nature as likely to lead to 'poor' function. In engineering science, natural systems are impossibly

Text Box 1.2 Teaching and learning tissue engineering is especially tricky

It is likely, even this far into the chapter, that you have been struck by the way some parts of the story seem to sound overly simple (or, dare we say, boring). This is *the* difficulty which is inherent in any text trying to describe the basics of tissue engineering, for reasons which are obvious (once pointed out). Researchers, from aspiring undergraduates to grizzled post-docs, tend to have a specialist's training and background within one of the single disciplines, the *tribes*, of TERM.

Also, and let us be clear about this, we are not talking about close relatives, such as cell biology and genetics or biochemistry on one hand, or mechanical with electrical or process engineering on the other. These are linkages between major cultural and philosophical divides, with seriously different *approaches*, as we have glimpsed already. Students trained in physical sciences tend to rely on precision and mathematical predictability (formulae) which are unnerving to biologists. However, they suffer similar insecurities as they grapple the seemingly infinite complexity, imprecision and variability of the biomedical sciences. Since the characteristic feature of TERM states that it *only* exists where traditional disciplines are crossed, then you, the reader, must be aspiring to cross those boundaries.

It then becomes inevitable that some parts of the story will sound very simple, even naive, to some readers. The problem is that these 'obvious, boring bits' are essential and not at all obvious to other pools of readers, such as no less genuine aspiring tissue engineers, who are coming from a different discipline. It is a truism that it is the 'simple and obvious' which divides and so retards good tissue engineers, as these are the parts of our own specialities that we are *least* likely to clarify in discussion.

So, where you start to feel bored and fidgety, console yourself in the knowledge that there is no alternative. In this case, *all* of the basics have to go into a single book. In other words, explanations of stress, strain and stiffness will appear, even though it is clear that some of you are mechanical engineers. If it is any consolation, though, the cell biologists will be treated to distinctions between epithelial and stromal cells and between tissues and organs.

It is fascinating how many biological colleagues believe that it must be easier for physical scientists to learn the descriptive world of biology than it is for biologists to cope with the mathematics in engineering and physics. Yet, in the very next room, you can meet as many engineers no less convinced that biology consists of an impenetrable range of fact-cliffs and concept-mountains – give them a good, long computer model any day. The truth is that it is difficult in both directions, and we just have to learn to live with 'going over the basics' as it is not basic to everyone. This chapter is only doing its job if the reader at one point feels bored and then at another suddenly anxious and informed. Clearly, though, preparing a book which is guaranteed to bore most of its readers at some point is itself a teaching challenge.

Exercise: Try swapping reading materials with colleagues from other disciplines (e.g. cell biologists with engineers, surgery and repair biology with materials scientists) and check out what sections *they* have highlighted, compared to you. The chances are that you will learn valuable lessons about the locations of your respective comfort and uncertainty zones. Identifying the location of your collaborator's uncertainty zones is a key part of learning to be a good tissue engineer.

complex to copy in the detail they seem to have. In blending these two cultures, it is inevitable that the tracks ahead will twist and turn from (i) to (iii) above, and so must be under continuous review. Paradoxically, there is a strong argument for working hard to preserve this process of oscillation, although it can seem to some like instability or indecision.

The pragmatic value of the oscillation here[4] comes because the systems *are* so complex and we *cannot* know, for any given application and

---

[4] As with life by the ocean; when you adapt to working with slow, powerful oscillations, you learn to be *very* certain when the *turning points* are coming (and equally confident in the tide-chart tools you use!).

technology, whether an apparently very simple solution will be sufficient for function (i). Conversely, how do members of the engineering tribes know when a technology is just too simple to work and stop trying to apply it (iii)? On the other side of the coin, there is a perfectly reasonable and correct default that in such complex bio-systems: 'we always need more basic knowledge' (ii: the bio-medical tribes).

Since this can easily become an open-ended problem with indefinite timelines, it needs to be tested to identify when such 'enabling' research is just too ambitious to be currently practical. The result may well be that we identify an effective system for progress without fully understanding why in the first place. After all, we have many examples of simple approaches that worked before we knew how they worked, because they happened to tap into natural bio-controls. When this happens, it can bring rapid and inexpensive solutions. Historical examples can be seen in immunization against viral infection or sunlight therapies for rickets. It might also come to include the injection of naked, cultured cells into tissue lesions (i.e. simple cell therapies).

However, some equally important successes have depended on long developments, involving many complex, knowledge-based technologies, more similar to fabricating complex tissues in 3D tissue-bioreactors.

1.3 What are the 'tribes' of tissue engineering?

One thing about tissue engineering which almost every author agrees on is that it is interdisciplinary (sometimes cross-disciplinary or multi-disciplinary). It is important to scratch the surface of that statement to find out exactly what each particular author really means by the term, but nevertheless this is a consensus that we can work from (more of interdisciplinarity later).

The practical truth of this was evident in the very earliest years of tissue engineering, when the subject was brand new, full of exciting possibilities and, above all, highly fundable. Suddenly, there

were many tissue engineers at meetings, publishing and securing grants. These many scientists were legitimate members of a growing community, and yet there had been no route to become officially trained in TERM, nor a reasonable lag period where expertise of the new field might have been gained.

The reason for this was simple. The community of tissue engineers (at least in Europe) had sprouted *directly* from expertise in its component disciplines. If you were a biomaterials scientist, a bio-mechanical engineer, a tissue repair or a cell biologist, or a surgeon with special interests in engineering a tissue, you could reasonably claim to be a tissue engineer. All of the traditional learned societies (to which the new tissue engineers also belonged) sprouted sessions on TERM, and then complete conferences focused on the subject. These societies still routinely have either tissue engineering or regenerative medicine as a default topic-for-invited-papers at their annual symposia. Importantly, each of these scientific and engineering communities tend to tackle and consider the TERM field in their own special way and from their own particular standpoint.

Working in the TERM field, then, feels like being within a loose federation of tribes. It is intellectually rich and behaviourally diverse, but it also can be slow to progress and prone to misunderstanding – even naivety. To the newcomer or trainee tissue engineer, the effect can be bewildering until this clarity emerges (Text Box 1.2).

In more familiar terms, its nature can be understood from parallels with the Scottish clans or the communities of the early settlements in New England. Before the advent of easy, rapid transport, the clans or extended family groups that lived near to the coast would prosper by fishing, boat-building, smuggling, etc. Those in the mountains would forage off wild game, cut timber or mug lost travellers, while others, living by rivers in the lowlands, would grow crops, establish banks, build roads or sell fraudulent maps for travellers going to the mountains. Each group would bring their own views and technical expertise to events which demanded common effort or joint defence against outside elements. That is, they would also work together for the wider 'national' needs and common interests of Scotland or New England.

Smaller and larger versions of this pooling of diverse experience can be found in the histories of China, Switzerland and modern South Africa. Examples of the opposite of this analogy (i.e. non-cooperation and competition at every scale) might be represented in the medieval city-states of Italy or the recent history of the Balkan states. In effect, the great strength of such cooperation springs from the very diversity of experience which makes the 'tribes' different. So it is with tissue engineering. Challenging as the habits of the different tribes/disciplines may seem (especially the smuggling, mugging and fraud), they are the source of joint progress.

Perhaps, then, homogeneity or consensus should be viewed as the main *enemy* of TERM. The key to cooperative success in such systems lies in the tension between opposing targets. On the one side is the need to get closer and to promote useful cooperation. On the other is the imperative to maintain the specialist tribal skills and contributions by keeping separate. In some cases, this tension can resolve into a balance between the demands of competition and cooperation. Because this is so central to the understanding of tissue engineering, yet so rarely analysed, it is a recurrent theme of this first chapter.

To understand a little of how the scientific and engineering specialities contribute to successful tissue engineering, we need to identify who they (the tribes) are. Indeed, this can be subtly very helpful for analysing the messages of lectures and papers in the field. Throughout the 1990s, many talks on tissue engineering started with a slide of the speaker's perception of its component disciplines (Figure 1.3a). In fact, the speaker's selection of these disciplines

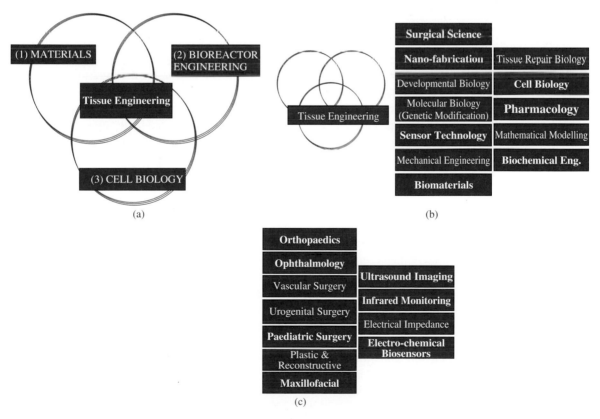

Figure 1.3 (a) One typical depiction of the essential intersecting quality of contributing TERM subjects. (b) Some other major component disciplines of TERM. (c) Some sub-disciplines possible within *only* surgical science and monitoring technology.

was actually more informative of his or her origins and intellectual approach than they would like us to know. It was, and still is, important to listen *very* closely to this introduction. It should, indeed, form the basis of any critical, academic analysis of the technology and scientific interpretations that follow, in the body of the lecture or review.

There are several major science and engineering disciplines (and many, many sub-specialties) which could contribute to the overlap which makes tissue engineering exist. Figure 1.3a shows one example, with tissue engineering being that part of the intersecting rings where at least two, and ideally all, are overlapping. Clearly, where there is no overlap, there is no reason to think that the subject is anything more or less than the traditional discipline itself. Equally, it is possible to have as many intersecting rings as you choose, but here correctness and increasing division has to be balanced by clarity and brevity. It is important, then, to ask the question, 'why did they make *this* particular choice of disciplines?'

Analysis of the example in Figure 1.3a indicates that the three example components put forward are not on an equal size-footing. Two are broadly based general disciplines (biomaterials and cell biology), while the third is more specific, that is, a sub-specialty. This indicates (correctly in this case) that the talk being introduced was going to draw on aspects of cell biology and biomaterials as they applied to the design, construction or operation (i.e. the engineering) of bioreactors.

Now we come to the scary bit. Can we make a list of those disciplines that tissue engineering should cross? The answer of course must be 'Yes – but . . .'. Figure 1.3b gives an illustration of what are, in fact, fairly major subjects found in real examples of tissue engineering research. By the time we have set out from surgical science and tissue repair biology and we reach mathematical modelling and biochemical engineering, it is becoming clear that this could include most of biomedicine and a sizeable chunk of physical sciences and engineering.

Where the component disciplines are not particularly diverse, it may be that the work fits better into a more traditional field and may not really be tissue engineering at all. For example, a main discipline of development biology teamed with cell culture technology and optical imaging might, in fact, be better described in terms of conventional research into mechanisms of embryo development. It is sometimes suggested that research themes are 'tissue engineering', based on a possibility that it *could* lead to important, if serendipitous, findings. This cannot be an acceptable criterion, as it could be said of almost any activity, but it erodes away the basic logic of the field. For example, research into polymer chemistry would be considered primarily *polymer chemistry* rather than biomaterials in nature, until a viable tissue engineering biomaterial is likely to be prepared. It is not the common convention to classify research fields based on their *possible* long-term outcomes.

Understanding which sub-disciplines are involved in any given tissue engineering approach can be helpful. It partially indicates where the work and its ideas are coming from, and it helps to inform on where particular logics, technologies or concept will be strong (and where others may be usefully inserted). For example, we may want to review work on the follow-up of the fate of clinical implants (say, engineered large vessel grafts) by use of new approaches in minimally invasive sensor technology. This would be a collaboration involving the overlap of surgical science, biomaterials and sensor physics. But what type of sensors? Electrical field, mechanical or optical sensor technologies would each give their own distinct set of approaches and capabilities. However, the application of, say, ultrasound or magnetic imaging technologies would carry completely different implications and limitations. In effect:

- Which parameters of implant performance are measured (e.g. vessel wall structure, blood flow rates, clot formation)?
- How often can measurements be made (infrequent for heavy, costly equipment versus regular for indwelling sensors)?
- What is the data quality (e.g. resolution limits for identifying fine features of the vessel wall or micro-thrombi formation; tendency for signals to

be obscured by overlying tissues; poor relevance of the data as a measure of implant performance)?

- What are the dangers to the patient, damage to the implant or surrounding tissues (e.g. heating effects of some ultrasound treatments or the need for injection of disclosing agents to the patient to make structures visible)?
- How much work is needed to adapt the monitoring technology to the demands of the implant system (i.e. how much research is needed)?

If the problem of spiralling numbers of permutations needs any reinforcing, then the addition of sub-disciplines will help. Figure 1.3c indicates some possible sub-discipline pairings which would be expected for tissue engineering based on *only* the primary disciplines of surgical science and monitoring technology. Clearly these lists can, and do, go on and on, with more examples of new subject-matches being added to the literature every year. It is equally clear that these are not useful as a basis for definitions, and that they most definitely are not for predicting future successful combinations. They simply form a historic record, indicating the extent to which ever greater diversity of approaches and discipline combinations can prove useful.

Despite the diversity, then, can we put a name to the *chief* tribes of tissue engineering? It would certainly be an unusual tissue engineering application, for example, if there were no cell or developmental biologists, surgical or tissue repair specialists, biomaterials scientists, (perhaps pharmacologists), biomechanical engineers or optical physicists/imaging specialists. So, while hard and fast rules continue to be rare members of the TERM community, we do have a group of familiar suspects who turn up regularly. To continue the Scottish clan analogy, while you might not expect to see *all* possible Highland names at the Town Fling (the Oliphants or Rosses might be out of town), it would be reasonable to suspect a phoney event if there were absolutely no Campbells, McGregors, MacIntyres or McDougalls.

So, perhaps we should conclude at this early stage that the subject of tissue engineering is not only enriched by the contribution of many differing specialities, but indeed that it cannot really exist (at least as 'tissue engineering') without the contribution of at least two of the more traditional component specialties. After all, there must be a tangible reason why a topic can be distinguished as tissue engineering rather than, for example, biomaterial science or optical bio-monitoring.

1.3.1 Special needs for special characteristics: why is networking essential for TERM?

The 'accepted wisdom', of at least a decade, implies that you can hardly be a real scientist unless you work across disciplines. However, since it is now hardly possible to challenge this idea, it is also becoming more difficult to understand what different specialists *mean* by it. Despite the obvious advantages to novelty and scientific vigour, it is alarming that there is so little critical examination of the costs, more particularly the downsides, to interdisciplinary collaboration. After all, it is only politicians who suggest that we have such a thing as 'cost-free benefit'. So, where is the downside to being multi-disciplinarity?

One clue to this can be found by looking closely at why, for so many years, career scientists (particularly bio-medical) chose to focus their life's work on a pinhead specialization. As an entomologist, already pretty specialized in the study of insect biology, it has been possible to make a major reputation in *just* moth migration (British only), or in the distribution of a particular group of parasitic wasp. We might look, then, at what good things the specialist must give up to become truly cross-disciplinary. This question does assume that the cross-discipline gap we are talking about is more along the lines of 'entomologist to aeronautical engineer' than 'moth to beetle biology'. A similar story can be seen in the clinical specialities. Plastic surgeons, for example (famously expected to operate in almost any body space), will rapidly find a special niche. This may not even just be in hand, nerve or breast reconstruction, but often (s)he will become known for a particular surgical technique, sometimes associated with a new stitch type.

The potential for loss here is that you become less excellent (less knowledgeable, expert, up to date) as you get further from your sphere of expertise. The risk of leaving that focused niche is that of making naive mistakes. In other words, the perceived danger is *loss of intellectual safety*.

Loss of intellectual safety has been a huge factor in traditional subjects, acting against the obvious benefits of working across big subject discipline gaps. In effect, remaining an 'expert' (excellent at an international level) requires a huge depth of understanding of the published literature in all of the topics surrounding 'your speciality'. Given the depth and complexity of the modern literature, this can be an immense task. While it is possible to read and critique the annual output of published material in (say) hand reconstruction or the secret life of Antarctic whales, the equation rapidly becomes impossible if it has to be blended with advances in tendon gene abnormalities or satellite tracking. And actually, these examples are relatively modest in modern tissue engineering collaborations. To help to understand this, Figure 1.3 illustrates how multi-disciplines can turn into a real nightmare for the would-be tissue engineer.

As you look through the basic texts in tissue engineering, you may notice how common it is for a research group or institute to describe itself as being within a 'type' of tissue engineering (see tribes and identities, above). This normally takes the form of a prefix to the group name, to designate their tissue focus. There can be bone, cartilage, skin, nerve, muscle. cardiac, liver, blood vessel (small and large) and bladder/urogenital tissue-engineers to name a few. Figure 1.4a lists only four examples of these in our matrix of expertise.

It is then also useful to focus on the engineering problems of a particular tissue. Even within one tissue group, there are usually many sub-groups or forms of tissue. There may, for example, be different clinical needs at different stages of tissue formation (e.g. for childhood disorders) or in different body sites. Cartilages, for example, are needed for joint repair, but also for reconstruction of facial tissues (nose and ears) or for reformation of trachea and inter-vertebral discs. Even for the replacement of joint surfaces (i.e. articular or 'hyaline' cartilage), the tissues which are needed can vary between joints and according to patient age and disease, while in some cases (e.g. meniscus) a quite different 'fibro-cartilage' is thought to be required. Consequently, engineering even single tissue types really can merit such levels of specialization.

However, contributing research knowledge to the pathology and physiology of your tissue (cartilage) is the job of the appropriate biomedical scientist – *not* the tissue engineer (a caveat to this comes later). Tissue engineers should be special, and so distinct from a tissue specialist, because they aspire to *fabricate* their specialist tissue. Consequently, they must also understand the elements of the main platform technologies which they would be expected to use to achieve that characteristic 'fabrication' aspiration. These may, for example, include 'cell

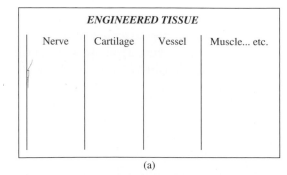

	ENGINEERED TISSUE		
Nerve	Cartilage	Vessel	Muscle... etc.

(a)

PLATFORM
Scaffold Materials
Bioreactors
Monitoring/Bio-sensing
Tissue Models & Modelling
Cell Sourcing/Bio-Processing

(b)

Figure 1.4 (a) Short example-list of tissue engineering 'tissue' specialities. (b) Short example-list of tissue engineering enabling or platform technologies.

sourcing', 'scaffold materials' or 'monitoring and bio-sensing'. A few of the many possible examples of these platforms are given in Figure 1.4b.

The special problem which comes inevitably with the *engineering* of tissues is that the plethora of platform technologies which are used across the range of tissues can have different levels of success (often surprising) *in different tissues*. For example, a breakthrough in polymer scaffold 'design' or 'controllability', as applied to cartilage tissue engineering, may well have enormous advantages for skin and vessel engineering. Similarly, if neural engineers identify a revolutionary way to process and monitor tube structures or promote cell guidance, then vascular or urothelial tissue engineers will also stand to benefit. The problem is that the research advantage goes to the group who pick up on the key enabling innovation *first*. Those who do not look across the wider tissue engineering spectrum will see it late, and so may end up appearing 'naive'*.

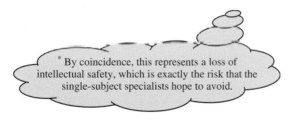

* By coincidence, this represents a loss of intellectual safety, which is exactly the risk that the single-subject specialists hope to avoid.

In fact, in the special case of tissue engineering, their very diligence in the use of traditional focus and specialization onto a narrow tissue-base can become a *major weakness*. This is illustrated in the exemplar matrix of cross-interests in Table 1.1. Clearly, the breadth of cross-over here between apparently distinct tissues and the enabling areas (often engineering) makes traditional monitoring of research progress inadequate. This effect is amplified by the tendency for tissue engineering tribes to publish in their favourite 'tribal', rather than TERM, journals.

For example, if your primary area is cell biology, how often do you scan titles such as *Advanced Functional Materials* or *The Journal of Biophysics*? Equally, if your work is primarily in scaffold biomaterials, would you catch original work from journals such as *Gut* or *Cartilage and Osteoarthritis*, where you might learn of specialist surgical and repair innovations, or from *Development* for new thinking on tissue regeneration. The telling factor here is that these are all successful, high-impact journals – attractive honey pots for excellent researchers *in their respective specializations*.

So, the central problem of identifying relevant innovations in time remains. It is implausible to cover the volume of knowledge and out-of-field innovation needed through literature scanning or by attending specialist conferences. This, then, is the source of the idea that tissue engineering, like no other discipline, depends on aggressive, continuous *networking*. Effective networking allows participants across very diverse areas to pick up early hints and indications of cross-disciplinary excitement or innovation which normally would be slow to traverse the divides.

There are now numerous organisations which perform this role, from the international tissue engineering and regenerative medicine societies (with acronyms like TERMIS and ICCE) to continental, national and even regional network organisations.

Table 1.1 Example of a full knowledge-matrix for tissue engineering, using only 4 × 5 tissue/platforms; derived from Figures 1.4a and 1.4b.

Platforms	Tissue 1	Tissue 2	Tissue 3	Tissue 4
	Nerve	Cartilage	Vessel	Muscle
Scaffold materials				
Bioreactor engineering				
Monitoring and sensing				
Tissue models and modelling				
Cell sourcing/bio-processing				

Figure 1.5 Bronze sculpture of a human tower (Tarragona, Spain), illustrating the principle of network cooperation. Note the many figures needed to support the base.

The nature and logic of many tissue engineering networks can be illustrated through analogy with the bronze sculpture shown in Figure 1.5. This depicts a human tower traditionally formed in some parts of Spain (this case celebrates a record-breaking tower from Tarragona). Notice that this tower supports at its summit a small boy. Members of the tower-layers may take turns to enjoy the position and its view. However, the key point is that the weight of the tower itself has to be supported by a surprisingly large number of collaborators pressing inwards to hold up its base. In other words, a few people at any one time can, in turn, get a clear long view from the tower because of the concerted support efforts of the 'many' participants.

For reasons which are mainly cultural and political, the European Union probably has the best example of integration and networking between groups. Figure 1.6 illustrates approximately the spread and distribution of major networking groups active in tissue engineering. Around the year 1999,

this could be drawn as an approximate north–south corridor (roughly the red box). Ten years later, and with considerable central EU support, the geographical (and intellectual) spread of this has expanded dramatically to fill two linked shapes (blue triangle plus crescent), covering western Europe, including the northern Mediterranean rim.

Interestingly, networking activities tend to be a balance between those collaborative activities which participant groupings consider to be beneficial and the natural tendency to consider other groups as competitors. In other words, there is commonly a tension between networking (sharing) and perfectly healthy competitive activities.

The very time and effort required for network formation and collaboration can be seen as potential loss of competitive edge, and it is important to understand this balance when comparing activities in different continental zones. For example, there has been a sustained and conscious effort to promote (and fund) collaborative and networking activities across the European Union (EU). This has given a characteristic style to cross-disciplinary EU science, which is arguably less well developed in the same fields in North America or Asia. Here, there may currently be a greater tendency to adopt competitive models, at least over wider geographical areas. The jury will be out for some time over the question of which of these models is closer to being the most effective balance.

The 'take home concept' from this section is that we have identified the first (of a series) of the tissue engineering tensions – between cross-disciplinary cooperation and essential competition. The key point, though, is that the existence of both elements of the tension is necessary for success in tissue engineering. There will be no Utopia using only one or the other, so our task must be to discover ways to 'ride the tension' – to find effective balance points, not to eliminate the tension.

1.4 Surprises from tissue engineering (Veselius to Vacanti)

It is probably true that followers (evangelists?) from almost every new wave of research, from

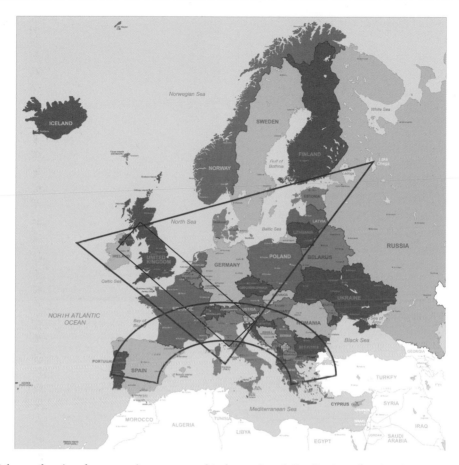

Figure 1.6 Scheme showing the approximate geographical growth and distribution of major tissue engineering groups around Europe. Key: ≈1999: red corridor (NW-SE central/western EU). ≈2010: contained in two blue areas. © iStockphoto.com/Pingebat.

biochemistry and molecular biology to nano-medicine and synthetic biology, have claimed that their emerging perspective will bring with it new revolutions and scientific dawns. Of course, tissue engineering and regenerative medicine is no exception.

Some of these are admittedly small dawns, but we can never tell in advance. The fact that a book such as this inevitably has just this bias does not mean we can avoid the idea that there really is something revolutionary about tissue engineering. The first stage must be to identify what this 'special feature' might be. Only then can we ask if it really is *so* special.

The proposition here is that something very new and surprising is already happening. In fact,

the ability to fabricate basic biomimetic structures (simple tissues) is turning upside-down the fundamental, traditional approaches used for centuries in biological scientific research.

To understand where this view comes from, it is necessary to appreciate that biological sciences and scientists attempt to understand some of the most seriously complex systems anywhere. This is perhaps not surprising, after a thousand million years of evolution refining the complexity of bio-systems. Complexity, versatility and diversity are, in fact, their defining characteristics. An equally characteristic approach to research in the field has developed over its long history, quite distinct from that in the physical sciences and engineering. This might reasonably be described as:

1. breaking down of the intact complex system into component parts;
2. description and classification of the parts; and then
3. hypothesis-driven deduction and analysis of how the parts *might* have functioned in the intact system or organism.

This has proved to be hugely successful over many hundreds of years, from the early anatomists such as Vesalius (Figure 1.7) to the present day.

Andreas Vesalius (1514–1564) was a Flemish surgeon living in Brussels. He insisted on basing the anatomy he taught to his students on original, systematic dissections of human cadavers, as opposed to animal dissections or artistic interpretations. Although this was initially illegal, a helpful judge in the Italian city of Padova eventually eased his task by supplying the cadavers of executed criminals. From such systematic dismantling of intact body structures, Vesalius was able to describe in detail the layout of both the nervous and blood systems. It is not unreasonable then to consider that this descriptive work in part enabled the British physician William Harvey (1578–1657; Figure 1.8) to deduce and demonstrate the mechanisms by which the heart pumped blood through a discrete circulatory system. This is a classic example of the approach of dissect-describe-hypothesize, spread over a long period.

Figure 1.8 Portrait of William Harvey (1578–1657), who is credited with the first detailed description of the blood circulation system. Harvey was a physician at St Bartholomew's Hospital, London and 'physician in ordinary' to King Charles I. The latter got him into significant trouble with Cromwell's Parliamentary (anti-monarchy) troops, and his great discovery did little to help Charles in the end.

The Swedish naturalist Carl Linnaeus (1707–1778; Figure 1.9) is considered to be the father of systematic taxonomy (classification of species). The work of Linnaeus and many other descriptive biologists after him provided the basic interrelationships between animals and plants which have enabled all branches of bio-science to deduce, test and refine our understanding of key biological mechanisms, from Kreb's cycle and respiration biochemistry to modern genetic shift and inheritance.

This same basic pattern of 'break down, describe and reassemble' is even visible in the present day, with the solving of the human genome. First, with genomics the human genetic code was discovered, broken down and progressively described. However, close on the heels of the full genome description came the predictable quest to understand the mechanisms by which these coded proteins operate, leading to proteomics. The remarkable constancy of approach is clear – though, interestingly, the time

 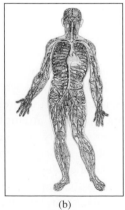

(a) (b)

Figure 1.7 Andreas Vesalius and his description of the human system of blood vessels. (a) Reproduced with permission © Getty Images; (b) Reproduced with permission © Royal Society of Medicine.

Figure 1.9 Carl Linnaeus (1707–1778). Portrait in oils, by L Pasch after A Roslin, 1775 copied for Sir Joseph Banks. With kind permission of the Linnean Society of London.

course for the cycle is shrinking from centuries to decades.

Through tissue engineering, however, it can be argued that a completely new pattern of progression could be emerging. In their pivotal review of tissue engineering, Langer & Vacanti (1993)[5] outlined the aim of *building up* tissues and generic biological structures from their basic units, rather like engineering would be expected to fabricate a human-designed device. In this case, the basic building blocks consisted of suitable cells, 3D support materials and suitable growth control signals (mechanics,

growth factors, nutrients). As we shall see later, having one of the basic building components as complex as 'a living cell' makes the assembly process rather more complex than one would normally choose for a fabrication process. Nevertheless, this is similar to the process that many engineers would immediately recognize as bottom-up logic.

Bottom-up approaches lead to intimate understandings of the operating mechanisms of the systems by virtue of their relative simplicity and the many iterative assembly cycles needed. These cycles are characteristic, first to make the system function, then function better, then faster, cleaner and cheaper, etc. In the automotive industry, developments have progressed for so long that much of what can be known about the basic process *is* known. Innovations now commonly come via other technologies which impact on the industry, such as computer-based engine management, aerodynamics, surface coatings or changing social pressures. In the case of assembling biological systems, the final target level is where it operates in exactly the same way that the original tissue does in nature.

In effect, almost 500 years post-Vesalius, we now have a pretty sophisticated idea of what any given tissue should look like and even how it should perform when intact. Tissue engineering effectively aims to make increasingly complex versions of simplified (reductionist) tissues, to assess how they perform and to keep reiterating the process to improve the functional result. In theory, we should know when our efforts are approaching functional and useful, as the tissue we fabricate will start to work more and more like the 'real thing'. Indeed, there is a case for the term 'biomimetic engineering' to cover this process.

The astute reader will see immediately that we are now, after almost half a millennium of tradition, peering the other way up the research avenue. This process promises to show us how biological mechanisms operate through progressively refining what we can *make to work*, rather than what we *think the parts should do*. Critically, each time we design and fabricate a tissue and it does *not* work, we can eliminate one more possible operating mechanism(s). Indeed, this view is already pointing the

[5]Langer, R. & Vacanti, J.P. (1993) Tissue engineering. *Science* **260**, 920–926.

way to completely new (often remarkably simple) understandings of how cells might assemble and refine tissue structures in nature. These will emerge periodically in later chapters.

However, perhaps the most persuasive glimpse of this mechanism inaction comes from the realization, through a number of reviews, that the engineering of tissues promises much more than the simple fabrication of body parts for clinical use. The driver of early stage clinical applications (initially accelerated by commercial forces) has tended to be premature and out of proportion. What we are seeing now is a whole segment of tissue engineering research dedicated to making biomimetic model tissues. These have a value in their own right as test platforms, 3D screening tools and diagnostic systems. But they are also the visible evidence that the 'make-assess-improve' iterative cycle is turning.

1.5 So, really, *is* there any difference between tissue engineering and regenerative medicine?

1.5.1 Questions never really asked: repair versus regeneration?

Many gardeners will have experienced that all too common, but slightly squirmy, moment when digging a flower bed – the 'earthworm incident'. They just get in the way. There you are: one worm, two halves. This may quickly be followed by reinterring the parts and a guilt-removing recollection that both (or was it just the head?) ends will regenerate into new animals. Have I done my bit for soil ecology, then? Rightly or wrongly, the idea that invertebrates can grow complete new, working parts even after complete and major losses seems relatively unsurprising.

Where this '*regeneration*' extends to fellow vertebrates, and in particular regeneration of amputated limbs in amphibians, such as newts and salamanders, it may seem a little more special. We can, after all, identify much more closely with the new limb and its movements – though not in any shape or form with the idea that this could happen to us. In fact, writers of comics and films frequently use this

idea in their plots and story lines for superheroes who heal as fast as they are injured. But no matter how good the film graphics or comic storyline, this is still just fiction and a dream for we mammals. Indeed, the mammalian reality is that our tissues *repair*, and this repair process is a pale shadow if the ideal, which is *regeneration*.

Although not all amphibians or their wounds regenerate quite this well, some (Figure 1.10) will go on to form complete new limbs at the site of amputation. This occurs by a type of growth resembling that of embryonic limb formation and

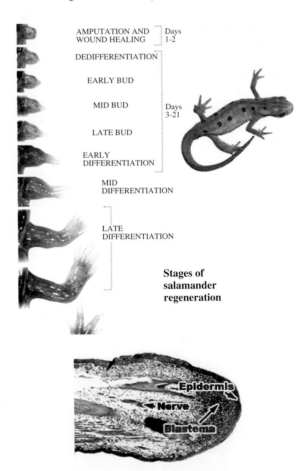

Figure 1.10 Serial images showing the regeneration of a salamander limb over 42 days. Inset: Histological section showing the growing point of the regenerating limb with its differentiated extensions. Including nerve and covering epidermis. From: Mescher, A.L. & Neff, A.W. (2004). Loss of Regenerative Capacity: A Trade-off for Immune Specificity? *Cellscience*. Reproduced by kind permission of A. L. Mescher.

some impression of it can be seen from the figure, in which the remaining stump of the lost limb starts to grow and extend. The growth and extension occurs predominantly at the extreme distal edge (i.e. most remote from the body). This growing strip, or edge, is a rather ill-defined (non-differentiate) cell mass which pushes away from the body wall. Behind it, component structures of the limb (blood vessels, nerve, bone etc) begin to form as cells progressively differentiate down each of the specific tissue lineages involved.

The same process of programmed differentiation, from generalist ('stem' or 'progenitor' cells) to specialist (tissue) cells, along a programmed series, is typical of limb development. This continues as the main bones of the limb grow, ending eventually with the formation of separate terminal digits (toes). Admittedly, the anatomy is simpler than it is in human limbs, but the regenerate has a new, normal skin, covering normal long and toe bones with tendons and joints, so they can bend and move.

Though less obvious, the appearance of another feature should be equally impressive (perhaps *magical* if we are considering how it might be engineered). This key step, so familiar to us that it can be almost invisible unless pointed out, is known as **integration.** After all, the new tissue could (in theory at least) have formed as an independent limb on the end of the stump with little connection to the body. However, this is not what happens; the limb bones and tendons organize and physically link to the rest of the salamander so that they move and function. This means that the new limb is not *just* physically joined to carry physical load, but is fed by nerves and blood vessels, which must grown *into* the new limb from the pre-existing body side of the stump (Text Box 1.3).

In addition, the size and geometry of the new limb are, mostly, similar to the original and they are a match to the size and needs of the animal. This dimensional (size) or spatial control is particularly intriguing and is poorly understood. By no means all of it can be explained by the idea that the cells 'know' (are genetically programmed) how to rebuild a salamander. One example of this is the enigma of the 'switch-off' control of new tissue formation, typical of repair, regeneration and normal tissue growth seen throughout the body, though (perhaps simplistically) not so for tumours. How does the new limb come to end up just the right size to match the others, and not three times or half the original length?

The process of limb regeneration in amphibia bears a strong resemblance to that of limb formation in the embryo. In other words, this form of natural 'tissue regeneration' is in the domain of developmental biology, which is significant, as we shall see later. The clue to this can be seen in Figure 1.10, and its inset showing a cross section through the regenerating limb-stump. At the growing tip is a plug of undifferentiated cells (stem cells), known in embryos as the blastema. These generate the forward outgrowth of tissue mass but, as this mass moves away (elongating the new leg), those cells behind start to differentiate into the structures we recognize as the layers and components of a new leg, skin, nerve, skeleton. Each gradually matures into their respective final parts of the leg, specialized to their individual functions.

So much, then, for our lower-vertebrate cousins. However, if we mere mammals are unfortunate enough to have a limb cut off in surgery or in an accident, it remains 'gone'. We, the victims, are left with the stump – that is, whatever (undamaged) tissue was left attached to the proximal, body side, of the injury site. Importantly, the stump will remain, non-sprouting, whatever we do and however long we try. Furthermore, the otherwise uninjured (distal) parts of the lost limb (e.g. fingers or toes) show absolutely no tendency to regenerate a new body (this one really *is* for the worms and sci-fi animators). Even though we are familiar, even resigned, to this and we take it for granted, our native abilities (relative to newts) are particularly modest and disappointing, because we do not even do particularly well with the stump end (or, to be more precise, the scar).

In the case of major limb loss such as this, human patients are understandably more concerned about the loss of the leg or arm function (or relieved to have survived at all). What sort of tissue reformation occurs at the stump is not, perhaps, the victim's main concern when you would prefer the stump not to end as it does at all. At least bleeding from

Text Box 1.3 Integration in tissue repair and regeneration: so familiar, it's almost invisible

Integration is a very small word for a critical part of both tissue repair and regeneration. It is the process by which any new tissue structures become attached and 'plumbed in' to the existing surrounding tissues. This ensures that the newly formed repaired or regenerate tissue is connected into the central systems of supply-and-control (i.e. is *systemically* linked).

The most basic of these linkages is the in-growth of new blood vessels (see (1) on Figure 1.11). This automatically brings 'connection to' (and control by) the host animal, in the shape of immune surveillance and inflammatory cells, hormones and growth factors, as well as coagulation, nutrients and oxygen. The in-growth of nerves (2), of course, brings its own control, where the nerve tips 'dock' with muscles or sensory endings. Finally, (3) a durable mechanical linkage, attaching the new tissue into the surrounding, parent tissues, is almost always essential. This is formed when connective tissue collagen fibres are woven, by fibroblasts, across the new-old join.

The importance of vascular integration is very widely appreciated (though mostly for nutrient and oxygen supply), but this is just the most colourful of the set. Without appropriate levels of integration in all three of these areas, the new tissue would either die or have very poor function. Indeed, the pattern of integration can affect the very nature of the new tissue; for example, in the cartilages, the entry/non-entry of vessels can determine when/where it is calcified (and forms bone), or where it forms articular or fibro-cartilages (as in the meniscus). Figure 1.11 summarizes these three elements. We can, then, consider there to be three distinct (but linked) integration processes:

1. Revascularization (normally angiogenesis, the in-growth of new vessels by sprouting of existing surrounding capillaries).
2. Re-innervation; outgrowth/sprouting of the injured surrounded nerve axons into the new tissues, establishing both sensory and motor controls to new tissues where appropriate.
3. Mechanical integration/attachment. This is probably the most fundamental, and so least noticed, element (i.e. the salamander's new limb does not drop off its stump when moved), and your repaired skin wound does not pop out to leave a hole when you wash (though, interestingly, the blood clot eventually does).

Rather like the work done by a team of skilled plumbers, electricians and plasterers who follow the builders of your new house or extension, these stages only become obvious where they *do not* work properly.

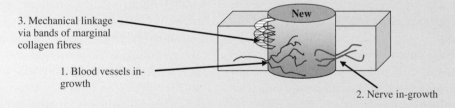

3. Mechanical linkage via bands of marginal collagen fibres

1. Blood vessels in-growth

2. Nerve in-growth

Figure 1.11 The three areas of integration.

the major arteries has been stopped in time and a covering of sorts has formed to keep out massive infections. Indeed, there is now a well accepted view that we mammals have evolved a system dedicated primarily to *survival*, and less to the quality of life afterwards.

However, some clues to the wider problem are perhaps evident when we hear of the 'phantom limb' effect (the failure of integration between injured nerves and the brain), necrosis or infection due to poor re-vascularization or skin contractures/deformation. These are features of a mammalian tissue *repair* process (i.e. not regeneration). In effect, evolution has favoured rapid, aggressive tissue in-filling, over the restoration of three-dimensional structure and spatial organisation which would restore function.

In short, higher vertebrates seem to have evolved to minimize the imminent and lethal dangers posed by rapid fluid loss out and equally rapid pathogen access through the wound. The process of generating bulk in-fill (i.e. repair tissue and scarring), as fast as possible, seems to apply to most parts of the body, even where they do not border the outside world as the skin does. The process of bulk-dumping of poorly organized connective tissue into repair sites is what we call 'scarring'.

It is worth emphasizing, though, that scarring is the *generic* loss of normal tissue architecture and function. In other words, it is the default, and it occurs to lesser or greater extents at most injury sites (not just the skin). The dermal scar in Figure 1.12 is poorly functional. It has different (stiffer) mechanical properties than the surroundings and it affects facial movement, social interactions and appearance (commonly with psychological impact). Notice, though, how the new tissue stands up from the surrounding skin. In effect, it did NOT stop growing at an appropriate point (unlike the salamander limb) and is too big to accurately replace the injury site. Spatial control in repair tissues is poor.

The uncomfortable truth is that in higher vertebrates, scarring is the normal, default process in response to injury. This is a key concept for tissue engineers to recall and hold close, as it means that we live with the constant possibility that we are engineering scars!

Scarring, comparable with the type we are familiar with in skin, also occurs in blood vessels, tendons, heart and other muscles, and all major organs from lung and kidney to guts and urogenital tract. The fact that scarring/repair is the default mammalian response to injury is rather poorly appreciated, as we only refer to repair sites as 'scar' where they are a problem. When the scar is not a problem (a mixture of luck and insensitivity to loss of function), we are happy and we call it tissue repair.

What determines whether a post-injury repair tissue (for example in Figure 1.12) impacts significantly on function is generally down to luck, location and injury size (with a smattering of genetics). However, scarring is the source of an enormous variety of major and minor forms of human suffering, proportional to the perceived impact of the lost function. It is also important to recall that scarring/repair is the *normal* response.

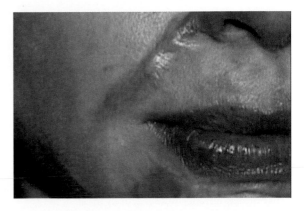

Figure 1.12 Image of a facial burn scar, long after injury. The new replacement tissue is the wrong size, geometry, colour and texture (material properties). Its function is seriously altered from the original. The fact that this *repair* tissue is on the face (and *after* surgical correction) helps us to understand the nature of the problem, but in fact this repair default occurs almost everywhere in the body. Reproduced with permission © R. K. Mishra.

There are, in fact, many abnormal or downright pathological forms of the process. The tip of this particular iceberg can be glimpsed in examples of pathological *dermal scarring* conditions such as hypertrophic and keloid scars. In these exaggerated repair tissues, the shape and material properties of the scars show signs of a failure to shut down at the appropriate time (in some cases ever), leading to oversized or physically deforming repair tissues. We are probably only now starting to understand how such faulty repair processing affects other internal tissues.

We have now identified the first plank in our understanding of how tissue engineering and regenerative medicine might be distinct – based on what each is trying to achieve (Text Box 1.4).

1.5.2 Understanding the full spectrum: tissue replacement, repair and regeneration

Is there *really* any difference between tissue engineering and regenerative medicine? The answer for most workers in the field is 'yes', though it is often less simple to explain why. In fact, both terms describe an aspiration which is as old as mankind – to restore previously lost function to body parts (whatever the cause). It is helpful to

Text Box 1.4 How tissue regeneration differs from repair

Tissue regeneration (whether in major part of a salamander or small layers of a human) is the replacement of lost or injured tissue structures by near-identical structures with the same function as the original.

   Tissue repair (the default process in mammals) is the replacement of lost, damaged or injured tissues with an approximation to the original tissue (sometimes, but not always, of the same shape and dimensions), which may or may not substitute effectively for all of the original tissue functions.

Exercise: Write a short analysis of the difference (especially spatial organisation) between regeneration of the salamander limb in Figure 1.10 and adult mammalian skin in Figure 1.12. Suggest one or more environmental cues which might help to explain how the size and shapes of mechanical tissues such as these are controlled (or fail to be controlled).

Tip: how is it that almost everyone's left leg grows to almost exactly the same length as their right, yet variation between even close relatives' leg lengths can be large?

know, though, that there are probably only three broad approaches to this vision. These are:

1. to *replace* the tissue (with a device providing *some* function);
2. to get back *better* function by enhancing the natural *repair* process;
3. to *regenerate* functionally *perfect* matching tissue to that which is defective.

Modern bio-medicine, with its development of advanced artificial prosthetics and its ability to suppress rejection of transplants and to re-connect microvasculature, has made major progress with item number 1 above (replacement). From this standpoint, tissue engineering and regenerative medicine can be seen as different approaches towards the same goal (i.e. restoration of function) using very different techniques.

   However, there is a vision among clinicians and researchers that we are moving (Figure 1.13) progressively along a left-to-right, past-to-future, time-line. This is moving away from the era of replacing defective body parts with metal/plastic implants or pieces of previously used tissues or organs. These are approaches which, paradoxically, have been and will continue to be enormously successful and important for real patient care well into the foreseeable future. This timeline ends at the point where we can achieve perfect restoration of function (i.e. in regenerative medicine). En route,

it passes through tissue engineering, in which the aim is to develop ever better levels of engineered biomimetic repair.

   According to this analysis, we now find ourselves at an intermediate stage (Figure 1.14), in which the dominant research questions and clinical objectives are designed to improve more on the natural *tissue repair* process than to achieve true *tissue regeneration*. In other words, tissue engineering (engineering enhanced repair) is characterized as using advanced bio-processing, monitoring and control technologies. This is currently a wide progression-front. Such technologies include those of biomaterials processing and tracking, bioreactor and monitoring/sensing systems, drug and growth factor-controlled release and biomimetic engineering of the extracellular matrix.

   In contrast, regenerative medicine, with its more distant, elevated target, is based largely on new and evolving fundamental concepts of stem cell biology and cell plasticity. While some cell therapies and clinical applications are in clinical trial and evaluation, these are at early stages and tend to be characterized by a heavy reliance on the native behaviours of certain, selected cells. The cells in question can be adult, differentiated (such as cartilage chondocytes) or adult progenitor or stem cells (for example derived from bone marrow, corneal limbus or fat) or, more commonly, uncertain combinations of the two. Embryonic and reprogrammed stem cells are as yet on the horizon.

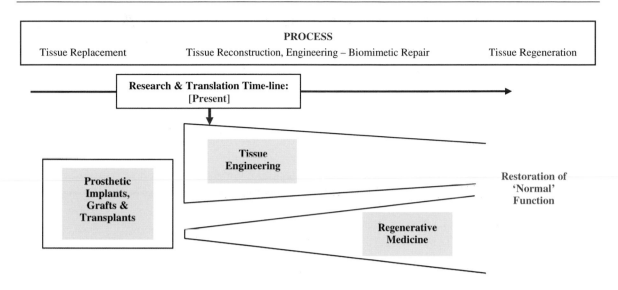

Figure 1.13 Diagram to illustrate how tissue engineering and regenerative medicine can be considered to relate to each other, based on the understanding of repair/regeneration biology. From this standpoint, they are moving towards the same goal (restoration of function) but are using different approaches. Importantly, they are running in parallel, though not necessarily at the same rate or using the same track or starting point. Unfortunately, to those outside the field, these two very distinct approaches *sound* the same, and they are often viewed as almost the same thing. The evolving approaches are indicated in the top box, while the general timeline of progression is indicated below it (left/past to right/future).

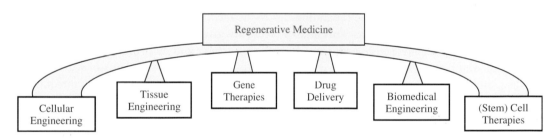

Figure 1.14 Diagram showing an alternative (aspirational) view of the position of regenerative medicine as an overarching, umbrella term. Here, it is supported by a clutch of related disciplines, using the idea that technologies which feed into the broad aspiration of regenerating perfect tissue function represent part of the same, target-defined structure. Such a technology-based structure, with regenerative medicine as its ultimate aim/aspiration, might also logically include developmental and tissue repair biology.

The opportunities for *engineering* what happens in such cell therapies are often fairly low, and there is frequently much less ability (even aspiration) to control what happens externally. This is an inevitable position, given our limited understanding of what these cells might be and how they might achieve the therapeutic ends we hope for. In the case of adult stem cells for therapy, the extent of this limited understanding is made clear by the complex and empirical immune-staining patterns that remain the only way to identify many populations of 'stem cells' and their early differentiation.

As a result, we can crudely distinguish between the two basic strategies (tissue engineering and regenerative medicine) in terms of how we approach them practically. Tissue engineering might be viewed as involving our best attempts to fabricate 3D biomimetic structures, i.e. tissues. It is generally biomimetic of tissue *repair*, using any biological process available to achieve or improve on natural repair. Clearly, this means that tissue engineering is not limited to using or mimicking regeneration processes, or even to having regeneration as its goal, however welcome it is when it happens.

Cell therapies, aiming at achieving tissue regeneration and regenerative medicine therapies, largely *do* aspire to achieving tissue regeneration, frequently by trying to identify special cells which will mimic regeneration when implanted. However, these approaches presently depend heavily on identifying, tuning and isolating cells with such innate regenerative capacity and behaviour. Technological manipulations in this field are at an early stage and are largely limited to the selection of promising cell populations (ideally enriched in stem or progenitor cells) and encouraging some of these to differentiate towards cells needed in the target (in short, cell farming). Such culture or farming-like technologies are designed primarily to expand populations of desirable cells, to drive unsuitable cells in the required direction or to eliminate cells with irrelevant or unhelpful activities.

This is illustrated in Figure 1.13 as the expanding, diverging bar of regenerative medicine. This analysis suggests that tissue engineering tackles much higher technical and process control targets within a rather less ambitious overall vision (i.e. tissue repair rather than regeneration). In contrast, regenerative medicine aspires to a highly ambitious end vision but is forced to set itself relatively low, empirical targets, chiefly because of the currently modest technology base.

Attentive readers who identified with the earlier part of this chapter and its description of the tribal nature of TERM should now be expecting the inevitable 'alternative' or caveat. In this case, the alternative view of the tissue engineering and regenerative medicine relationship avoids reference to goals and relies on technological interdependencies.

Figure 1.14 illustrates this apparently regeneration-centric view, in which a number of existing research areas are considered to be supporting the overall, umbrella vision of regenerative medicine. In fact, it is possible to see this as a derivative form of the previous model, with the time dimension omitted. In this structure, regenerative

medicine takes on an overarching position precisely because it is such a distant and high aspiration. In fact, regenerative medicine then becomes the vision or aspiration of those other (component) disciplines which do have sound, technically definable foundations (i.e. not requiring an open ended basic research commitment). Of course, where and/or if it turns out that clinically useful regeneration can be achieved *without* detailed technical understanding of how it occurs, then such empirical approaches will come to dominate.

So the answer to the question we started with, on the difference between tissue engineering and regenerative medicine, seems to come down to the *quality* of the function which it is hoped to restore. In previous decades (even centuries) we have concentrated on *replacing* tissues with the modest ambition of giving back some function (or for limited periods of time). In some areas, this approach (e.g. wooden legs to artificial hip replacements, eye-patches to contact lenses and corneal transplants) is perfectly adequate for the problem and, indeed, continues to improve the lot of patients.

We have also been trying to improve on natural tissue *repair* for centuries, though mainly by tackling its grossest, most acute failings of infection, haemorrhage and deformity (Figure 1.15).

Research into the problems of scarring, unstable repair tissue material or poor integration and organisation have only really become mainstream within the past 25–30 years. Simple patient survival is now not enough. We are now far more fussy about what is considered the 'functional' quality of

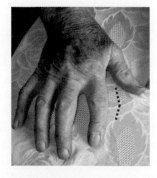

Figure 1.15 A bending finger, known as a fixed flexion deformity (in this case as a consequence of rheumatoid arthritis), can also follow from Dupuytren's disease or scarring after a tendon injury and adhesion.

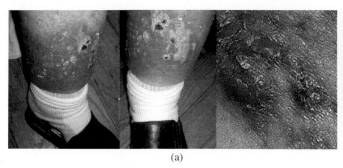
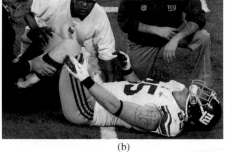

(a)   (b)

Figure 1.16 Modern society sets higher and higher demands on the traditional concept of restoring function. These have gone from just getting a wound to heal over (a) to current needs for it to 'look good' as well, or from previously getting back some walking/running function after a foot injury to now returning to first team performance (b). Photo in (b) © http://deadspin.com/nfl-roundtable/.

the repair tissue at the end of our treatments. These new aspirations to functional restoration are far higher than before and clearly are unlikely to be satisfied by the typical properties of either natural tissue repair or prosthetic implants (Figure 1.16).

Such solutions tend to leave:

- large red or inelastic scars;
- poor ranges of body motion;
- no feeling;
- poor circulation;
- limited implant survival expectation.

In fact, 'functional' has most recently migrated to include:

- convenience (too slow, too many stages – e.g. slow-healing or chronic wounds); and
- cosmetic/aesthetic (including concepts of *desirable* – rather than necessary – shape, colour or size).

With this has come an increasing understanding that many of these aims might be achieved by learning how to better enhance the native *repair* process. Examples have included:

- injections of growth factors (e.g. PDGF or TGFβ1);
- alteration of cell type or activity;
- manipulation of the mechanical forces acting; and
- fabrication of 3D physical repair-templates; or
- fabrication of graft tissues (ideally off-the-shelf).

Here we can recognize the emergence of tissue engineering approaches (though closely entwined

with other therapeutic strands of repair biology). Only when the outcome of these treatments is sufficiently effective to produce a 'regenerate' (as good as the original), as opposed to a repair tissue (i.e. not bad but not perfect either), will we be over the border into the kingdom of regenerative medicine.

1.6 Conclusions

A great deal of this chapter has been spent dissecting the nature and aspirations of tissue engineering and regenerative medicine. If the student wishes to take this subject seriously and make any new contributions, it is important to have a grasp of these concepts and basic understandings. Initially, this can seem surprising, but on closer examination it is clear that we are not dealing with a conventional field of research. This makes it unusually important to understand how the 'cogs and pulleys' operate in this case and, more particularly, what makes them different.

Critically, TERM has had a very different evolution to that followed in the past by other major initiatives. It was not born out of a revolution in technology, nor a breakthrough in scientific understanding (e.g. molecular biology to genomics/proteomics, or histology/optical microscopy to electron microscopy to magnetic resonance and computer tomography). Rather, it was generated from a fusion of technologies and concepts which were actually rather well known. It is this *fusion* aspect, drawing on three or four very different but major subject areas, which has moulded tissue engineering. Where it can, this chapter has explored:

- how these ideas and technologies start to fit together (or where they have trouble fitting);
- what they really *can* do together; and (*more than ever before*)
- how the emphasis is on getting the most out of the crossovers and novelties which are generated every time an idea or technique migrates across a traditional discipline interface.

In chasing this particular rabbit to ground, we have tackled the question of which disciplines tissue engineers come from (and why it matters). We have looked back at our origins and the way that one's 'home discipline' might explain why definitions tend to be either bland or partisan. We have identified the defining factors embedded in the greater aspirations – the vision – of tissue engineering (namely whether we aim to replace, repair or regenerate a tissue). Finally, we have begun an initial sketch of how different approaches, understandings and requirements of the bio-science and physical science/engineering communities can generate their own form of scientific revolution. Indeed, we may now only just be starting to recognize the nature of this revolution.

As for defining the field, the message perhaps should be one of continuing to modify this, as the field continues to evolve, expand and subdivide (see Text Box 1.1). We, in fact, may just be too soon at the dance to know exactly what style the band will be playing. Critically, students of tissue engineering will always be a risk from Peter Cook's all too plausible warning* on the danger of weak

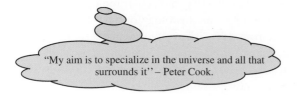

"My aim is to specialize in the universe and all that surrounds it" – Peter Cook.

self restraint in the face of the apparently endless diversity of options discussed here. As we have seen, this can be resisted by cold and critical analysis of what we propose to do and how we plan to reach our extreme tissue engineering goals.

1.7 Summarizing definitions

Definitions are, at least at some levels, an essential part of most intellectual activities. However, along with their obvious value can come many (well known) difficulties. This is especially true in the case of tissue engineering and regenerative medicine (TERM). Not least, the whole idea can be regarded as a fusion of already well-established concepts and fields of research. It might be, then, that we have the best result we can expect – a series of retrospective representations each looking at the topic from its own standpoint (e.g. biomaterials, surgery, cell and repair biology, bioengineering, etc.).

However balanced the author tries to be, it must come from one or other bias, because of our training. This is also true of the version you are reading now, though others may have more or less critical analysis. Listed here is a small collection of the more widely published efforts. Interestingly, while it is most common to use the entire string (TERM), published definitions generally focus separately on tissue engineering and regenerative medicine, usually without trying to explaining the difference. This may be that there is currently no clear idea of how and why they are different (so it is safer to lump the two together). Alternatively, it may be that the two terms are truly synonymous, covering essentially the same ground (with only subtle differences). This author's view is that they not only *are* quite distinct, but that it is important for our comprehension that we are clear about the differences. Below is a small collection of published definitions, for reference, starting with the pivotal 'Science' description/review of tissue engineering from Langer & Vacanti in 1993.

Annex 1 Other people's definitions of tissue engineering

Definitions are, at least on some levels, an essential part of most intellectual activities. However, along with this come many (well known) difficulties. In the case of tissue engineering and regenerative medicine (TERM) this is especially true, not least as the original idea might be said to be a fusion of already well established concepts, which almost immediately began to sprout retrospective definitions in every direction.

As a result of this it might be, then, that we have a series of perspectives on what the field really represents, each looking from the standpoint of one of the component traditional disciplines – exactly as I have done, now! This annex provides a small collection of the more widely published efforts, as an instruction of what might be and how the field has evolved. Interestingly, these pretty well all define either 'tissue engineering' or 'regenerative medicine', despite the fact that most writers will use the entire string (TERM). This may be that authors do not have a clear idea of the differences (so err on the safe side), or that they are truly synonymous, as sometimes claimed (almost certainly wrongly). Here we provide a small collection for reference. Since Langer and Vacanti are often (though far from universally) credited with coining the term, this starts with a definition of 'tissue engineering' from their widely cited *Science* review article of 1993:

'Tissue engineering is an interdisciplinary field that applies the principles of engineering and life sciences toward the development of biological substitutes that restore, maintain, or improve tissue function or a whole organ.'

Langer, R. & Vacanti, J.P. (1993). Tissue engineering. *Science* **260**, 920–6.

'The term regenerative medicine is often used synonymously with tissue engineering, although those involved in regenerative medicine place more emphasis on the use of stem cells to produce tissues.'

Addendum from current entry (2009) in Wikipedia (i.e. popular definition).

Many authorities (particularly those dependent on US funding) might choose the NIH definition to be the most useful:

'Tissue engineering is an emerging multidisciplinary field involving biology, medicine and engineering that is likely to revolutionize the ways we improve the health and quality of life for millions of people worldwide by restoring, maintaining or enhancing tissue and organ function. In addition to having a therapeutic application, where the tissue is either grown in a patient or outside the patient and transplanted, tissue engineering can have diagnostic applications where the tissue is made in vitro *and used for testing drug metabolism and uptake, toxicity and pathogenicity. The foundation of tissue engineering for either therapeutic or diagnostic applications is the ability to exploit living cells in a variety of ways. Tissue engineering research includes biomaterials, cells, biomolecules, engineering design aspects, biomechanics, informatics to support tissue engineering and stem cell research.'*

NIH definition of tissue engineering.

A decade on from Langer and Vacanti, in 2004, amidst the many other 'perspectives' that had emerged came this example of tissue engineering from a biomaterials viewpoint:

'There is an inherent, virtuous logic to tissue engineering that sounds too good to be true. By my definition, tissue engineering is the persuasion of the body to heal itself, achieved by the delivery to the appropriate site of cells, biomolecules, and supporting structures. It specifically involves the regeneration of new tissue to replace that which has become diseased or injured, the significance of which is that we, as adult humans, do not normally possess this ability. We may repair ourselves under some very limited circumstances (for example, bone fractures and injured skin may undergo repair) but, even when this does occur, this usually involves nonspecific reparative tissue (i.e. scar tissue) rather than the regeneration of the specific functional tissue that has been affected.'

Williams, D.F. (2004). Benefit and risk in tissue engineering. *Materials Today* **7**, 24–29.

Annex 2 Other people's definitions of regenerative medicine

The current NIH working definition states:

'Regenerative medicine/tissue engineering is a rapidly growing multidisciplinary field involving the life, physical and engineering sciences that seeks to develop functional cell, tissue and organ substitutes to repair, replace or enhance biological function that has been lost due to congenital abnormalities, injury, disease or aging. It includes both the regeneration of tissues in vitro *for subsequent implantation* in vivo *as well as regeneration directly* in vivo. *In addition to having a therapeutic application, tissue engineering can have a diagnostic application where the engineered tissue is used as a biosensor. Engineered tissues can also be used for the development of drugs, including screening for novel drug candidates, identifying novel genes as drug targets and testing for drug metabolism, uptake and toxicity.'*

This can be qualified by the addendum from the Pittsburgh Tissue Engineering Initiative:

'A distinguishing characteristic of regenerative medicine is that it has the potential to cure disease through repair or replacement of damaged or failing tissues.'

The NIH Facts sheet goes on to state:

'Regenerative medicine is the process of creating living, functional tissues to repair or replace tissue or organ function lost due to age, disease, damage or congenital defects. This field holds the promise of regenerating damaged tissues and organs in the body by stimulating previously irreparable organs to heal themselves. Regenerative medicine also empowers scientists to grow tissues and organs in the laboratory and safely implant them when the body cannot heal itself. Importantly, regenerative medicine has the potential to solve the problem of the shortage of organs available for donation compared to the number of patients that require life-saving organ transplantation.'

Though it is perhaps unusual for a definition to include the 'potential' or 'promise' of its subject matter rather than what they actually are now, this aspirational quality may, itself, be a defining characteristic of the new field.

After almost a decade of use a new definition, based on its brevity, comes from the journal *Regenerative Medicine*:

'Regenerative Medicine replaces or regenerates human cells, tissue or organs, to restore or establish normal function.'

Mason, C. & Dunnill, P. (2008). A brief definition of regenerative medicine. *Regenerative Medicine* **3**(1), 1–5.

Further reading

1. Vacanti, J. & Vacanti C. (2007). The history and scope of tissue engineering. In: Lanza, R., Langer, R. & Vacanti, J. P. (eds.) *Principles of Tissue Engineering*, pp. 3–6. Academic Press, Burlington, MA.
[A good start to the origins of general tissue engineering from a combined biomaterials and surgical perspective.]
2. van Blitterswijk, C. A., Moroni, L., Rouwkema, J., Siddappa, R. & Sohier, J. (2008). Tissue Engineering – an Introduction. In: van Blitterswijk, C. A. (ed.) *Tissue Engineering*, pp. 14–36. Academic Press, London.
[Modern introduction to basics, with biomaterial leanings.]

3. Clark, R. A. F. (1996). *The Molecular and Cellular Biology of Wound Repair*, pp. 611. Plenum Press, New York.
[Classic view of the basic teachings of the wound repair biology tribe, excellent analysis – still widely referenced.]
4. Grinnell, F. (2009). *Everyday Practice of Science*, pp. 230. Oxford University Press, Oxford.
5. Stocum, D. L. (2006). *Regenerative Biology and Medicine*, pp. 464. Academic Press, Burlington, MA.
6. Langer, R. & Vacanti, J. P. (1993). Tissue Engineering. *Science* 260, 920–926.
[The classic review which acted as the starting-pistol for the international expansion of effort to engineer tissues.]

A massive wooden tower in Lausanne, Switzerland (La Tour de Sauvabelin).
Is it bio-mimetic, or just good mechanics? It comprises an integrated spiral staircase and walls made in wood. Spiral structures occur in nature, for example in snails and collagen molecules, but they are just a geometric form with particular mechanical strengths. They are uncommon, at least at the gross scale, in human anatomy. Wood is clearly bio-mimetic in composition but, strictly speaking, only where we are mimicking advanced plants such as trees. Just because something looks natural, it does not automatically make it bio-mimetic.

2

Checking Out the Tissue Groupings and the Small Print

or: Avoiding the low aim that still misses

2.1 Checking the small print: what did we agree to engineer?

I have never worked for Airbus or Boeing. But you can be sure that before their huge design and engineering teams so much as reach for a pen to sketch a new aeroplane, they have a *very* clear analysis of what they are being asked to make. This is most definitely not the case (yet) in tissue engineering, although – for reasons even more pressing than those of the aerospace industry – that *should* be our aim. After all, if Airbus Industries need to switch from a metal alloy fuel line to plastic in the final prototype, they just fabricate the replacement, insert and verify that the new one performs as their model predicts. The additional costs should be modest. No one would reasonably expect them

Extreme Tissue Engineering: Concepts and Strategies for Tissue Fabrication, First Edition. Robert A. Brown.
© 2013 John Wiley & Sons, Ltd. Published 2013 by John Wiley & Sons, Ltd.

to smash up their previous prototype planes and start again!

However, this is not the case (again at present) for an engineered tissue therapy, real or imagined. Our ability to predict, analyse and model the tissue systems we aspire to fabricate is minimal in comparison to non-biological, manufactured devices. The consequences of getting things wrong can be just as disastrous for both types of engineering, though perhaps more spectacular and immediate for passenger planes.

The regulatory authorities responsible for civil aircraft would ask for data and validation of predictions on the performance of the new fuel pipe in the existing device. This is most obvious when we remember that the odd Boeing or Airbus has crashed in the past. Occasionally this has been due to control software, engine parts or structural surface failure. Yet identical aircraft types remain in service; they have been checked out, replacement parts fitted and are back, better than before. No one normally expects all examples of that plane to disappear overnight to the breaker's yard.

Not so with tissue engineered implant devices. It is likely that regulators responsible for the quality/ safety of human implants would send the producer right back to the beginning of the development, in the process incinerating all examples of the failed design. The simple explanation, of course, is that we do not yet understand enough about tissues, and exactly what they need to do, to make them as predictable as parts for modern aircraft. However, it cannot be accurate to describe our example structure here of the Airbus airliner as a 'simple' device. These 100-tonne machines routinely bullet around at 200 metres/sec on the edge of space, where outside temperatures are good for freezing your blood. Still, the risk to passenger life and limb is judged to be negligible.

This can be illustrated by a closer look at the aeroplane engines next time you fly. Many Airbus types, for example, are made to take one of perhaps two or three completely different engine types (i.e. produced by the main jet engine manufacturers, Rolls Royce, General Electric or Pratt and Whitney: see Figure 2.1). Yet still each aircraft version performs predictably and safely within tolerances which would make biologists weep with envy.

In contrast, when we change only single minor process steps or components in biological tissues

Figure 2.1 Airbus 320 can be fitted with either Rolls-Royce or General Electric jet engines. Spot the different cowlings.

and implantable materials, we really do lose all confidence in its subsequent performance. Until we 'try' the experiment, it is hard to predict whether performance will change totally, not at all or somewhere in between. Commonly, the extent or even the direction of change cannot be predicted – sometimes not even as basically as saying it is likely to be better or worse!

We can perhaps glimpse the extent of this difference from the history of knock-out mice and their informative surprises. In preparing a knock-out mouse, the biologist takes out just *one* gene (so eliminating one protein component from the entire mouse). But far from sitting back, confident that this modest, focused removal will elicit a single change in function, one shift in behaviour or a unique block in a signalling pathway, the biologist investigates the entire animal. Every tissue, every habit and every metabolic pathway is catalogued to look for 'the phenotype' – that *pattern* of changes which characterizes the deletion (but could not be predicted). Sometimes these changes can be so great or so numerous that the animals cannot breed or develop beyond embryonic stages. But sometimes, the complexity of cross-support systems or the process of duplication and protein redundancy means that no 'phenotype' is immediately detectable. On occasions, where repair/remodelling mechanisms are affected, the knock-out mice must actually be wounded before any effect is apparent.

In fact, the difference with aircraft systems is not at all surprising, however complex and interlinked they may be. The aircraft systems were built up, from the bottom, by engineers. They have a full understanding of how each component part works, both alone and in combination with its co-parts. After all, they *made* the components. This is not true of the knock-out biologist, who is working by making alterations, top-down, on an already highly complex system, not knowing but guessing at the workings of the whole mouse.

This difference is reflected in how we modify and regulate the bio-fabrication of engineered tissues. When we 'grow' a tissue implant that produces a good result (e.g. satisfactory to both patient and surgeon), that is *it* – cast in stone! Changing pretty well

any component (e.g. the sequence, timing, sometimes even reagent suppliers), fills us (and the government regulators) with a profound insecurity – so much so that we are sent right back to the start of our designing, testing and proving-what-to-expect process. It is as if our aircraft manufacturers did not know how to make devices fly in general, just how this type works, and even then only as long as it is an Airbus A320, serial number A320-000417-D, with Rolls-Royce engines, tuned for Shell kerosene. Getting the design of an engineered clinical tissue even slightly wrong can be, and often has been, disastrous because of this 'return-to-go' principle.

There may, then, be an opportunity for us eventually to fabricate tissues as if we were engineers. However, if this is our claim, then the non-engineering tribes will need to adjust to the reality of working like engineers. This means understanding what quantifiable functions we want to produce *and* how they can be measured once they are assembled. Ideally this should apply to our basic components as well as the finished article, so that we can change or improve components without 'surprises'. The problem is that this is not really a typical approach for biological scientists, and this how the gulf between aspiration and reality has been excavated in some areas. In brief, biological methods alone are rarely ideal for making structures that are expected to perform as if they were engineered.

Hence, it is critical to accept the implications of adopting the 'engineering' word. It will be interpreted (e.g. by engineers) in the manner that aircraft manufacturers **engineer** large planes. This involves understanding the operation of the wings, fuselage and engines to a high level of mathematical accuracy. Such mechanistic understanding allows them to compute (using their predictive models) that they want to fly 200 passengers for 3,500 km at 550 km/h, with tolerances for extremes of wind speed. The model predicts the ideal patterns of wing shape, dimensions, engine power, fuel consumption and maintenance intervals. The engineers make the plane and then identify *exactly* where there is the slightest deviation in performance from their prediction. If a parameter goes outside its performance range, alternative structures or materials can be substituted and key adjustments made to other factors, which will be altered as a consequence.

For perfectly good reasons, this is not the understanding of the biological community:

- Bio-systems are extremely complex and integrated (so the mechanistic understanding is still simplistic) and mostly not quantitative or even fully reproducible.
- The properties of construction materials which are available to modern tissue engineers are largely uncertain under most biological conditions.

In fact, the biomaterials part of the tissue engineering partnership might feel more comfortable with the analogy shown in Figure 2.2, of early aircraft designs by A.V. Roe and Anthony Fokker. These early aviators knew roughly the tricks that should get a heavier-than-air-machine into the air – but only just. They sadly knew rather less about the tricks needed to get down in one piece, with the result that they did not always take flight and remain airborne for the required periods or in the intended direction. They made informed, and sometimes inspired, guesses, but all too often these were based more on emotional feelings than a knowledge of the material strengths, forces and durability of their creations. For a considerable period of the evolution of early aircraft, the plan focused on investigating the crashes! For example, with the luxury of modern retrospection, we can look at Figure 2.2 and question the wisdom of the rear, strapped-down fuel tank in (b) and the close-set pram wheels in (a). Neither of these was even likely to catch on.

It is interesting to reflect on just how much of this analogy (including the wording) rings true of recent tissue engineering. Indeed, it is possible to extend the analogy one more step, to include the recent biological drive to use stem/progenitor cells to tackle our limitations in engineering tissues. This might be seen as an abandonment of 'wing design' altogether, in exchange for a completely different form of flight without wings (i.e. we cannot understand

(a)

(b)

Figure 2.2 Early (pre-1918) aircraft designs by Anthony Fokker (top panel) and A.V. Roe (lower panel), showing contemporary mono- and biplane formats, with traction (pulling forward (a)) and pusher (b) propellers, respectively.

the heavier-than-air engineering mechanism, so we dodge it: see Figure 2.3).

It might be worth a passing thought at this stage, that we could re-examine the need for such regulatory rigidity if we ever hit on a way to fabricate biological structures from the bottom up, as we do with aeroplanes. When we can fabricate tissues from well-defined components that work together predictably through well-understood processes which can be mathematically modelled, *then* we can tune our systems and products. Perhaps a good target here is the pharmaceutical industry. Once it is established that a chemical compound has a series of desirable clinical effects, then that compound can be formulated in many different

Figure 2.3 Airship R34: flight without wings, demanding alternative, lighter-than-air technologies. © CSG CIC Glasgow Museums Collection.

ways and combinations. This often attracts only modest regulation, provided the chemical purity, concentration and sterility can be assured. We know what pitfalls to look for and what really matters for success and safety. Tissues are like that, except with spatial-mechanical complexities **and cells**!

The reader may wonder if we are shuffling off topic here, but it is an important analogy as it gives the biological part of our community some valuable context (and modesty). It should also help inform engineering partners about the void which we must bridge in concepts and expectations. Tissue engineering is well known for the tradition of progressively 'talking-up' its vision, often to attract valuable industrial and public support. But this positive, upbeat impression can sometimes lead new recruits to miss the enormity of our task ahead. This is *not* good for strategic thinking. When you perceive that you are 'nearly there', the plans you formulate are very different from those you make when you have a long journey before you. The aim here is to take a fresh look at the 'small print' of the tissue engineering contract we are signing so that we know exactly where we are.

2.2 Identifying special tissue needs, problems and opportunities

Each tissue type carries with it special requirements which represent its 'problems and opportunities'. Building up a rational and detailed profile

of these is the key starting point for *engineering* that tissue. In practice, this involves a bottom-up or minimalist approach. It would be counter-productive – especially at the outset – to aim to engineer, say, a **left carotid artery**. This is both too variable (patient-specific) in its detailed anatomy and too specialized in its application to be a useful design starting point. Perhaps a better description of the target would be 'a visco-elastic vascular tube carrying clottable liquid under pulsatile pressure, with minimal turbulence' (Figure 2.4). It is certainly a sufficiently high hurdle. The important phrase to remember here is '**Key Functional Properties**' (KFP for short).

Notice that by denying ourselves the shorthand of using the anatomical name and instead identifying the KFP, we have been forced to list the *real* properties that we *really* need. This is a great start, particularly as it is likely to demand extensive discussions with the end user of your construct – perhaps a surgeon – rather than quick look at a textbook. It is also likely that these (KFP) properties will be useful across many vessels other than the left carotid artery. This can become a 'platform construct'. In other words, it may be possible to adapt it for use at all sorts of anatomical sites, simply by changing its shape.

Indeed, we can go further and start to put numbers to the KFPs, as allowable ranges for each property. In many cases, this will allow us to define

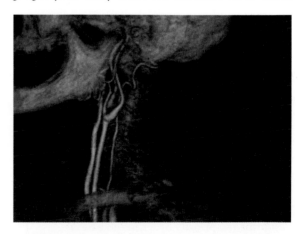

Figure 2.4 Left carotid artery, or 'a visco-elastic vascular tube carrying clottable liquid under pulsatile pressure, with minimal turbulence' to describe it using its KFP. © www.rime.pt.

those vessels, or families of tissues, for which our design will be appropriate. Equally important, it then becomes simple to exclude those for which it is unsuitable. Careful selection of KFPs at the outset will pay huge dividends later. Indeed, its iterations can make it one of the longest of stages.

We now have the skeleton of *good practice* for the planning stage:

1. Specify the key functional properties (KFPs), prioritized as 1st, 2nd, 3rd, etc., from the most to least important (i.e. critical to beneficial, or essential to desired).
2. Identify the range of values for which the performance of that tissue function would be:
   (i) ideal;
   (ii) acceptable;
   (iii) absolutely disastrous.
3. Specify (even in general) how the KFPs would be measured. Start by deciding which units (e.g. cells/ml; % live cells; MPa material stiffness; ml/min fluid flow rate) would be most useful to describe the function you need to measure.
4. List any caveats to this analysis, in particular at which stage(s) of the process these KFPs need to apply. The stages can go from initial assembly of the tissue components (e.g. cells, temporary scaffold, extra cellular matrix) to the end of bioreactor culture and on to post implantation. This can mean that for each caveat, a new set of slightly different KFP ranges need to be specified (e.g. range 1 = post-bioreactor culture stage; range 2 = one-month post-implantation stage).

These targets (KFPs), then, should ideally be the *first* section of a target application described or discussed by a tissue engineering partnership, as opposed to the anatomical site or surgical problem, which is more commonly the opening. Obviously, both the site and the problem play important parts in shaping decisions while the KFPs are being assembled, but their importance need not go far beyond assigning the priority order in the KFP list.

Once the tissue construct has its top rank KFPs, within their acceptable ranges, refinement of this generic construct can easily follow. To continue our analogy, we might hope that these specifications could become as easy to change as it is for Airbus to switch from Pratt and Whitney to Rolls-Royce engines to meet divergent airline needs. Throughout such iterations of design and testing, the aim is that KFP ranges come closer to their 'ideal' as they are better understood.

Clearly then, where the KFP model is used, constructs not only improve in function but also provide new knowledge of the factors which *control* that performance. This is a major, hidden opportunity as it represents a refining database which can grow into a model system for accurate prediction of future designs. It is a direct parallel of systems which now allow aircraft manufacturers to predict wing performance in a way that could never have happened if aeronautics had set out, say, simply, to mimic the wings of a crow, then of a vulture, then a swallow . . . an albatross . . . a dodo . . . ?

In some cases, the KFP progression may take the form of an evolution from an ultra-simple, almost embryonic tissue to a fully adult-functional, mature tissue. Alternatively, they can form a series of increasing complexity:

- 1st stage: fabrication of a model tissue (e.g. *animal sparing* for testing and research).
- 2nd (clinical) stage: as simple generic *spare part* surgical implants.
- 3rd stage: implants designed for a single *specific clinical* problem (such as the carotid artery in our initial example).

To summarize, describing our construct tissue targets as a rational series of KFPs, as opposed to naming the bio-anatomy, significantly shifts the early (and later) processing towards the 'engineering' benefits we want. It also lays a solid foundation of *function-based* design and generates a database of measureable, important factors to reduce the later incidence of 'surprises'. At least as important, it enhances the chance that our designs will have wider applications, based on the 'performance envelope' that we shall become able to produce for each key function. These should be applicable to similar sites or types of tissue or related defects. Quantifying how the engineered structures/components perform, once they are made, makes it possible for future iterations to identify additional tissues or injury sites where they can be applied.

2.3 When is 'aiming high' just 'over the top'?

It is difficult to point to many forms of '**non-living** heavier-than-air' flight (excluding the short duration, seasonal (though passive) migrations of Oklahoma trailer homes in the tornado season). On this basis, it might be argued that human airliners are a form of bio-mimesis, though in this case of bird rather than human function. Oddly enough, this is an interesting thought that is relevant to contemporary tissue engineering, as it demonstrates how society's attitude has, over the last 100 years, already come to accommodate *de facto* 'bio-mimetic engineering'.

The point here is we are perfectly able to be pragmatic about 'how bio-mimetic' our flying devices need to be. We all can now visit colleagues or relatives a continent or two away, in Shanghai, Boston, Mumbai or Sydney. Yet we never blink an eye at the sight of the featherless Boeing that will fly us there. What could go more to the core of bird-flight-function than feathers? But ever since Icarus* and

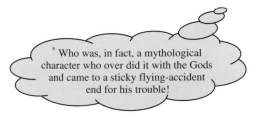

* Who was, in fact, a mythological character who over did it with the Gods and came to a sticky flying-accident end for his trouble!

the earliest glider pioneers, it has been obvious that we **do not need** this particular bio-mimetic component (see Chapter 9); instead, we cross the oceans in a casing of sheet metal and rivets! So, by what good logical reason can we automatically assume which part of any given 'bio' we need to 'mimic' in our target tissue?

This is a fascinating question in contemporary tissue engineering and regenerative medicine (TERM). How close a 'copy' or mimic of the target tissue does our construct need to be? Equally, how can we assess

when the 'imaginative and visionary' has drifted into 'wacky' dreams, especially where hyper-focused enthusiasts are the dreamers? The question has its roots in the tension (probably essential to TERM) between two contradictory needs. Let's call this the '*safe-hype*' tension. The first is for forward-looking analyses of potential applications which have imagination and vision. But the second requires that these same analyses are scientifically balanced, prudent and defensible.

Of course, pretty well anything *may* be possible if we work long enough at it, but sometime-never is not a permissible time frame. The *imaginative-vision* side of the strategic planning must have a more critical analysis of time scales than a Hollywood Sci-Fi movie. Hence, there is a need to balance this with the *prudent and defensible*. The question of how long it will take us to acquire the key understandings we are lacking is the pivot-point for assessing this 'reality' balance. Who can tell? Maybe a feathered fuselage will *eventually* improve our flight . . . but in the meantime . . .

Arguably more than many modern scientific fields, TERM has a reputation for hype (over-selling or exaggeration of its objectives). This matters, as the public are particularly interested in the prospects of having new parts of their body made painlessly available when vital bits are injured, decay or drop off. Indeed, they have every right to be interested, and realistically informed, as their taxes pay for the research. This interest is evident from the most casual glance at national newspapers and TV channels, with their seemingly endless series of upcoming miracle-systems for new hearts, eyes or skin. Clearly, TERM has more than its share of optimistic 'amazing-but-true' stories. But aside from feeding scientists' dreams and filling newspaper columns, this is one important half of valuable *safe-hype* tension – the half that drives us on.

Like all essential tensions, though, there is never a fixed stable balance point – no such thing as 'safe hype'. Scientists must inform and inspire their paymaster-sponsors but, at the same time, they must be balanced and cautious. This is particularly tricky where the subject matter detail is so complex and uncertain but the overall idea *seems* so simple. One of the most important points where this tension can be effectively balanced is that between the optimism of building *biology-as-it-was* and the prudent-pragmatism that knows we can often get *function-without-perfect-recapitulation* (Text Box 2.1).

By looking for a timely appropriate balance to the *safe-hype* tension, we are moving to the possibility of putting stages or graded-milestones against our application targets from:

(i) modest-achievable (success soon); to
(ii) difficult-with-hard-grind (success in the mid-term); and
(iii) that-might-take-a-while (success for our descendants).

Listening and reading about the 'in process' progress in regenerative medicine generally can leave the impression that most initiatives are in the (i) tending to (ii) category. This is the effect of the *positive vision*. Though experienced individuals may learn to recognize these levels, there is a guideline which can make it easier for the newcomer. This uses the principle that some long journeys bring smaller benefits en route *before* we reach the great destination. So we can judge the balance of time-risk against end-point benefit of any engineered construct, based on what smaller or earlier outputs will emerge en route. There are three such application targets (matching the milestones in the list above):

1. Model 3D tissues (research and screening lab tools).
2. Simple spare parts for general surgical reconstruction/repair.
3. Fully integrated (regenerated) tissue and organ replacements.

Text Box 2.1 Looking for the 'functional' compromise?

Crisp, defining lines of logic are hard to find in this area, as there is a tendency for all things to overlap. However, one way to plot research progress or the evolution of strategies is to identify where the requirement for improved function really is. This can vary in any given decade and at any one tissue or lesion site. For example, early prosthetic hips were valuable *replacements* where no alternatives were available. Later versions have concentrated on longer life and simpler surgical fitting. These functions are now all so good that research focuses on making them easier to replace once they wear out (which they must, eventually!) or avoiding them completely with engineered tissues.

Tissue *repair* was initially revolutionized by 'simple' technologies (at least they are now) to prevent massive infection or bleeding. These improve repair by allowing the patient to survive long enough to mount a repair response at all. Approaches to 'engineer' the natural repair process have evolved subsequently to include the full spectrum of approaches, from genetic engineering of repair cells to supporting temporary scaffolds and manipulation of local growth factor levels. Out from this spectrum has emerged the idea that we can use classical engineering and bio-engineering processing to improve the final *repair tissue function*. This leaves tissue engineering as one (particularly appropriate) approach amongst a number of others.

Finally, then, we reach the pinnacle of functional restoration embedded in the idea of tissue *regeneration*. This involves producing an exact replica of the failed tissue. Consequently, its function will, by definition, be perfectly matched to the target tissue. For example, an advance/future target for loss of a patch of eyelid would be to restore it with new eyelid skin, as opposed to forearm, buttock or a generic/average skin. These would be the last generation solutions, OK for tissue replacement or repair strategies – it would, of course, be a Utopian vision at present. At its extreme, this vision is rooted in the concept that it will one day be possible to recapitulate embryological growth and development in order to 'regenerate' *perfect function*, in the way that seems to occur in some amphibia (see Chapter 1).

This illustrates the full spectrum of aspirations, ranging from the pragmatic baseline of replacing *some* function (with all its implied compromises) to the Utopian end-stop of perfect regeneration, with a near-infinite variety of fine-functional matching. That would seem to be no compromise at all – except for the time compromise! So, we may be tacitly accepting the biggest compromise of all where we have no idea *when* these ideals will be achieved.

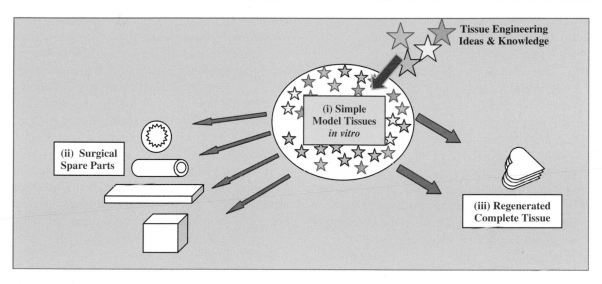

Figure 2.5 Scheme of the route map leading away from basic tissue engineering and the inter-relationship of: (i) tissue models (experimental & testing systems); (ii) spare parts (surgical repair kits); and (iii) complete tissue systems (regenerated implants).

To summarize this section, it is a characteristic of our subject that it must constantly balance between, on one hand, a vision of the promised benefits, and on the other, a sound assessment of what *is* possible. The safe-hype balance is shorthand for this tension, between the conservative drive to make tissues exactly as they are in nature and the vision of quick, low-cost tissues rolling off a production line. But there are questions that help us to balance this tension. At one end, we can ask how we *really* know that this molecule or that pattern of channels is essential for function. At the other end, we must ask ourselves how long it will realistically take to make the key process work.

2.4 Opportunities, risks and problems

The focus in this section will be on comparing the special needs and opportunities offered where we aim to produce model tissues. The corresponding analysis of engineering clinical implant tissues can be brief as it is largely the mirror image of, and the comparator for, model tissues.

2.4.1 Experimental model tissues (as distinct from spare-parts and fully regenerated tissues)

In its early days, tissue engineering was forged with a strong focus on clinical application targets (replacing specific high demand/high value tissues). As a result, the idea of preparing model (not-for-clinic) tissues was for many years overlooked, and it has only recently become a popular aim. However, the potential benefits and undoubted logic of using tissue models means that this is a growing field of activity. In effect, the very process of fabricating and assembling model tissues provides a natural spin-off of biological knowledge, as outlined in Chapter 1, Section 1.4. Consequently, while there are many sub-plots and alternative high- and low-risk clinical targets, the *en route* production of model tissues is a distinct and early target (Figure 2.5).

The special feature of **tissue models** is that they exist only *in vitro* (in our case probably 3D; culture), as simplified but strongly biomimetic forms of the tissue being modelled. We can cut rapidly to explain just what makes tissue models so attractive through two central points:

1. There is a major unmet need for 3D models for:
   (a) pharmaceutical and other screening
   (b) toxicity testing
   (c) clinical diagnosis/research.

   Simple but well-defined 3D *in vitro* model tissues have major potential uses in their own right, by providing alternatives to *in vivo* testing (i.e. they offer animal-sparing alternatives).

2. Much (though not all) of the work involved in generating 3D model tissues is essentially the same as that needed for clinical implant tissues, so they can represent a translational payback stage (i.e. a valuable output) on the way to conventional engineered tissue implants. In effect, it is essential to understand the performance, tolerances and limits of model tissues for testing, and these provide an excellent foundation for more complex clinical implant technology. Also, regulatory demands for testing systems have a far lighter (different?) touch than for clinical uses.

2.4.2 The pressing need for 3D model tissues

A well-documented and mature example is that of skin equivalent engineering. The huge and unmet demand for skin implants and grafts for patients is only too well known. But the demands on pharmaceutical, cosmetic and chemical industries for rapid, accurate, reproducible (and ideally humanized) test systems is equally pressing, though perhaps less well appreciated. In particular, under current European legislation on animal procedures, the need for alternative test models to replace or reduce animal use is becoming still more pressing, particularly in the cosmetics sector. Without new and well-characterized 3D tissue models, particularly of barrier functions, industrial progress will be hampered. What is more, where the new model tissues succeed, they can generate new industries of their own. It is a double win.

The argument against using of animal testing models has important lessons for tissue engineering, once they are understood. First, there is the obvious moral aspect of testing on live animals, especially in the numbers currently used. In order to guarantee that this, at least, involves the minimum

of suffering, governments have instituted many demanding requirements and control processes. This makes animal testing both an expensive and a time-hungry option, dramatically increasing the cost of drug research. Finally, though, it is becoming clear just how *wrong*, as well as variable, results from animal models can really be when we extrapolate them to human physiology, drug responses and diseases.

This means that engineered model tissues present particular and special opportunities. As we might expect, they are relatively inexpensive, not least because of the minimal bureaucracy involved. Their relative simplicity tends to make them more rapid and reproducible to use and easier to interpret. Beyond this, though, they have the potential to be tailored to mimic exactly the site or function we require (i.e. customization). Finally, the potential in many cases to use human cells makes it feasible to fabricate human test tissues. In other words, the small print in this particular tissue engineering contract points to the potential for an absolutely enormous radiation of highly specialized applications. This depends, however, on us (the extreme tissue engineers) actually delivering on the clauses headed 'reproducible', 'customized' and 'humanized' – in that order!

Typical targets and testing applications include:

- drug access through the skin, gut and blood-brain barrier;
- agent toxicity in the liver, lung and kidneys;
- prediction of pathological/age-related changes in the joint surfaces (cartilage) and bone;
- drug responsiveness of tumours, blood vessel wall and skin.
- testing if, or how an individual patient will respond to a drug, i.e. which version of the drug is best (customized medicine)? (Text Box 2.2)

2.4.3 Tissue models can be useful spin-offs on the way to implants

Engineered model tissues for testing do not have the baggage of having to function *in vivo* after implantation. They do not, for example, need to

Text Box 2.2 Non-animal testing

Topicality of this field can be judged from the near disappearance of animal testing in the European cosmetics industry (as of 2008/9). Paris-based L'Oreal is the largest cosmetics company in the world (the total value of the European cosmetics industry sector was estimated to be €35 billion in 2007).

L'Oreal ceased to rely on traditional animal-based testing over a decade ago and emphasised this recently by acquiring the tissue model (epithelial engineering)

company, Skin Ethic Laboratories. Along with other model epithelial cell structures, Skin Ethic manufactures a model skin, comparable to that of MatTek, around which much of the parent company bases its product development.

In yet another sector, the multinational company Unilever, better known for soaps and foods, has had over two decades of research experience in producing and applying skin equivalent models of various levels of complexity.

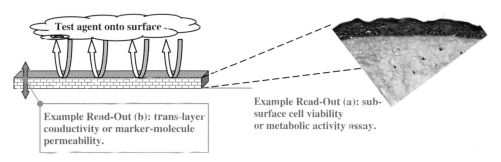

Figure 2.6 This summarises example testing strategies for one commercially available epidermal (MatTek (Inc.) EpiDerm skin-substitute: http://www.mattek.com). This consists of a multi-layer of epidermal skin cells (keratinocytes) grown in culture to form a sheet, using original technology of Reinwald and Green. The inset micrograph (right) shows the appearance of a dermal-epidermal skin 3D model with a similar epithelial layer, but grown on a compressed collagen-fibroblast layer. The spatial organisation provides the capacity for a 3D readout, in this case measuring the rates at which agents pass through the epithelial barrier to cells and matrix of the lower dermal layers. Agents can be drugs, toxins, chemicals, cosmetic components, UV or other radiation, applied specifically to the outer (upper) surface. Responses measured at deeper cell levels over a time course as a range of measures, from cell viability to specific gene expression or growth factor release and apoptosis. Penetration rates of drugs can be monitored fluorescently. In some cases it is important to confirm that a low permeability physiological epithelium has formed before estimating transcutaneous permeability. This is typically measured as a function of the transepithelial electrical resistance (TEER).

incorporate features to regulate complex integration into the recipient host site. In effect, they represent a 'lower bar' of application (i.e. achievable more easily and sooner), particularly in terms of the level of regulatory approval needed (Figure 2.6). They can be regarded commercially as the 'low hanging fruit' of tissue engineering. For example:

- Skin-equivalent models discussed already need to be sterile at the point of use (or they could not be used reliably for assay purposes). However, with professional handling, it is not as urgent as it is

for implants to show they are free from human pathogens (e.g. viruses) or modified genes.

- They are not going to be rejected or provoke an inflammatory response, so they can be made using animal products with no difficulties. These present significant regulatory and safety problems for human therapeutic implants.

- A major regulatory hurdle is that therapeutic agents must be shown, in controlled trials, to be effective in exactly the manner (and to the limits) claimed. Clearly, if a model tissue did *not* perform a useful function it would not be used, but this is not the same as facing a direct barrier to production.

2.5 Special needs for model tissues

2.5.1 Cell selection: constancy versus correctness

Sadly, there is no 'free lunch', and model tissues developed for testing purposes *do* have their own special needs and demands. These are inescapable consequences of their very specific 'function'. Function in this case is defined as: to give reliable, accurate reproducible responses which can be unambiguously interpreted in a way which is a reasonable reflection of the target tissue.

While an implant must provide a reliable benefit to the recipient patient, this will always contain a substantial degree of variability in its detail (e.g. rate of integration, strength, physical appearance). This is inevitable, not least as each recipient/patient is different from the next. The implant surgeon (e.g. plastic, orthopaedic, maxillo-facial) may well measure key patient performance indicators. This might be, say, the range of movement, pull strength and joint rotation angle after reconstruction of hand tendons. However, such measures of clinical success are commonly expressed as wide ranges of values, reflecting the spread of patient responses and injury type. In effect, it is *relative* improvement that is the key for patients. As long as the patient becomes substantially better than before surgery, the reconstruction was a success. Absolute or precise values for improving performance are hard to find and often not so appropriate.

This is not true for test-bed systems or a drug screening assay, including 3D model tissues. Here it is the norm to expect a numerical readout of the test response, expressed in absolute terms, or at least with very tight ranges, relative to a time-zero or zero concentration baseline. In short, this demands levels of reproducibility and precision which would not normally be expected of therapeutic systems. The most acute consequence of this pressure is evident

in the type of cells selected to seed such 3D model constructs. "Where do we get our cells from?" could almost be considered the tissue engineer's mantra. In this case, for model tissue and implants the answer is swung around by 180° from that for clinical implants.

Therapeutically desirable cells tend to be synthetically active, non-immunogenic (or as close to the recipient as possible) and free from pathogens such as viruses. In contrast, for building an assay or test-bed based on a 3D construct, almost *none* of these previously indispensible requirements are particularly important. Viral agents and immunogenicity are marginal factors. The use of human cells is desirable, but not really essential. Interestingly, the gold standard aim for therapeutics, using cells from one particular human (i.e. autologous) becomes a 'no-no' for screening and test models, where pooled or 'average' cell responses are a benefit. Animal cells are quite acceptable. In testing, the *central* demand is 'Reproducibility, Reproducibility, Reproducibility'. But reproducibility is not the strong point of primary cells, freshly extracted from a tissue – especially human tissue.

The key tension-balance underlying cell selection for clinical implants is around how close we can get to actually taking cells from the individual patient, economically and without causing extra harm. In complete contrast, for testing it is how much *reproducibility* we can afford to give up while still keeping the cells that are relevant to the system we want to test. This is because the most consistent, constant cells which would give the most reproducible cultures are transformed cell lines (similar to, and in some cases derived from, cancer cells). While these are, indeed, in wide use for conventional cell-based assays, they have frequently lost many of the properties of the parent native-tissue, or primary cells (Text Box 2.3).

To fall back on analogy, if the various forms and types of cell from a tissue are thought of as a cell 'family', then transformed cells would be, at best, the eccentric cousin who went to sea amid shady rumours. At worst they might be the mad uncle who has to be watched carefully on days out and has difficulty with everyday social interactions. On the plus side, you know just where you are with these cells, and their behaviours are normally very well documented. As a result, we can just look them up and work out whether the properties they retain will do the job we have in mind for them.

The pragmatic tension or compromise, then, is between tolerating these cell eccentricities and enjoying the fact that they are constant and predictable. The trick is to ensure that the eccentricities do not interfere with the main parameters under test and that the madness *really is* as constant as we think. They should perform much the same month after month, passage after passage, without aging, developing into other cell types or differentiating new, imaginative features to surprise us. In some cases, cells with tumour-like properties should ring some alarm bells where we are designing test systems with a strong spatial element – spatial organisation and attachment is often not their strong point.

However, we are at a necessary, and still useful, staging position from where progressively more sophisticated gene modified cell lines may be produced. To paraphrase an old saying; no one is likely to produce this particular 'horse' (i.e. develop the 'ideal' cell line) before the demand rises for effective 3D test systems (i.e. the 'cart' for it to pull).

2.5.2 Support matrices – can synthetics fake it?

In later sections, there will be much comparison between the benefits and drawbacks of synthetic scaffolds versus native protein cell support materials for engineering of tissues. In fact, this forms one of the defining differences between tracks towards implantable and model tissues. In brief, synthetic polymers are used to support cell growth in 3D, with the aim that the polymer slowly degrades as the cells deposit a native extracellular matrix replacement. For clinical implants, this strategy has distinct advantages and a sound, long-term rationale. In effect, the early 'tissue' made of cells and synthetic polymer scaffold (plus optional, small amounts of native matrix) is implanted with the aim of maturing, *in the body*, to become a functional tissue. Host tissue in-growth, vascularization and local factors would help the transformation, often over a period

---

Text Box 2.3   Transformed cell lines

When cells are freshly isolated from a tissue, either by disrupting the tissue or by tempting its cells to migrate out onto a culture dish, this is called a ***primary culture***. Depending on where it is grown from and how, primary cultures can contain seriously mixed (heterogeneous) populations of cells. However, this is commonly considered to be a reasonable representation of cells in the original tissue.

As primary cultures expand, they are sub-cultured and this produces a cell line – those cells which survive on plastic and divide fastest. Individual cells can be can be selected out and cloned to give more homogeneous populations, but such cell lines tend to divide rather slowly and this rate reduces continuously with time and further sub-culturing. These are 'finite' cell lines, which gradually run out of proliferative steam.

However, such cell lines can give rise to continuous or transformed cell lines where growth and cell division continues rapidly in an unregulated, undiminished manner, rather like *in vitro* tumours. This can happen spontaneously in some cells, or due to the action of viruses, radiation, transfection or chemicals. Transformed cell lines can also be derived from tumours but, although they can share some features, normal transformed and malignant cells are not necessarily the same.

Clearly, the fast division rate and consistency of transformed cells is really handy, especially for routine testing. The downside, though, is that in acquiring such happy characteristics, some can seem to become the cell version of 'bonkers'. In losing their constraints on division, they also lose some of the basic properties and behaviours which made them typical of their tissue of origin. However, enough of these are retained to be useful, and a wide range of continuous cell lines are available with well described properties, including examples of epithelial, fibroblast and neural tissue properties.

Source: Freshney, R.I. (2005). *Culture of animal cells: a manual of basic technique*. Wiley-Liss, Hoboken, NJ. (ISBN 0471453293).

of months, whereby the artificial 3D 'matrix' is replaced by one which is natural.

While this is plausible for a patient, it is not feasible for mass production of model tissues for screening pharmacological candidates.

Present synthetic polymer supports effectively fall between two adjacent hard places! First, they are not particularly biomimetic in composition, patterns of biodegradation or (often) 3D μ-structure, so they make a poor, even negative contribution to the modelling of real tissues, especially those with matrix (i.e. all connective and many cell-rich tissues). On the other hand, waiting weeks in culture for the synthetic-natural transformation to occur is not practical or economic, even if the cells used are capable of that transformation.

Consequently, engineered constructs based on the most widely used synthetic polymer supports are generally poor candidates for model test tissues. New forms of synthetic, biomimetic support materials may change this by using components which are sufficiently biomimetic, *without cell action*, to act immediately as model 3D matrices. These are the hybrid matrix types which will be analyzed in Chapter 4. This dual problem does not normally apply to support materials made from natural proteins, as these can be rapidly fabricated in forms which mimic natural extracellular matrix, from the start and without cell action.

This, then, is another version/example of the tension we must balance between building in too much and too little biomimesis, and at what stage. For tissue models, the biomimetic bar can be low – but it should be reached very quickly.

2.5.3 Tissue dimensions: when size does matter!

Another significant difference between the inherent aims of engineered tissues for clinical versus test uses is size. Certainly for the development of mass testing and screening, it is a high priority that the test constructs are small and plentiful, to satisfy the need for many tests, many replicates and small volumes. Presently this is envisaged as systems which use 12, 24 or 96 well plates, maintained in conventional culture. This is important, as it favours rapid test times and low reagent consumption, some of which can be very costly. It is fortunate, then, that many existing output measures – molecular, optical and electrical – are minimally invasive and collect data rapidly from low tissue volumes. In contrast, clinical implants commonly need to be of much larger dimensions.

On the whole, tissue defects of the size envisaged for a 96-well plate (few mm) heal themselves reasonable well. Many surgical applications need much larger grafts, in the multi-gram to kilogram range. While small constructs with μ-scale structure can be challenging to fabricate, because of the scale of structures and the range of hierarchies involved, fabricating a mass of tissue and keeping it alive and functional is a separate problem.

Again, we see that there can be a clear segregation of options which is implicit in our choice to engineer either model tissues or clinical implants.

2.6 Opportunities and sub-divisions for engineering clinical implant tissues

So, what sort of tissue-making opportunities are out there? Actually a great deal more than we might first imagine – even within the overall groupings of *models* and *implants*.

It is not really necessary to go into detail at this stage, but it is important to understand at least some of the general options and directions. In fact, new approaches and imaginative forms of target tissue-functions are still appearing as our understanding and technologies grow. Some of the general families and groupings are listed below.

1. Implantable, physiological tissue copies:
   (a) MATURE
   (b) IMMATURE/temporary, repair templates.
2. Model tissues or copies of:
   (i) NORMAL tissues
   (ii) ABNORMAL/pathological tissues
   (iii) biological PROCESSES (e.g. integration).

Within these general groupings, we can distinguish opposing categories:

- Large (>mm scale), versus small tissues (e.g. 10s–100s of μm in scale).
- **Matrix-rich** (commonly connective) tissues versus **cell-rich** tissues (organs or epithelia).
- Hard versus soft.
- Random, symmetric (non-directional) and asymmetric or anisotropic tissues.
- Vascular versus avascular tissues.
- Mechanically fixed interface tissues versus gliding interfaces.
- Biologically active versus bio-inert.
- Permeable versus barrier.
- Tissues which operate as defined mechanical units versus those which operate as metabolic units (e.g. glands producing hormones; filtration organs).
- Aphysiological tissues, i.e. copies of natural tissue, but used in new ways or unnatural locations. These include constructs which copy tissue function for non-natural reasons (e.g. controlled drug release, cell carrier devices).

2.6.1 Making physiological implants: spare parts or complete replacement?

This is a distinction between tissue targets where, on the one hand, the surgeon aims to plug in the whole functional component (like a garage might fit your car with a complete new engine) or where, in some cases, the preferred surgical approach is just to repair a key defective component using spare parts and surgical skill (back to the garage, a skilled mechanic might make your engine as good as new by replacing its pistons and valves).

The 'making a whole heart' approach mirrors conventional cadaveric transplantation logic (but without the cadaver). Spare part engineering in the same area might be seen as engineering heart valves, chordae tendineae (so called 'heart strings') or micro-vascular patches. Clearly, engineering the whole functional unit is a big call for the tissue engineer, but easier for the surgeon. Spare part tissue engineering inverts this approach, requiring simpler tissue implants but

placing greater demands on the surgeon: another tissue-engineering-tension.

2.6.2 Making pathological and aphysiological constructs: inventing new parts and new uses

Making pathological tissues sounds like a contradiction in terms until we think of the many uses of model (non-therapeutic) constructs (Text Box 2.4). In this case, once we have normal tissue mimics for measuring drugs effects (or the poisonous potential of this wonder-fertilizer or that baby shampoo), it is logical to want to make them go wrong! When we can make a model cartilage-bone tissue fibrillate and break down, perhaps we will also understand why it happens in osteoarthritis. If we can engineer a replica kidney, perhaps it will be possible to injure it and understand what causes it to fail – and perhaps to screen a candidate drug to reverse the pathology.

Aphysiological tissue targets provide us with a very different and fascinating view of tissue engineering logic. There are a number of ways of viewing this concept. One illustration builds on the idea that some therapeutically useful tissues really never existed, even though we *can* make them. For example, consider the need to improve the quality of life for spinal injury patients. The tissues and organs down-stream of the spinal lesion are, of course, initially fine (though muscle function gradually degrades), but they are no longer under the patient's control. Restoration of a few key functions can be tantalizingly close. In this case, if we were able to engineer nerve conduit tissues to guide nerve re-growth efficiently over significant distances, it might be possible to re-innervate critical muscles *below* the injury. In such patients, it is clear that the ability to cough is surprisingly important and can be a significant functional loss. Nerve redirection from above the spinal lesion to the diaphragm muscle responsible for this function would be a major benefit. No such nerve exists naturally in the human body yet, were it to be achieved, it would behave as a 'natural' nerve, but with an aphysiological anatomy.

Text Box 2.4 Examples of 3D model tissues

Engineered tissue models can also be used in *research* to define new elements of well known pathologies (or disease processes), or to identify the basic elements of normal tissue physiology (especially cell responses in systems which are otherwise too complex to dissect apart). Examples include:

(i) 3D models of articular cartilage. These include chondrocytes embedded and cultured within weak agarose gels and:
   (a) provided with growth factors to alter cell-matrix metabolism, making it possible to understand how cartilage matrix is/is not rebuilt after damage or degeneration, for example in osteoarthritis[1];
   (b) monitored under confocal microscopy while applying controlled compressive loading, to understand how cells and cell nuclei are distorted and so help explain common pathways in cartilage mechano-biology[2].
(ii) 3D models to test the mechanisms underlying hypoxia-driven angiogenesis (new sprouting and growth of blood vessels from existing capillaries towards sites of low oxygen). This involves growth of various cell types (fibroblasts, vascular smooth muscle or bone marrow stromal stem cells) in large-diameter 3D dense collagen matrices with an embedded (core) oxygen monitoring probe. This has allowed direct correlation of the dynamics of cell-induced hypoxia on the production and release of growth factors stimulating angiogenesis. Key to this is that, unlike previous *in vivo* systems, the key determining factors of O_2 consumption – diffusion transport through the matrix and delivery (via vessels) – can be modelled, calculated and correlated with actual O_2 levels[3].

Sources:

1. Jenniskens, Y. M., Koevoet, W., de Bart, A. C., Weinans, H., Jahr, H., Verhaar, J. A., DeGroot, J. & van Osch, G. J. (2006). Biochemical and functional modulation of the cartilage collagen network by IGF1, TGFbeta2 and FGF2. *Osteoarthritis and Cartilage* **14**, 1136–1146.
2. Knight, M. M., Toyoda, T., Lee, D. A. & Bader, D. L. (2006). Mechanical compression and hydrostatic pressure induce reversible changes in actin cytoskeletal organisation in chondrocytes in agarose. *Journal of Biomechanics* **39**, 1547–1551.
3. Hadjipanayi, E., Cheema, U., Mudera, V., Deng, D., Liu, W. & Brown, R. A. (2011). First implantable device for hypoxia-mediated angiogenic induction. *Journal of Controlled Release* **153**, 217–224.

By taking the same idea in another direction, we can move towards tissue engineering of controlled release devices. It is true that if we aim to make, say, pancreatic islet glands for diabetics or adrenal glands, these would, at the same time, be conventional engineered tissues. They also, incidentally, teach us a great deal about the special controlled perfusion properties needed if we aspire to make implantable controlled-release depots. Many such applications are under development, normally towards achieving prolonged or controlled rates of release of entirely unnatural drug agents or therapeutic proteins as they have never been delivered before. These, then, can be considered as forms of aphysiological construct, performing non-native jobs but using physiological mechanisms and tissue-mimetic structures.

2.6.3 Learning to use the plethora of tissue requirements as an opportunity

Having visited a few of the target threads of tissue engineering, the reader might now recognize the huge radiation of possibilities that has evolved in tissue engineering logic. As each tissue brings its own opportunities and demands, so these are multiplied by the different possible implant locations and the ways in which they can be degraded. This spreads further where our vision takes us beyond 'normal' physiology. Imagine for a moment designing a tissue engineering approach to enhance 'repair' of joint articular cartilage (which does not normally repair much at all) by controlled growth factor delivery. This would involve controlled release, chronological delivery and diffusion gradients, under controlled and incremental compression-shear loading. How different these demands are to those required for a blood vessel construct, using a bio-resorbable polymer scaffold, seeded with cells and growth factors. The neo-vessel must produce an appropriately elastic extracellular matrix and host cell in-growth over the lumen to prevent blood clot formation, while resisting peeling off under fluid shear.

This is tissue engineering. It aspires to enhance and improve on natural tissue repair, but through a huge variety of routes applied to almost any tissue, situation and failure state. The variants are almost infinite. Not to be daunted, this means that opportunities for adaptation of our available technologies, are also limitless. However, to take advantage of this huge opportunity we must accept the responsibility to be:

- selective in the extreme;
- strategically imaginative; and
- logically robust, in the tissues and targets we choose.

This will be a recurrent theme in later chapters. Interestingly, this series of traits was thin on the ground in the early stages of (not-so-extreme) tissue engineering. The joy of 20:20 hindsight allows us to understand how '*high intensity, low attention-span*' commercial support for research dictated the selection of tissue and application targets. It is now quite clear that some of the tissue applications are *not* such 'low-hanging fruit' as industry initially imagined. In view of our current learning, it will be useful to identify where the fruit was:

- not as low as we imagined,
- not a desirable fruit at all,
- or just an example of boardroom-wordsmanship.

What is now clear is that the need to solve society's greatest injuries 'in a single bound', using untried technologies from poorly connected disciplines, was a Superman quest.

2.7 Overall summary

Given the view of this treasure chest of opportunities which hindsight supplies us, we might also suspect that our problem has been one of *aiming low and still missing*. But perhaps the first target we *really* had to hit was simply to generate the motivation for the biological-, engineering-, surgical- and materials-tribes to meet up, talk and work together, so that we could properly understand the problem. This, for sure, was successful and worthwhile.

Further reading

1. Yannas, I. V. (2001). *Tissue and Organ Regeneration in Adults*, pp. 383. Springer-Verlag, New York. [Examples of engineering and regeneration of tissues worked out in great detail.]
2. Brown, R. A. (2000). Bioartificial implants: design and tissue engineering. In Elices, M. (ed.) *Structural Biological Materials: Design and Structure-Property Relationships*, pp. 107–160. Pergamon, Oxford. [Expanded examples of identification and use of the key functional parameters for target tissues.]
3. Abbott, A. (2005). Animal testing: More than a cosmetic change *Nature* **438**, 144–146.
4. Holmes, A., Shaksheff, K. & Brown, R. A. (2009). Engineering tissue alternatives to animals: applying tissue engineering to basic research and safety testing. *Regenerative Medicine* **4**, 579–92. [References 3 & 4: A view of the need and possible future tracks towards making 3D tissue models to replace animals for testing.]
5. Vemuri, M. C., Chase, L. G. & Rao, M. S. (2011). Mesenchymal stem cell assays and applications (a review). *Methods In Molecular Biology* **698**, 3–8.
6. Salim, S. & Ali, S. A. (2011). Vertebrate melanophores as potential model for drug discovery and development: a review. *Cellular and Molecular Biology Letters* **16**, 162–200. [References 5 & 6: Examples of routes for acquiring cells to make tissues for testing.]
7. Warwick, R. M. & Kearney, J. N. (2002). Safety of human tissues and cells for transplantation. In: Polak, J. M., Hench, L. L. & Kemp, P. (eds.) *Future Strategies for Tissue and Organ Replacement*, pp. 381–419. Imperial College Press, London. [Strategies for sourcing cells and tissues: the beginnings of regulation.]

This chapter is all about spatial perception and cells. Understanding another person's perceptions is tricky even when we speak the same language. Understanding how animals sense and perceive the environment is more difficult still, even where we can recognize similarities in the structures and sensing machinery. Even then, much of it is a guess. Understanding how spatial signals are processed and interpreted in single cells or cell clusters is a whole different ball game – not least at the cell-scale of 10–50 μm, where many of the basic assumptions do not apply, or work differently. Perhaps, then, we should give much more time to our concepts of how cells detect direction and movement.

As an example, we understand the spatial perceptions of the dog – his forward-facing eyes give stereo 3D images of the world he is pointing into. In contrast, the chicken seems to have two non-overlapping, independent and sideways views of the world, useful for all-round danger-warning but rather less so for precision 3D perception – we think! We struggle as much with chicken vision as with the dog's tongue-perception, because we cannot experience what they do.

Working out the basics of how cells measure asymmetry in their μ-environment is similar if we can try thinking at the cell scale and eliminating the implausible. For example, it is likely that cells rely much more on a 'tongue and whisker' type of sensing (i.e. chemical and mechanical) than the light/sound systems of our two multi-cellular chums above.

Chicken photograph reproduced with permission © iStockphoto.com/Peter Seager.

3

What Cells 'Hear' When We Say '3D'

or: How do you know you are moving when you close your eyes?

3.1 Sensing your environment in three dimensions: seeing the cues

In many European cities, the walkways and crossings now have a bewildering collection of accessories to help blind people. It is instructive to take a walk with a stick and learn how to read the information available from these. All the time, extra clues are being provided about what is coming up or how you might move between static objects (e.g. walls, edgings, barriers) or moving projectiles (avoiding cycle-ways (Rotterdam), finding gaps between cars (Barcelona) or buses (London)). These are designed to give instant, functional mechanical information (touch through a stick or vibration through sonic pedestrian crossings) about the static and dynamic space surrounding the footpath.

Although there are a few exceptions, there is far less useful and reliable information to be gained by sniffing or tasting the air. So it must be in the micrometre (μm) scale 3D physical space which cells inhabit. *Physical* measurements (e.g. of

Extreme Tissue Engineering: Concepts and Strategies for Tissue Fabrication, First Edition. Robert A. Brown.
© 2013 John Wiley & Sons, Ltd. Published 2013 by John Wiley & Sons, Ltd.

mechanical contacts or reflectance of sound or light) must provide much better information streams on the physical 3D space than (bio)chemical signals. Chemical signals are good at telling us we are near to doughnuts or cattle but less good at indicating position, size or rapid movement (shop, farmyard or delivery truck). For this analysis, we shall assume that most cells are functionally 'blind'.

In this chapter, it is useful if you can imagine what it is like to 'be' a cell within its 3D space, and so identify what types of basic spatial information it needs and whether this information is available (Text Box 3.1). In particular, the key question is how cells gather information about their immediate and neighbouring 3D environment – their location within that space.

The first obvious factor here is scale: that space is very small to us. However, while this space measures, at most, tens of microns in any direction, it operates under many of the same physical laws as our human space. The monitoring of physical signals and cues represents the main source of this type of data for any system. Since direct use of sound (i.e. significant pulse frequencies), optical or gravitational attraction by cells is considered specialist and rare, much physical information will come from monitoring of changes in the mechanical environment.

Clearly, if we aspire to control how, when and in what form cells build 3D tissue structures, it is essential to understand the mechanisms by which they 'find out' where they are and what is within their space. In other words, in which language must we communicate spatial instructions to cells?

As with any complex question, it is always good to start by describing the question in the simplest terms possible, focusing on the most dominant factors. In the case of spatial matters, orientation and material mechanics are good starting points (for example, which way is up, down, left and right and which is harder and softer?). For this, there is a particularly potent analogy with the human scale (Figures 3.1a to 3.1c).

The activity of sea travel illustrates how it is essential to adapt to the **two basic positions**: on and in. Each requires profoundly different forms of structure and environmental monitoring. The submarine (Figure 3.1a) is the sort of craft we are familiar with for travel *through* the bulk of

Text Box 3.1 What does '3D' really mean for different cell types and (why) does this matter for tissue engineering?

If cells could 'feel', then, how would they feel about 3D? In biology, we fully expect that cells can build and maintain their home tissues with exquisite fine structure in a way which is only plausible if they are able to detect complex orientations and forces within their 3D space. Because space and 3D structure (morphology) are so central to our aims in tissue engineering, this is not a bullet that it is possible for us to dodge.

Unfortunately, despite increasing recent interest, we are working from a low base and presently we have only a sketchy understanding of how cells collect, process and use physical information about their space. In contrast, we have a far more complex and sophisticated understanding of molecular control mechanisms. In this chapter we shall explore how this is made more difficult by nomenclature-logic clashes in traditional (*in vitro*) cell biology. This is visible in the very idea that 3D cell culture is a special state, when to engineers and physical scientists, cells in practice *always* operate in a 3D environment.

To hunt out some truths around this paradox, we must burrow into what is really meant by the 'physical cell space', with the aim of identifying what information different cell types *may* be able to collect and use. Important conclusions are that the answers depend on the types of cell in use. This analysis draws as little as possible on chemical/molecular signalling, primarily because many familiar pathways have already been elaborated, but also because monitoring of the physical space must be dominated by data about its physical properties. Even at the human scale, if you are wearing a blindfold and earplugs, how much help is it to sniff and taste the air for getting across the street or making a bed?

What is the use of this understanding? The hope is that by knowing which physical cues and information cells use to make and maintain their space, we will be able to design rational systems to control growth, both in the body and in culture. In a way, we are trying to learn the body language of cells.

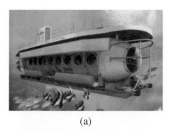

(a)

(b)

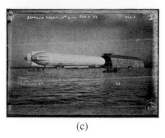

(c)

Figure 3.1 (a) Submarines are designed to work within a relatively homogeneous, single phase (fluid) environment. Clues as to where you are, especially orientation in 3D within that medium, are few, and are non-trivial to collect. © Balicruises.com. (b) In surface ships, we know so much more about our 3D spatial location from the obvious fact that we are working at an air-water fluid *interface* (we assume mariners find 'sinking' obvious). This is particularly clear where the air-liquid interactions are used for propulsion and steering. © Europeanbarging.com. (c) Once this airship takes off, it is in a similar homogeneous environment to the submarine, except the fluid is now a gas-air. Like the submariners, the airship's pilot must measure how high/low the vessel is and how close to horizontal or vertical it is. © US Government, Library of Congress.

the medium (in this case, the bulk-water phase). Interestingly, we call it 'underwater travel'. The second type is the surface ship (more familiar to us only due to the accident of our own primary habitat). This operates at the air-sea interface (Figure 3.1b).

Both the craft construction and the sort of information collected by their crews as they move around 'their 3D space' are very different. The submarine is characterized by symmetrical structure, mechanically equally robust in all planes, with streamlining/elongation in the direction of principle axis of movement. The surface vessel is highly asymmetric, with profound differences in its upper and lower shape, structure and material properties (strength). Characteristically, this leads to a great deal more variety (asymmetry) in the overall shape of interface/ship than we see with undersea vessels.

Incidentally, there is a third means of ocean travel, namely over the water phase (e.g. the airship, Figure 3.1c). Because this again involves motion through the bulk of a fluid (in this case, air) it has more structural similarity with the submarine – tubular and symmetrical. The point here is that these systems are adapted not to the water or the air fluids themselves but either (i) to being surrounded (symmetrically) by the fluid, or (ii) to operation at the (asymmetric) interfaces between different fluids or fluid-solid interfaces.[6]

The earthworm (Figure 3.2a) is clearly well adapted to life embedded in a relatively uniform bulk material (moist soil), but much less so to the stark, symmetrical contrast of life at the surface – the air-soil interface (Figure 3.2b). In contrast, the limpet is another invertebrate with the alternate space adaptation, having evolved to a tough life at the rock-sea interface of tidal zones (Figure 3.2c). Here it is harshly reminded of 'up and down' twice every day.

---

[6] It might be assumed that heavier-than-air aeroplanes represent an exception to the rule of symmetry and within-fluid travel, but this would be a mistake. Planes are a special case as they are *forced* to use interface effects (lift-generating surfaces) to overcome their weight. On the other hand, rockets use raw blast energy to dodge this, and so they are again made as tube-shapes.

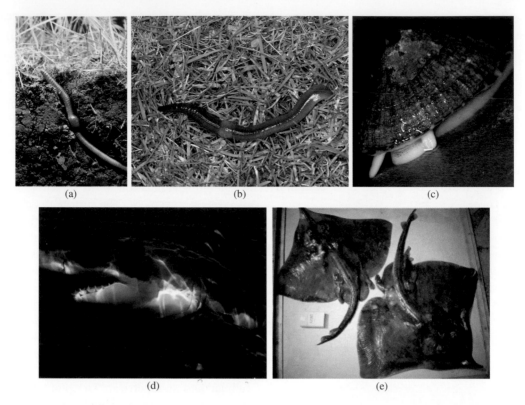

Figure 3.2 Animal adaptation to life in bulk, symmetric surroundings versus asymmetric surfaces (i.e. interfaces). Illustrated: An earthworm in its burrow (a) versus on the surface with the air (b); and (c) a limpet on a rock-sea interface surface; (d) a shark in mid-level swimming, and (e) a bottom-adapted skate. Credits: (b) © Steve Hopkin/ardea.com; (c) reproduced with permission from John Banks, Caithness Biodiversity Collection; (d) © US Government.

To bring the analogy back to vertebrates, sharks and skates are closely related but extremely adapted forms of fish. While sharks (Figure 3.2d) are well known to range freely through the bulk of ocean depths – again a largely symmetric aquatic space – their cousins the skates are adapted to an asymmetric sea-bottom life. Sharks, are characteristically round-bodied (with roughly radial symmetry but axially elongate and streamlined. Skates are flattened in structure to hug the interface. They are axially more symmetrical, with their eyes on top and mouth underneath (Figure 3.2e). Neither sharks nor skates are particularly happy in the other's space – in Darwinian language, they are poorly adapted to compete and survive there.

These examples illustrate just how important certain features of our 3D space can be – particularly its symmetry – and how the information collected from these spaces can be taken for granted.

3.2 What is this 3D cell culture thing?

There has, for some time, been a growing awareness in the biomedical and biotech worlds—accelerated by tissue engineering – that future concepts of cell physiology must take more account of the 3D cell living space. At the same time, there has been a rather superficial view that moving from '2D' to '3D' culture models will glide us gently into these new and fruitful waters. Why, then, has this apparently gentle gradient of logic given so much trouble and so little new understanding? While all we now know shouts that the basic premise of '3D' is good, could it be that the simplicity of the idea is concealing a complexity of gargantuan and shocking proportions. Only a few hardy souls seem to have begun to grapple with what is now revealed to be a multi-armed monster at anything like the level it demands. In fact, the beast lurking in this deceptive Pandora's box may have been made more forbidding because of our own naming habits, and particularly in the 'cell biology' philosophy of what 2D and 3D really mean.

It is often good practice, where we are 'surprised', to initiate a paradox hunt, since it is common for our surprise to be rooted in contradictions or definition-anomalies. In this case, there is a clue within the seemingly fatuous question of 'what do we *really* mean by 2D', as this will elicit different answers from the cell biologist and the physicist. In cell biology, the terms '2D' and 'cell monolayer' or 'cell sheet' have become synonymous. This is potentially a dangerous definition slip because, in the present world of nanotechnology and molecular-scale surfacing, there are many tribes who do not understand how such a fundamental concept as spatial dimensions can become 'cell-dependent'. Once this problem is pointed out, we can probably all agree that two-dimensionality or 2D is either:

1. (practical usage; *effective* 2D) where the dimension of a structure in its z plane is functionally insignificant relative to its two other dimensions (x and y, or length and width, i.e. for practicalities of the system, the thickness is negligible); or
2. (absolute terms) a theoretical state in which a structure has substance in the x and y planes but non in its 3rd (z) plane. To illustrate: Hawking has pointed out that a truly 2D dog would fall into quite separate top and bottom, bisected by its gut (see further reading).

Panels (a) and (b) of Figure 3.3 try to express this graphically and, although they seem at first to be drawing out the completely obvious, it is important, as we approach any paradox, to be crystal clear. A layer with large x–y surface area (i) but insignificant thickness (z) still has a finite thickness, shown in the (ii) plane. Panel (b), shows the same cross section

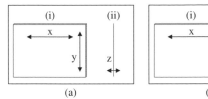

Figure 3.3 Diagram to illustrate the distinction between functional (a) and absolute 2D (b) in terms of x, y and z plane dimension. In absolute terms, 2D has absolutely *no* z but, practically, it can be just 'insignificant' – **but insignificant relative to what?**

plane (i) but with no thickness at all (ii). Although it is simple to draw (b) and to talk about it, it is quite rare to experience it (outside of theoretical physics). Indeed, it will not exist in our practical world. The point of this illustration is that it highlights the heavy burden which we must accept (in claiming '2D') to fully define and explain why z is not significant *in the system we are working with*.

Perhaps most telling – as we shall see – is the question, 'at what stage, then, does a stack of 2D layers become in reality 3D?'

3.3 Is 3D, for cells, more than a stack of 2Ds?

The idea that cells have insignificant thickness (even when seriously flattened onto a plastic culture dish) would definitely not be a happy position for even the most traditional cell biologist. They quite clearly have significant aspect ratios (length to thickness) – normally many thousands of nanometres or hundreds of molecular diameters. In addition, there are entire texts and journals dedicated to the study of trans-cellular transport of drugs and proteins across (the thickness of) single cell layers which line our organs.

It is equally well established that some of the most essential (i.e. significant) cell functions take place at scales far below that of the cell itself. These are dependent on complex membrane structures, surface and membrane receptor proteins, enzyme systems and critical molecular-scale control systems, all operating in the nano-scale – orders of magnitude smaller than the thickness of cells themselves. No, the real source of the paradox here is not that cell scientists are unaware of the *normal* meaning of 2D/3D. Rather, they have (almost unconsciously) invented *another* meaning which only has cellular significance. This now lives in a parallel cytological universe, as convenient shorthand. In fact, it more correctly relates (like the sharks and skates) to the stark difference between life in symmetric and asymmetric spaces.

'2D' is in fact being used in place of the term 'monolayer' (Figure 3.4), in a way which damages the concepts of our students. In this 2D world, cells are attached to a solid surface, though only on one surface, normally the basal or 'dorsal' surface (Figure 3.4). Cells in 3D have other 'things' all around (to which they may or may not attach).

This concept of '2D' is made more remarkable (paradoxical?), not by the fact that cells in monolayer culture *do* in reality have material on their non-attached (dorsal) surface, but that it is a clear fluid (i.e. different phase, with very different mechanical properties; more of this later). Neither is it that this form of cell growth (with all solid attachment on *only* one surface) is typical of most cells in their natural life. In fact, only a

small proportion of specialist metazoan cell types are adapted to living at fluid interfaces.

The real paradox is that these are the cells in nature which have the *least* problem in obtaining simple, accurate spatial and mechanical data about their 3D environment. Interface cells, in fact, live in a crystal-clear 3D world where the difference between 'up', 'down' and side-to-side could not be clearer or easier to monitor. It is reasonable to assume that the bulk of the animal cell universe is adapted in the struggle of detecting and interpreting far more complex and ambiguous 3D clues which hint at the changing nature of their native space. We can only guess, then, how such cells might be 'blinded' and their 3D responses completely modified when dropped onto a fluid-plastic interface, where information is stark and glaring, in the same way we imagine that bats must struggle with light-information overload when they are in full daylight (Figure 3.5).

Epithelial cells characteristically 'line' many tissues, forming the interface with other tissues or external fluids. These include the gut, bronchial and corneal epithelia, urothelial linings (bladder, urethra, etc.) and the many varieties of vascular endothelial cells. These cell types are adapted to a general positional existence where (in nature) they grip tightly to some form of (fairly) solid underlayer and to all of their adjacent neighbouring cells

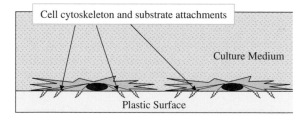

Figure 3.4 Although it is common to *call* cells grown in monolayer '2D cultures', the reality could not be more misplaced. The cell thickness (z-plane dimension) cannot be regarded as negligible in absolute terms, being around 5–10 per cent of the total cell width and more than 1,000 times larger than the thickness of essential components which hold the cells onto the plastic, such as integrin cell-substrate receptors or tethering rods of the actin cell skeleton (arrowed below). Paradoxically, though, cells grown in monolayer on planar surfaces (especially epithelial cells adapted to fluid surfaces) do have unambiguous cues available about their 3D space. These include both the presence (down) and absence (up) of attachment sites and the mechanical properties (e.g. compliance) of the two surfaces.

Figure 3.5 Bats and daylight: A well-known example of sensory confusion – deprivation by overload. Eric Isselée/Shutterstock.com.

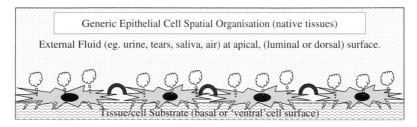

Figure 3.6 Epithelial cells (including endothelial cells lining the vascular system) are adapted to life at very special 3D locations, on surfaces and interfaces. The general pattern of 3D arrangement of these cells is remarkably consistent. They attach down to the underlying solid substrate (lower cells or extracellular matrix) by receptors (integrins) which pass through the cell membrane. At the same time such cells attach laterally to their closest neighbours within the sheet, via different receptors (frequently cadherins). In this way, they form fluid-tight sheet coverings. Again, contrasting with the integrin receptors over their basal surfaces, these cells express and segregate yet other forms of receptor on their outer or luminal surfaces (normally adapted to recognising key, soluble components in the overlying fluid). These cells define the term '*direct* environmental contact' with a fiercely bipolar adaptation, leaving little scope for confusion between up and down.

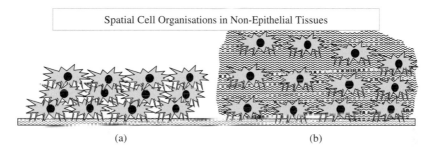

Figure 3.7 Non-epithelial cells, particularly cells of the matrix stroma, are characteristic of the bulk 3D mass of tissues. Some tissues or stages may grow to form dense cellular masses (a), where the emphasis is on cell-cell contact. Such configurations would contain little extracellular matrix material. However, where such cells are matrix-producing, connective tissue (stromal) cells such as fibroblasts, chondrocytes or bone cells, the organisation in (a) quickly gives rise to that in (b), as each cell produces more extracellular matrix material around itself, like a 3D coat. Since these matrix coats are effectively trapped in space, the neighbouring cells are progressively pushed further and further apart. This produces the typical stromal cell (matrix-rich) 3D organisation, characterized by variable cell densities within and attached to the hydrated (gel-like) extracellular matrix, whose density is inversely proportionate to that of the cells.

(Figure 3.6). At the same time, they must expose their unattached (upper or dorsal surface) to liquid or air, with all that this brings with it in terms of fluid shear forces, random passing contacts, unstable gradients and, most of all, asymmetric mechanics.

Most other cell types (e.g. those living within extracellular matrices or stroma – stromal cells) do not have this natural extreme of polarized attachment in their native tissues. They are variably connected to cells or extra-cellular matrix components of many forms, all with distinct mechanical properties (Figure 3.7). This is particularly true in the connective tissues. Not only can the nature (strength/stiffness), spatial pattern and number

(density) of these attachments be unpredictable, but they inevitably change with time as the cells and matrix move or are remodelled/reshaped. These, then, are the starkly contrasting lifestyles of epithelial and non-epithelial cells in nature.

The answer to our question (is 3D for cells more than a stack of 2Ds ?) now becomes a little easier to predict, accurately, through the use of biological and time caveats. Obviously, in absolute terms, if we make a stack of a sufficient number of sheets which are 'effectively' 2D sheets (i.e. thickness is functionally minimal), that stack will eventually become functionally 3D. But here we come to glimpse the flaw in the question that is so informative. In fact, it never *was* the simple 'thickness' which distinguished the cell biology term '2D'. It was the asymmetry of being at an interface. Consequently, the answer to our question is 'No'. As far as cells are concerned, a stack of layers still provides a cell-space which remains a *stack of interfaces*, no matter how many there are. Functional thickness was never the issue anyway, as illustrated by 'epithelial stratification'.

However, there is still that 'time-caveat'. While a stack of layers does represent a series of parallel, '2D' interfaces at set-up time (time zero), the resident cells will remodel that structure during culture or after implantation. One outcome of remodelling can be new physical attachments formed between the adjacent layers, so producing a more symmetrical 3D bulk structure.

When, where and how fast this transition occurs will depend on the resident cells in question. Epithelial cells would initially see this as a familiar series of adjacent interfaces, but non-epithelial cells will routinely attach to both available surfaces between the layers, so degrading the asymmetry. The 'time caveat', then, as so often in tissue engineering, becomes part of the process. Indeed, 'time' is where biological (cell-based) activities of our tissue engineering will have their strongest role (see Chapter 9). Cellular activities can generate huge functional diversity (i.e. the tissue detail), with all that this implies for the versatility of our clinical applications. But it needs time.

3.4 On, in and between tissues: what is it like to be a cell?

It can be intellectually risky to 'humanize' your cells (i.e. imagine that they have complex sentient attributes). It is classic to hear, even at major conferences, how cells in this culture or that system are 'happy' or 'looking' for receptors. However, it can often be very helpful, for example in understanding the basics of environmental sampling and data handling at the cell level, to think of ourselves as cells. The trick here is to imagine what basic information they would need in any particular situation.

The first cell type example, epithelial/endothelial cells, live in a thin layer (Figure 3.6). Information on the properties of their narrow 3D space (at most a few cells thick) comes to them directly in a glaringly unambiguous manner. This might be like sampling the kaleidoscope of changing sights and sounds from the sun-deck of a cruise liner. You can tell pretty well immediately which way is up, down, left and right, which way the wind-shear and sun-heat are coming from. In fact, spatial information can be monitored *directly* on a minute-by-minute time base.

At the other extreme, stromal cells have an embedded existence, deep in their surrounding extracellular matrix (Figure 3.7). They might be imagined as living deep in the midst of a 'muted cacophony' of complex mechanical hints about their spatial location. The effects of external loads (e.g. strain magnitudes and vectors) are altered, deflected and reduced as they pass through surrounding materials (the ship's hull). These surrounding, modifying materials comprise neighbouring cells and the visco-elastic extracellular matrix in which they are embedded – their extracellular matrix.

If we use the human analogy of monitoring your 3D environment from the sounds you hear, it is almost as if the stromal cells have adapted to detect the groans or creaks that would come to you deep within the hull of an old steamship. They respond to such noises and vibrations, based on what they imply is happening to the ship and their micro-environment within the ship (e.g. a storm, docking, collisions, etc.). Unfortunately, such signals would inevitably come with complex, confusing echoes and harmonics – a far cry from the clear bangs, hoots and whistles available to the surface epithelial cells living on the metaphorical top-deck.

It is possible to get an impression of this from the diagrams within Figure 3.7. Figure 3.7a illustrates a cell-rich mass (typical of growing, repairing tissues, contractile, glandular and neural tissues), with little matrix and mainly cell-cell connections. Environmental physical cues, in this instance, are extremely variable and presumably among the most difficult to interpret usefully. The mechanical properties of surrounding cells are inherently soft (or compliant).

However, there will be additional and conflicting information streams from the high and low stiffness layers, which are in other deeper planes (e.g. tough extracellular matrix sheets, bone or natural inter-tissue gliding surfaces). Detection of these planes seems to depend on movements of the cell against its surroundings using the stiff rods and contractile filaments of its cytoskeleton. Furthermore, since the cytoskeleton can and *does* change rapidly in the surrounding cells, these stiffness signals will also vary rather unpredictably in direction and time.

Again, some cell-types and cell-stages are adapted to relatively active, constant motion – for example, many stromal cells and cells within a repair site. Others, such as mature epithelial cells and stromal cells deep within healthy, adult connective tissues, are relatively sloth-like, tending to stay in one place – and for all we know, they just contemplate their happiness-factor (see above).

In many cell-types, this very motion generates its own localized mechanical feedback signals by applying tensile forces which pull and distort both the neighbouring cells and surrounding matrix, independent of external loads. These contractions are generated by the inherent cell cytoskeletal motor activity, based on actin-myosin fibres within the cell cytoplasm. To continue our humanized analogy, we might imagine the problem of gathering spatial information again from deep in the steamer's hull, but now surrounded by a cargo of assorted farm livestock. This situation can be understood a little better by studying the scale and structure of life as a cow in the old type cattle-transport ship (Figure 3.8a). In this case, cattle squeezed together below decks would pick up clues about sudden jolts as the ship was docking, or about the appearance of storms, or even the direction of large waves. However, they would be vague and confusing clues, affected by the size/position of the adjacent cows on each side and whether one or more of these neighbours started moving themselves.

However, in the case of stromal cells, this position of being packed in between lots of other squishy, moving cells (or cows) changes as they synthesize and deposit extracellular matrix proteins. This matrix provides the resident cells with a relatively

(a)

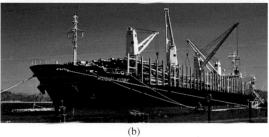

(b)

Figure 3.8 (a) Historic picture of cattle being transported by ship; loading them deep into the hold. (b) A coastal freighter loading with timber. Small animals living between the wood during the voyage will have a simpler time interpreting what is happening when the ship rolls or turns than they would if they were all packed together like the cattle in Figure 3.8a. Photo in (b) © Atlas Marine Services.

stiff, predictable support material to which they can attach, and so gather more reliable spatial information (Figure 3.8b).

This might occur, for example, during scar formation after a skin/dermal wound. At this point, the monitoring situation of resident cells will improve dramatically. This embedding material is mechanically *far* more stable than animated, moving cell bodies, both in terms of its lack of change over time and in their overall material stiffness. For any given, fixed volume of tissue, there is an inescapable inverse correlation between this increasing accumulation of extracellular matrix material and the overall cell density (expressed as cells/mm^3 of tissue). In other words, more matrix = less cells. As a result, the progression from cell-rich to matrix-rich composition brings with it both greater biological (improved space monitoring) and mechanical stability, in proportion to the reduction in cell density. Importantly, that cell-rich : matrix-rich progression is almost inevitable in connective tissue growth and repair, and also during connective tissue engineering. In our shipping analogy it is as if the steamer trades some of its cattle for timber at each port of call, progressively improving and simplifying the collection and interpretation of spatial information as it goes.

A key factor here, then, is that our considerable understanding of the different general cell types (e.g. **epithelial** cell sheets and extracellular matrix-rich **stroma**) suggests that they are adapted to living in (and gathering information from) their very different natural locations. Therefore, just as we are

comfortable with the idea that the sensitive hearing of whales may be disorientated by unnatural and random shipping sonar (Figure 3.9a), and night-flying moths are confused by electric lights, it should not be surprising that the monitoring systems of cells adapted to complex, low intensity information can be overwhelmed by excess spatial information.

To extend the previous shark analogy, bottom dwelling flatfish such as rays or plaice (Figure 3.9b) are poor at comprehending or processing visual cues, particularly those related to perspective and distance, such as complex 3D motion. This is not a little affected by the location and structure of their eyes. After all, their needs are adapted to utilize a spatially distinct sea-earth interface. They make short movements – quick bursts of shimmying,

(a)

(b)

(c)

Figure 3.9 (a) Whales may sometimes become spatially confused and beached due to the effects of additional man-made sonar information. (b) Flatfish have evolved to live *on their sides* on the sea-bottom. One of their eyes has migrated round the head so as to be on the 'top-side' and the jaw has swivelled – to look ugly. Now that is spatial adaptation to life at an interface! (c) A bird of prey coping with life in the bulk 3D. In this case, it is rapidly computing subtle movements of its landing branch, while falling forward and down against a gusting swirling air-flow; now that is complex spatial adaptation. Credits: (a) © iStockphoto.com/Alan Drummond; (b) Photo by Tim Nicholson (timnicholson@manx.net); (c) Reproduced with permission © Russ Kerr.

inconspicuous movement as close as possible to the water-sand interface.

By contrast, hawks are clearly adapted to move through and sense a spatial environment where almost everything in its surroundings moves constantly and independently (Figure 3.9c). Such bulk media-adapted animals gather and use the information they need as it becomes available in their 3D space. By analogy, cells which live *within* a relatively homogeneous bulk material must have similarly adapted systems for gathering the spatial information they need, e.g. to distinguish between stiff, impenetrable areas and crevices where movement is possible. This adaptation is likely to have a profound influence on how they interpret the 3D environment in which they operate.

Unlike the cases of hawks and flatfish, though, it is now common for many cell types to find themselves deep within spaces made by tissue engineers, where their monitoring systems are rather poorly adapted. As with any biological adaptation, any advantages the cells would normally have can become a very mixed blessing if the environment changes. However, despite obviously being unnatural, our tissue engineering culture systems can be sufficiently supportive to eliminate the most damaging aspects of the mismatches. In other words, we can say that our culture conditions 'support' cell survival/division. It is far from clear, though, that they are based on any substantive analysis of how the cells are adapted to monitor and use the 3D space.

This does not mean that our tissue engineering systems should try to perfectly mimic these adaptations – far from it. It does mean as with any good engineering design, that it is critical for us to understand the demands of that space-sensing adaptation and *match* this to the functions we hope to produce from this or that combination of cells.

The conclusion, here, is that epithelial cell types grown in monolayer culture may be as near to biomimetic heaven as it can get; comfortably monitoring and remodelling their 3D space, *next to the surface*. By epithelial standards, this monolayer is actually a good 3D culture! Similarly, if and where we can grow stromal cells, such as skin fibroblasts, deep within an appropriate, dense material, they will also find this to be a biomimetic 3D system.

However, each cell type will find the alternative environment variably 'confusing' (here we go again with that anthropomorphism) in terms of spatial and mechanical signalling. Under these circumstances, it would not be unreasonable for such cells to show extremes of behaviour, such as escape (migration), proliferation or even death (apoptosis being a form of programmed cell-suicide). These tend to be exactly the sort of undesirable, disappointing responses so often reported for tissue engineering systems. The epithelial/stromal cell example used here is simply the most clearly understood, because of the stark differences in their 3D space, but it seems certain that this will be true to some extent of any 3D engineered cell system.

It is, then, increasingly the task of serious tissue engineers to understand the nature of the cells we intend to use and the space in which they are being expected to work for us. In fact, this has been the hallmark of excellence in traditional engineering by humans for at least two centuries. So, when next you fly, do ponder how well Airbus or Boeing have analysed and engineered their systems. After all, *our* earth-adaptations do not react well to the temperatures, wind pressures and oxygen levels found when travelling near to the speed of sound on the edge of space. Yet we, as passengers, happily pull out our laptops to concentrate on that last minute report or exam revision without a thought for the proximity of the jet-stream. The air industry has engineered our environment.

To return to our original analysis, then, it is simply no longer reasonable to consider monolayer culture or culture in solid (3D) scaffolds as '2D or 3D'. It is even less relevant to label them in any given system as 'less or more physiological'. They really are neither of these. They are either appropriate, or not, to the cells under culture and the task and the bio-mimesis that we have set as our target. After all, why else would it be called tissue *engineering*?

As we shall see in the next section, this simple concept is both seriously liberating and hugely illuminating.

3.5 Different forms of cell-space: 2D, 3D, pseudo-3D and 4D cell culture

3.5.1 What has '3D' ever done for me?

Although the analysis so far might sound like harmless eccentricity, it begins to reveal why the language we use around '2D-3D' (monolayer-multi-layer or surface and embedded cultures) is at best unhelpful shorthand, and at worst the source of confusion between cell biologists and their students and collaborators in other disciplines. This, in general, is an ever-present problem in tissue engineering and regenerative medicine. Where ideas cut across disciplines, concepts can have different meanings (notably between engineers and biologists).

Cells in monolayer ('2D') culture are conveniently anchored down to a flat surface or interface through their basal surfaces. Anchorage as a single sheet also has major implications for access to the culture fluid, for supply of nutrients, oxygen and control-proteins and egress of wastes and synthetic products:

(i) maximal (subject to surface fluid mixing); and
(ii) equally accessible (i.e. same concentration) to all cells in the culture (more of this later).

Over the decades, the monolayer configuration has been used mostly to understand and test the many, many effects of soluble or exchangeable molecular agents as they pass into or out from the free, unattached upper surface of a cell sheet. Cell anchorage is convenient for cell handling and is frequently essential for division of the cells of interest. The directional attachment involves the bonding of cell membrane receptors (in this case mainly integrins) which come into contact with the culture surface. This integrin-binding more correctly links the cell membrane to other proteins which are attached, mainly by charge attraction, to the plastic culture surface. In other words, integrin receptors bind indirectly, through structural proteins – *not* directly to the plastic.

Direct cell-plastic attachment seems to be minimal in conventional cell culture. These structural matrix attachment proteins commonly derive from the serum in culture medium and, after a time, they are produced by the cells themselves. More recently, the culture surfaces have been pre-coated with such proteins (e.g. fibronectin, laminin, collagens) to avoid the lottery of how and when attachment progresses. This basal-surface-only attachment, then, is the source of perhaps the cell's most unambiguous spatial signal for 'up and down'.

Where cells attach to each other (either side to side or in forming multi-layers), this often uses another type of membrane attachment receptor, most often the cadherin receptor family. Occupation of both integrin and cadherin receptors (by matrix protein sites or cells, respectively) generates complex intracellular signalling cascades which may provide data to the cell – not only that it *is* attached, but also on the density of attachment sites, their cell surface location and substrate mechanical strength. Clearly, and by definition, cells in a monolayer will use their basal-surface, integrin-mediated attachments for simple anchorage.

At low densities, cells will have less ability or opportunity to form cell-cell (side to side) cadherin-based attachments, so basal (integrin) receptors will be the main source of signalling. As cells divide or migrate to form clusters on the flat culture surface, cell-cell contacts will become more and more common as a statistical inevitability, so changing the incoming receptor signalling data to participating cells. This shifts from being predominantly single (basal) surface, relatively stiff (protein-plastic) and integrin-mediated, to an increasing proportion of other surface (lateral), less stiff (another cell surface) and cadherin-mediated binding sites.

The fact that this shift represents an increasing ratio, or shift in *proportion*, of receptor signals makes this an inherently powerful, graded form of monitoring mechanism (Figure 3.10). As long as the cell is able to 'count' in some way, this represents a data stream indicating, with some precision, where and (in general) to what it is attached. In fact, the most plausible cell equivalent to 'counting' is likely to be their ability to react differently to different or

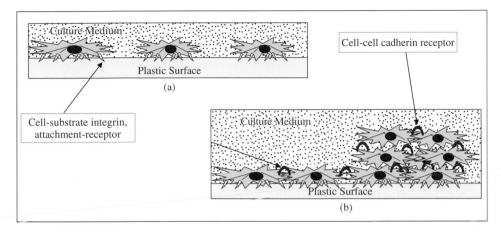

Figure 3.10 Cells have membrane attachment receptors (mainly integrins and cadherins) which pass through the cell membrane to attach, on the exterior, to either the matrix/substrate (for integrins) or adjacent cells (for cadherins). (a) At sparse cell densities cell-matrix, integrin-substrate binding predominates. (b) As cells divide or clump to higher densities cell-cell, cadherin interactions will increase in proportion and distribution. Receptor binding produces both physical links from *external* surfaces to the *internal* cell skeleton (cytoskeleton) and triggers internal cell-signalling pathways. The first of these is 'in-to-out' mechanical signalling. The second is biochemical signalling. Where clustering forms cell multi-layers ((b), right hand side), integrin attachment may be lost completely, leaving only cell-cell cadherin binding on all surfaces for some cells in the cluster (i.e. loss of direction signalling).

changing levels of chemical metabolites, generated either by receptor occupancy or clustering – i.e. as biochemical concentrations. Cells are, after all, first and foremost, sophisticated biochemical-mechano processing units.

As a result, we can glimpse possible mechanisms by which cells detect (and so react to):

(i) basal surface attachment to a planar surface, i.e. anchorage-dependent or substrate-mediated responses;

(ii) lateral edge, cell-to-cell attachments, i.e. confluence or density-dependent responses.

These responses have clear equivalents in native tissues and for certain cell-types.

The involvement of different *proportions* of receptor types (cadherin or integrin-attachment), along with distinct mechanical properties of these attachments, can be the source of information likely to push cells down one of a number of programmed responses (Figure 3.10). These responses are 'programmed' by the particular lineage and stage of differentiation (or adaptation) in which the cell finds itself when it receives the signal pattern. This cell state is, in turn, a function of its pattern of gene expression at that point in time.

We have many examples where even subtle changes to the mechanical properties of these attachments lead to dramatic shifts in both gene and protein expression by cells. This brings about visible changes in cell behaviour and so leads to the conclusion that such signalling will, itself, cause changes in cell differentiation or adaptation state. In this sequence, then, we can glimpse from the cell perspective how incoming spatial information can shift or deflect how that cell behaves. However, for any set of spatial properties, the cell reaction will be heavily dependent on cell type or stage, in just the same way that people from different walks of life will hear different messages in the same speech.

3.5.2 Introducing extracellular matrix

Let us, then, be more critical about why it is useful anyway to culture cells deep within a porous support material (i.e. '3D'), as opposed to asymmetrically on a flat surface ('2D'). The aim here is to avoid the rather weak justification (see the case above) that '3D' is 'more physiological' than '2D'.

How, then, do the cells 'see' this situation? Two key factors inevitably assume much greater, even front stage, importance as cells find themselves living deeper and deeper within a material. Paradoxically, both tend to take the form of lost or diminished signals. As we have seen, the first is a profound change of spatial mechanical signals to the cell. This is chiefly a loss of asymmetric attachment, or at least loss of major directional differences in attachment stiffness at different parts of the cell.

Cells shown diagrammatically in Figure 3.11a in a matrix-rich culture scaffold are surrounded by non-living material. This extracellular material (be it natural connective tissue matrix or synthetic polymer scaffold) will have stiff or compliant mechanical properties, which will dominate the cell space similarly in all planes. This means that, as cells develop their **internal forces** and pull on their support

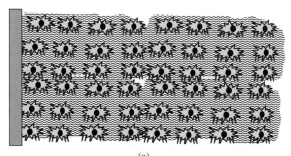

(a)

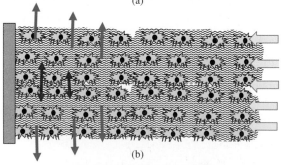

(b)

Figure 3.11 (a) Cells seeded and living throughout a material. Note: their environment is already made asymmetric by the stiff support on the left.
(b) Asymmetric external loads on cell-seeded materials generate complex strains. Note: even apparently *symmetric* loads (e.g. the tension shown in the green arrows) act asymmetrical because of the asymmetric material with its stiff left-hand edge.

material, it will resist those forces from all directions, in much the same way that water resists and moves similarly on all sides of the tail of a shark swimming through its symmetrical bulk environment. Exceptions to this relative uniformity will develop at the edges of the matrix material, where cells once again meet other 'surface' materials.

External forces, applied to the support matrix, commonly act as indicated in Figure 3.11b. The blue arrows are acting on the left hand (softer) surface of the material-and-cells construct to compress against the rigid support to its right. In this example, cells are compressed by the applied load (from the left), but there is an equal and opposite reaction from the fixed plate (right). Since the remaining faces – top and bottom – are unrestrained, this uniaxial compression will generate internal *tensile* forces acting at right angles (up and down, indicated by green arrows) as the soft construct bulges up and down.

Furthermore, where the cell-support material bulges and deforms (i.e. is strained (green arrows)), this generates shear forces around core cells (red arrows), tending to split between 'shear planes' (Figure 3.11b). It is clear, then that apparently simple external loads will affect many levels of cells in an apparently complex manner. The details of force/strain magnitude and direction (vectors) will vary dramatically, depending on exactly where in the support material each individual cell lies. This is totally different to that for cell monolayers.

We shall return to the main caveats to this model (Figure 3.11b) later – namely, its implication, from the cell perspective, that the support matrix is equally stiff/compliant in all planes and all regions. Happily, engineers and materials scientists can predict/quantify at least some of these changes, in detail, providing we can tackle the caveats.

However, this does illustrate (if we needed it) just how much more complex and *dynamic* (therefore harder to interpret) mechanical signals can be to cells-in-a-support-material than they seem to us when we 'just apply a compression'. The take-home message from this illustration should be, 'the support-material-matrix rules, OK!' If the reader has is any lingering uncertainty as to the *huge* increase in complexity, perhaps I should point out that you have deliberately been given *only* a 2D diagram. Imagine what happens when the system above is extended down into the page.

3.5.3 Diffusion and mass transport

The second critical difference felt by cells growing deep within a material is a profound change in the manner (rates) by which they can collect or get rid of nutrients, waste substances/metabolites, molecular messages and export products. Such molecular movement is also known as mass transport to and from the cells, and it is critical to cell survival and function. Clearly, even where mass transport to deep cells is rapid, there is again a loss of polarity or 'direction', relative to monolayer culture where, in one plane, these distances are negligible.

It is equally clear that the rates of mass transport will be significantly reduced by the presence of natural or artificial matrices or the presence of other cells. From the cell's standpoint, then, we can predict a loss of the rapid and highly directional molecular exchange which is inherent in surface monolayer cultures. Embedded in support materials, this will give way to an increased number and complexity (i.e. direction) of nutrient/waste *gradients*. The extent of depletion/excess of any particular molecule and the gradient direction will depend on cell position and density relative to the support material and neighbouring cells.

Before we concern ourselves with the damage benefits of these gradients on cells (a bio-tendency exaggerated by the habit of humanizing cells), we should first remind ourselves that they will primarily result in new, more complex signals. Interestingly, but far less well studied, there will be a parallel but inverse effect on the export of macromolecules, secreted by deep cells, which will be slowed or blocked by their surrounding matrix. This effect will be governed by the size of the cell products and diffusion properties (e.g. average pore diameter) of the surrounding matrix.

The factors affecting mass transport in model materials are well understood from basic physics and engineering. Key factors (Figure 3.12) are the diffusion path-length, diffusion properties of the matrix

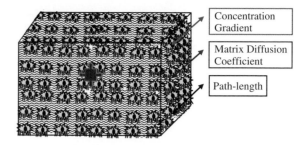

Figure 3.12 Illustration of idealized 3D tissue culture construct at t_0 indicating three of the key physical factors controlling mass transport, which will change substantially with time.

along that path and the concentration gradients of the molecules in question from one end of the path to the other (the diffusion gradient). Unfortunately, the caveat here – '*in model materials*'–dominates. We immediately introduce complexity to the basic model by seeding the materials with significant numbers of cells, distorting both diffusion gradients and properties of the material.

While the effects of cell seeding can be incorporated into defined culture systems *at time-zero* (t_0), each cell is a dynamic bio-factory. This means that the starting conditions will break down – and sooner rather than later for active systems. This key factor is often not incorporated into TE strategies. The very fact that our culture systems are designed to generate new tissue material structures makes it *inevitable* that t_0 'design' conditions for mass transport will alter. The only point in question is how fast we reach a point where we no longer understand, or even superficially control, the 3D culture. In fact, the very properties of bio-mimetic tissue structure that we aim to generate though cell action will change initial mass transport and cell activity. Consequently, many 3D culture systems tend to go out of operator control relatively rapidly and move towards cell 'autopilot'. Increasing the culture period, probably over just a few days for many cell systems, is likely to take this far from its design objectives.

This effect, then, helps to explain the strong tendency in this field for serendipitous advance as opposed to design prediction, i.e. conventional engineering. More importantly, it also makes the control of batch variation and industrial or production scale-up into major headaches (see Chapter 9). In such circumstances, poor reproducibility can only be tackled by the strictest possible technical rigour. In short, without complex, inbuilt adaptations over time, '3D' culture TE systems have an inherent tendency to go out of control – paradoxically, as a consequence of their own success.

The challenges which spring from this analysis are illustrated in Figure 3.12, representing a homogeneous material construct, seeded evenly with a homogenous cell population (a t_0 ideal in itself). Mass transport of nutrients and export proteins will rapidly produce diffusion gradients across the thickness of such constructs, proportional to (i) average cell activity/density and (ii) matrix density/diffusion coefficient. Both of these factors will, by definition, change over time, so altering the 3D cell space and feeding into the spiral of change.

Over culture periods of days (and especially of weeks), the very aim of the culture process requires that cells will:

(i) divide or die (altering cell density/distribution);
(ii) deposit or remove (i.e. remodel) the extracellular matrix differently in different directions to give anisotropic material properties (different in different planes) as they are in nature;
(iii) adapt (or differentiate?) to conditions, e.g. altering metabolic activity or behaviour patterns;
(iv) migrate within the matrix or artificial material.

All of these cell responses are *exactly* the events which are commonly assumed will occur in a culture (e.g. a 3D tissue bioreactor) as the target tissues such as skin, tendon or blood vessel begin to grow. But equally, all four will profoundly affect path-length, diffusion characteristics and key concentration gradients, not to mention matrix mechanical signals.

Perhaps the most basic law of system-design states that it is essential that your process can cope with the consequences of its own intended success. Consequently, great care is needed in setting the criteria for 3D culture success.

To sum up, then, from this analysis we can reliably predict that there will be a time-dependent increase in complexity in the resident cell 3D environment. The current challenge is that the nature of this complexity and the speed at which it forms leaves the tissue engineer unable to control the system. Next-generation tissue bioreactor science will need to wrest a greater degree of control back from the cells themselves, which is only likely with our current level of understanding, by detailed real-time monitoring of the properties of the cell space over the culture period. This will allow dynamic feedback-control to be fed into the 3D culture conditions.

This analysis indicates that long term culture systems will prove disproportionately problematic (even aside from economic constraints). As a result, it is clear that minimal culture duration needs to be a first-order design target, far higher than it is at present. Also, to minimize the rate of change of complexity and so simplify the monitoring, it is important to avoid cells at a 'volatile' stage of their development or adaptation (i.e. prone to undergo rapid shifts in behaviour). Such cell types will generate rapid changes in 3D space, amplifying system uncertainty. This suggests, for example, that stem or progenitor cells are best used in 3D cultures only after pre-processing to a more stable, committed state. Good practice would aim to expand and differentiate cells *prior* to seeding them into a bio-mimetic 3D space. In effect, this is the logic of segregating critical process stages to avoid compromise conditions.

3.5.4 Oxygen mass transport and gradients in 3D engineered tissues: scaling Mount Doom

Ask any cell biologist (non-3D specialist) what they think is the big deal (and big threat) about having cells in 3D and the answer is likely to come in two parts. Firstly, with positive enthusiasm, " . . . it's so much more '*physiological*'!" – but then, with an increasingly clouded face, they check themselves with the assertion, " . . . but all the cells in the middle will die". Put the same question to an engineer and (s)he will want to know the path length, diffusion coefficients and cell density/consumption rates.

In cell biology, there is a deeply engrained dogma (which seems not to need qualification or calculation) that cells in 3D, deeper than a few layers (or variously 200–500 μm) deep to an O_2 source, are *doomed*! We can trace its origins back to the early days of angiogenesis research, and tumour cell proximity to the nearest blood vessels.

This figure and concept of 3D cell life may well be realistic for tumours, liver, muscle and kidney, where cells in 3D are packed together into a cell-rich mass (see next section). However, it is certainly worth revisiting how valid this view is where cells are living in a 3D fibre-and-water-rich mesh – that is, your average connective tissue. In such matrix-rich tissues, the dense 3D mass of cells is dispersed throughout large volumes of watery material which is highly permeable to oxygen and other small nutrient molecules. Incidentally, this difference is exaggerated, since these cells (often fibroblast types) tend to be slow consumers and often have low cell activities relative to those of cell-rich tissues. All of this adds up to a completely different calculation and makes 3D living, in matrix-rich structures, a completely different proposition – and definitely *not* fingernail-dangling over the edge of Mount Doom!

Figure 3.13 shows what happens if we actually measure the real levels of oxygen at the core of a dense, collagen-engineered tissue filled with cells. In other words, what do we see where we monitor real oxygen transport across a living (if simple and defined) 'tissue'?

First, we find that even for relatively dense, tissue-like collagens (a gel of $\approx$12 per cent w/v in this example), *full* passive re-equilibration of

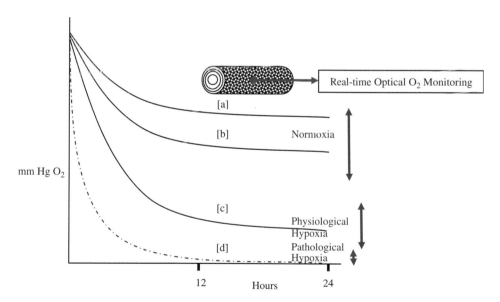

Figure 3.13 Real time plot of O_2 tension in the core of a (22 × 2.2 mm) rolled solid rod of dense, cell-seeded collagen (inset diagram, showing the indwelling fibre optic oxygen probe.) Lines (top to bottom) [a], [b] and [c] represent the fall-off of O_2 over time in culture with low, medium and high densities of dermal fibroblasts (respectively). Note: at the high cell density ($\approx$20 million cells/ml of tissue), cells are densely packed and difficult to see through by microscopy. Line [d] represents the response for pulmonary smooth muscle cells, at the same density as in [c]. Even after 24 hours, the highest fibroblast density had only fallen to levels of physiological hypoxia (this is the same as we find in normal tissues). Only the smooth muscle cells reached pathological hypoxia. Adapted from: Cheema, U., Brown, R.A., Alp, B. & MacRobert, A.J. (2008). Spatially defined oxygen gradients and vascular endothelial growth factor expression in an engineered 3D cell model. *Cellular and Molecular Life Sciences* **65**, 177–186; and Cheema, U., Hadjipanayi, E., Tammi, N., Alp, B., Mudera, V. & Brown, R.A. (2009). Identification of key factors in deep O_2 cell perfusion for vascular tissue engineering. *International Journal of Artificial Organs* **32**, 318–328.

oxygen occurred across a 1.1 mm diffusion distance in only 20–30 minutes (in other words, if the core is depleted to zero O_2 and the surface (path-length = 1.1 mm) placed at 120 mm Hg O_2, (cell incubator levels), the core becomes fully oxygenated in 20–30 minutes). This means that the apparently 'tissue density' collagen material is not the diffusion barrier to O_2 that it is often made out to be.

The second surprise from Figure 3.13 is that the core cells do not die, at least in the short- to mid-term. There is no detectable excess core-fibroblast death (i.e. over and above that at the surface) after 24 hours culture, even at extremely high cell densities (plot [c]). Indeed, core cell death was only just around 20 per cent between days 3–7, at which stage cell division and other factors complicate the interpretation of cumulative cell viability. However, in taking a closer look at the data, we should perhaps be less surprised by this, as the minimum level reached for core oxygen (plot [c]) is well within normal tissue levels (paradoxically known as 'physiological hypoxia')*.

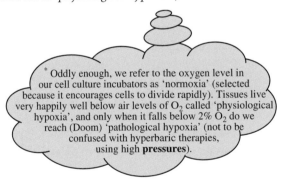

* Oddly enough, we refer to the oxygen level in our cell culture incubators as 'normoxia' (selected because it encourages cells to divide rapidly). Tissues live very happily well below air levels of O_2 called 'physiological hypoxia', and only when it falls below 2% O_2 do we reach (Doom) 'pathological hypoxia' (not to be confused with hyperbaric therapies, using high **pressures**).

Some of this effect is down to the cell type, as fibroblasts seem to be good at holding their collective breaths. For example, smooth muscle cells are much more energetic, and they behave as shown in Figure 3.13 by plot [d]. At the same cell density, core O_2 drops (like a stone) straight down to pathological hypoxic levels. Also, more of these cells do die, though still only gradually over the following days in culture.

Incidentally, bone marrow stromal stem cells behave in much the same way as the smooth muscle cells, though again only when cell densities are high. All three cell types have plenty of time to steadily

up-regulate their expression of angiogenic factors, as they would *in vivo*, to attract new vessels in growth (we discuss this – and stem cells – as a means to produce angiogenic depots in later chapters).

There is a very big 'however', here. It should be clear that we are no longer discussing the consequences of transport in a dense collagen material, acting as a barrier to oxygen diffusion (mass transport). Rather, we are balancing the consequences of O_2 *consumption* by other cells in the construct (i.e. each cell 'layer' between the surface and the core extracts what it can from what passes by). Even the great Colorado River was reduced to a steam at its outflow as a result of water withdrawal by thousands of farms and industries (e.g. Las Vegas) along its 2400 km track.

Now we can see why our great biological obsession with 'oxygen diffusion and 3D' can be a flawed overreaction. In native tissue matrices it is more likely to be a consumption, not a transport effect. Hence, cell density (cell-rich/matrix-rich) and cell type are critical. Neither is this a piece of academic hair-splitting. Put simply, if mass transport is generally as good as it is likely to get, then heroic efforts to improve it are going to make only a small difference to the 3D cell experience. On the other hand, applying our thinking to the **consumption** side of the equation can have a much greater impact. For example, what *are* all those cells needed for? Do they all have to work *so* hard, in such a dense mass?

So, are we getting unnecessarily worried about cells in 3D culture by fears of hypoxia and cell death? More than likely, the answer is *yes*, if we do not *measure* and *understand* the system we are using (system = material-density; path-lengths; cell type and density). This is made starkly obvious by the *final* part of the Figure 3.13 story. This oxygen transport-consumption example was measured in response to the 'obvious problem' that core fibroblasts were more than 1 mm from the surface, and so were '*sure to die*'.

As we can see, they did not die (and, in fact, they thrived for five weeks at the core!), because they did not consume enough oxygen to deplete below tissue levels. However, by that time a system had been developed for inserting many fine channels (of blood capillary diameter) through the length of the collagen rod. Even though these contain only static culture fluid, not flowing blood, the core O_2 levels were pushed back up to those of normoxia. The 'non-problem' had been 'over-fixed'. In effect, the channels just minimize the number of consuming cells acting on any one region (the core) by providing a tube full of undepleted medium *with no cells in it*.

The take-home moral of this fable is that in biological tissues (and good mimics), it is cell *consumption* of oxygen and nutrients that we should consider (and even worry about) as the potential wolf of the 3D cell experience. Certainly, we are unwise to allocate 'bad-guy' status to imaginary diffusion 'barriers' until the transport characteristics of our 3D system have been *measured*. This is a major Extreme Tissue Engineering lesson because, figuratively, it saves us from either taking long detours out of our way or investing in serious breathing and climbing gear, when 'Mount Doom' is really just a 300 metre high family ramble trail*.

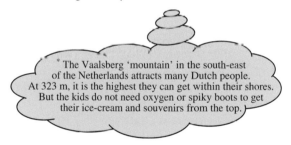

*The Vaalsberg 'mountain' in the south-east of the Netherlands attracts many Dutch people. At 323 m, it is the highest they can get within their shores. But the kids do not need oxygen or spiky boots to get their ice-cream and souvenirs from the top.

3.6 Matrix-rich, cell-rich and pseudo-3D cell cultures

In considering the theme of this chapter, the cell 'experience' in 3D culture, it is instructive to lift the lid on another key assumption which was embedded in the preceding discussion. This is the matter of biomaterials design and cell perfusion as they link to 3D space (see Chapter 5). A largely unintended consequence has emerged from the development of 3D cell support biomaterials/scaffolds with multi-micron diameter pores (i.e. greater than cell diameter).

The need for large pore scaffolds was a linchpin of early attempts to engineer tissues. It was based on the perceived need (a) for cells to be seeded deep into the material and (b) for 'rapid nutrient exchange' with the culture medium – i.e. mass transport). To maximize the direct contact of resident cells with culture medium, this class of scaffold materials was also designed with high interconnectivity between the pores. This structure means that the pores and cell-supporting struts which form the '3D culture surface' are many times larger than the cells they carry. The spatial relationship between attached cells and the walls of such porous materials, as opposed to conventional flat-plastic culture plates, is illustrated diagrammatically in Figures 3.14a and 3.14b.

However, this normally means that attached cells will experience an environment closer to that of monolayer ('2D') culture. Such cells at low, typical seeding densities will be anchored only at one (ventral) surface, as they are in monolayer culture (Figure 3.14a), with all that this implies. In effect, cells cannot easily distinguish between flat and slightly curved surfaces, which effectively eliminates the key distinguishing feature of 3D culture (Figure 3.14b), namely the shift from a strongly asymmetrical ('2D') to a more symmetrical cell space. As a result, systems where cells are seeded into large-pore materials have been termed 'pseudo-3D cultures'. In reality, this probably gives spatial cues to cells which are closer to conventional monolayer culture.

This situation will not change until cell proliferation begins to increase the cell density. Cell density in the pores increases to the point where lack of surface area forces cells to grow as multi-layers. This is where a new form of 3D culture gradually and locally develops (Figure 3.14c). Similar structures can also be seen in types of cells where they reach super-confluent densities and go from mono- to multi-layers.

Unfortunately, a central design feature of many porous scaffolds is a huge surface area for cell growth. This requires much cell division for complete coating, and there is no good reason why this cell saturation should occur evenly throughout. Hence, transitions from pseudo-3D configuration to true, cell-rich '3D' cultures will tend to be slow, local and ill-defined. Indeed, such culture systems

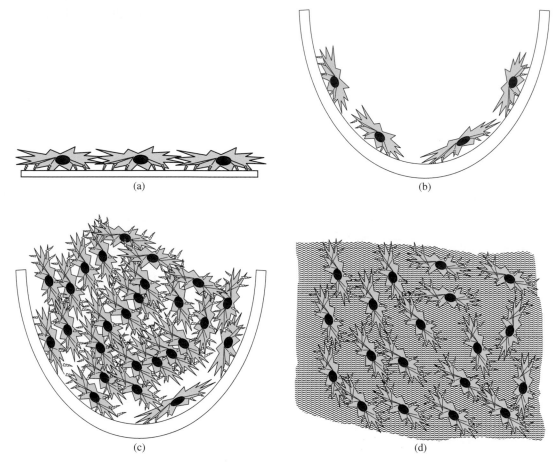

Figure 3.14 (a) Classic monolayer or '2D' culture of cells on a flat culture surface: basal attachment and cell skeleton (red). (b) 'Pseudo-3D' cells attached (basal surface only) onto the non-flat surface of large pores in a deep biomaterial scaffold. (c) Cell division in pseudo-3D culture eventually produces cell-rich true 3D culture within the large-pore ($>100\,\mu$m) biomaterial. Cell interaction is predominantly cell-cell, tending to cell-matrix with further culture time required for matrix deposition. (d) 3D matrix-rich culture without the pseudo-3D stage, in which cells attach mainly to matrix on all surfaces from the start (t_0); e.g. collagen gels, see Chapter 4.

would be *expected* to generate big differences in cell growth in deep and superficial zones, as we shall see later (see Chapters 5 and 8), forming unintended tissue layers.

By definition, the point at which such an indirect spatial transition occurs will be dependent on culture time, cell density, average pore surface area and cell type. Cell type, in particular, is critical here. It seems reasonable, then, to be cautious in suggesting that such configurations provide spatial cell cues resembling native tissues, as resident cells will only *gradually* receive 3D signals from their matrix and neighbours. Even then, this configuration is

predicted to develop at different rates in local patches throughout the substrate.

As discussed in the previous section, such cell masses tend to be characterized by cell-cell adhesions (dependent on cadherin membrane receptors). Only when cells within this mass start to secrete and accumulate a collagen-rich, extracellular matrix will this construct begin to resemble a native connective tissue. As shown also in Figure 3.10, spatial cues in connective tissues are dominated by symmetric interaction with protein fibres, i.e. with predominantly cell-matrix adhesion receptors (integrins), with all that this implies for cell signalling (Figure 3.14d).

Understanding the basis of this spatial transition, from pseudo-3D to cell-rich and eventually to matrix-rich 3D culture, is critical in the field of cell-scale spatial cuing. Its key contribution is in highlighting the difference between growing cells *on* large pore synthetic support materials, as opposed to *within* nano-fibrous meshes. The latter provides spatial and receptor-based cues which immediately resemble those of native connective tissues (Figure 3.14d). The former systems more closely resemble high-surface-area monolayer cultures in the first instance, giving way gradually to matrix-rich 3D system in a highly cell-dependent manner (Figures 3.14a–c).

3.7 4D cultures – or cultures with a 4th dimension?

It should be clear, from the many references so far to 'cell adaptation', that this is a significant factor in considering spatial cues in culture. Unfortunately, cell adaptation represents one of the most complex and least understood aspects of cell biology. Particularly in recent years, there have emerged new tracts of biological understanding which have revealed the extensive levels of cell plasticity possible both *in vivo* and in culture.

Some of these are better described than others, such as the dramatic shift towards fibroblastic synthetic characteristics when chondrocytes (cartilage cells) are grown for long periods in monolayer culture. In this classical example of adaptation to culture conditions, chondrocytes, in monolayer for one or two passages, shift from synthesizing cartilage-specific to skin/tendon collagen types (types II to I), with reduced proteoglycan synthesis. This can be prevented by growing in suspension or agarose gel culture, or by using a micro-mass culture with ultra-high chondrocyte seeding densities to give cell-rich 3D conditions. This chondrocyte to fibroblast shift is at least partly reversible. Cell phenotypic shifts such as this are often sensitive to poorly understood environmental cues, making

them major problems rather than opportunities for tissue engineering processes.

Unlike the characteristics of differentiated cell types present at time-zero, such adaptations – by definition – take time to develop. This emphasizes the role of 'time', or culture period, as a key factor in spatial control systems – and so we reach the place of 4D culture systems.

As we have discussed already (Figure 3.14), systems set up at t_0 as monolayer ('2D') or pseudo-3D cultures will gradually develop characteristics of true 3D with the loss of basal-only attachment. With more time still, the right cells, under the right conditions, may go on to deposit significant amounts of intercellular collagen, mimicking connective tissue 3D biology. This is the basis of a whole branch of (scaffold-free) engineered connective tissues.

However, even this short description of the potential difficulties of *cell adaptive changes* hints at the potential for control problems, where we cannot control or understand this aspect over time. In general, then, the tendency is for this time-dependence to introduce non-ideal conditions unless the dynamics, rate constants and efficiencies of such adaptations can be worked out. This certainly tends to be the case for systems using highly plastic stem/progenitor cells in 3D bioreactors, and where the aim is to deposit lots of extracellular matrix over extended time periods.

We can illustrate just how uncertain this effect can be, and thus just how much undesirable variability it can generate, by expanding on our previous 3D culture-progression (Figure 3.15). There is absolutely no good reason we should expect that any two or three cell types will progress *at the same rate* from pseudo-3D to cell-rich 3D (Figure 3.14a,b). The same is true for the next transition, from cell-rich to matrix-rich 3D culture – (b) to (c). The trouble is, these transitions are governed by two very different cell behaviours.

The driver for cell-rich 3D cultures (notional function 1 in Figure 3.15) is clearly cell density, so determined by proliferation and cell-death rates. But transition to matrix-rich 3D culture must be dependent on cell deposition of a collagen extracellular matrix – function 2. In other words, two

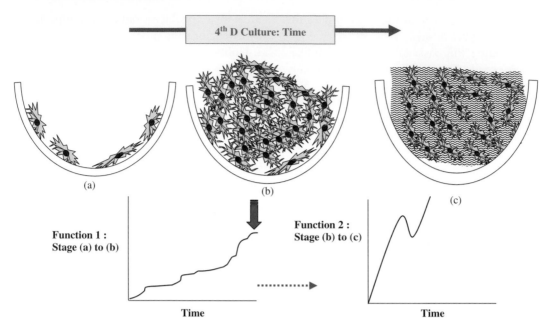

Figure 3.15 Example of '4D' culture, with transition stages, arrow (a to b and b to c), that trigger the start of very different but important functions (in this example, function 1 = cell division; function 2 = extracellular matrix deposition).

consequential processes are needed to give the 3D tissue structure: first, cell division; second, synthesis/deposition of a bulk matrix. This combination needs to be well understood and closely controlled, in order to prevent the system generating spatially variable heterogeneities. In other words, unexpected patchy tissue structure can be due to faster proliferation or matrix deposition rates in some, rather than other, scaffold pores. Such local rate differences, over just a few hundred μm in radius, will be almost inevitable in the absence of fine-scale control of mass transfer and of the 3D support material structure, as discussed above. Without this, increasing culture time will generate structural randomness in the final construct.

Uncontrolled variability of structure and composition of the newly deposited tissue has the potential to cause serious disappointments in the dream of fabricating tissues, and it leaves us with a less than ideal target. We are now left hoping only that constructs will be predictably 'average' in performance, over wider spatial volumes. This implies that some

zones of our constructs will perform differently from others, in a manner beyond our control.

It is very doubtful if this remarkably modest (and diluted) aim was even in the finest of fine print of the tissue engineering dream contract we bought into. Certainly, it would be hard to believe that the prospect of making 'metabolically active, average-function-tissue-blobs' would have set alight 1990s biotech as it did!

Given the nature of the troublesome fine print we have been looking at here, it is tempting to wonder if the original thinking did not go this far because it was hoped that '*the cells*' would take care of their own 3D organisation and micro-environment. The current hunt for 'special (stem) cells' then looks suspiciously like the same aversion to tackling the big problems which were always in the small print anyway. In fact, maybe it was not even a 'small print' issue. Perhaps we just missed the *neon-sign-obvious* part of the contract – the word 'engineering'. This is an idea we return to repeatedly on the track of extreme tissue engineering.

The take-home conclusion from this example is that it is critical to identify, analyze and minimize the consequences of 4D culture on the spatial behaviour of cells. At the moment, this means either:

(i) minimizing the culture time; or
(ii) reducing the tendency for our cells to adapt to the spatial cues provided.

This situation improves in relation to our ability to design, engineer and monitor tight control of the cell space (at the 100 µm scale) over time (see Chapter 9). It seems unlikely that it will be possible to eliminate cell-matrix adaptation from many of the biological systems we design. This makes it all the more important to mitigate their uncertainties.

Time-based monitoring of cultures, then, becomes a key requirement. This suggests that the appearance of functional stages in the culture will need to be monitored *in real time*, with corrective control steps as required. The time dimension of the 3D culture hinges on the use of minimally invasive, real-time monitoring of culture progression in a way which is far less critical than for '2D' cultures.

It should now be clear to the reader that time/sequence (culture period) is at least as important as more traditional parameters in controlling 3D space. **Real-time monitoring is, therefore, a major future requirement**.

3.8 Building our own personal understanding of cell position in its 3D space

The main take-home message of this chapter is that, for 3D cultures, it is critical to understand the cell space from a cell perspective. To do this effectively, it is essential to get right down (mentally at least) to the size of individual cells to understand just what that space consists of. Hand in hand with this is the need to understand which types of information are most important to each cell type (i.e. for any given tissue engineering application). In human terms, this translates to 'what *really* matters' to your circumstances and needs? These two principles work as an inseparable pair and are best applied together in constructing hypotheses to explain how cells will grow in any system, or in designing culture system modifications.

Again, let us try this dangerous trick of humanising cells. Take a look at Figure 3.16. This is a photograph of commuters travelling on a crowded London underground, the Tube, during the rush hour. If you *can* ride the Tube for real, do have a go; if not, then use your imagination. First, close your eyes and then stand, holding onto a secure position, and take in all the information that you can from the surroundings. Work out *how you know* what the train is doing as it moves. As you travel and get thrown about by the train, try to identify where you get your spatial information from. After a while, ask yourself how can you tell which is up and down, left and right? Is the train moving, or at a station, accelerating, slowing down or going round a bend?

It should become clear that many of your reference points come from your basic body asymmetry (see Chapter 6), coupled with an intimate knowledge of that asymmetry. You are standing upright, balanced against gravity, so 'up' and 'down' come from any tendencies you have to fall over against

Figure 3.16 Photograph of commuters travelling on the London Underground (the Tube) in rush hour. First, close your eyes and work out where you get your spatial information from? Second, what is important about that information to *you*? Reproduced with permission © Ryo Hirosawa.

the surfaces you are touching. These might be other passengers, seats, support poles or doors. More actively, these may be the uprights, straps and handles that you have selected as your full-time anchorage points (i.e. you are 'holding on tightly' as the guard recommends!). Similarly, you have a well-developed understanding of your left-right asymmetries. Again, as you lean one way or the other or start to pull harder on one of your anchor points, you can immediately calculate which way you are falling.

Once the train moves off and gathers speed between stations, it lurches and sways, turns and brakes, and these movements will throw you against soft or hard supports. Once again, you will lean on your anchor points or (by now) your adjacent fellow passengers. This simultaneously provides you with both for support (minimizing the embarrassment factor) and as sources of information. By deduction, you almost instantly compute that you are being moved by acceleration/deceleration and changes in direction/momentum, which you can relate to the movements of the train almost without thinking.

However, if you are to be able to use these as reliable sources of information, you must *already* know (or more dangerously, have made assumptions about) the mechanical properties of the structures now support you. This you might consciously consider as 'their ability' to resist your movements. For example, a lurch to the left may push you into contact with an apparently compliant, retreating surface. But is this a handrail, moving away from you *even faster* than you are leaning towards it, or is it just a shy passenger? The data ambiguity in this case is particularly stark as it could be suggesting two equally plausible interpretations; both dangerous in very different ways:

(i) The train is turning over: panic now? ... or ...
(ii) Careful, are you ready for a serious socio-legal incident?

Each requires more information (and quickly) to clarify the ambiguity. Similarly; is that a leather hand-strap you just grabbed for, or another passenger's hat? To compute the most likely meaning

of these ambiguities, you will need a clear understanding of the material properties of these 3D support structures – whether they are standard anchor points or parts of fellow passengers' anatomy.

Secondly, to understand the various ways that different people interpret this stream of positional information (i.e. 'individual sensitivities'), it is also important to understand how you routinely filter the significance of the data. We can see this as comparable with the adaptations of different cell-types. For example, people unfamiliar with big city habits – those who are comfortable with more personal distance, such as older people or those who are just vulnerable souls and fear muggers and nasty infections – will treat this monitoring system quiet differently from hardened city commuters and New Year revellers. Just like cells, then, we are *adapted*, and this makes a major difference to the assumptions through which we interpret the spatial monitoring data we collect.

The translation of these human-centric observations to understand how cells *might* use cues about space and position is clear, though we should be cautious about over-extrapolation. For example, it seems unlikely that gravity is a major *direct* cue to most cells, as their own mass is so small. However, this will also tend to be true for our commuter who is packed tightly between other passengers. We can modify our analogy here to imagine what would happen if we lost the gravity-cue, perhaps under the sardine-packing conditions of the Tokyo underground. In this case, gravity stops being too important as our weight is supported by the mass of passengers squashed between the carriage walls. However, cells (and commuters) can compensate for this by using (monitoring) the change of momentum, during sharp turns or accelerations, of the whole mass of passengers to which they are attached. For the commuter, this is a people-mass; for cells, it might be the large mass of extracellular matrix (e.g. dermis) into which they are meshed. In either case, the inertia of the relatively large masses involved can provide detectable strains on individual commuters or cells, indicating how the whole train is moving.

3.9 Conclusion

Different forms of cells in culture inevitably collect and use cues from their 3D space in very different ways. The major differences in spatial behaviour between epithelial and stromal cells (e.g. skin keratinocytes and dermal fibroblasts) form an unambiguous example of this.

However, we must expect to find a spectrum of smaller, less obvious or transient adaptations to spatial monitoring in the many different cell types employed to engineer tissues.

Further reading

1. Hawking, S. W. (1988). The Unification of Physics. In: *A Brief History of Time*, pp. 155–169. Transworld Ltd, London.
   [Check-out the 'Hawking' (basic physics) spin on what 2D means to biology–demonstrated using a dog.]

2. McGrouther, D. A. & Higgins, P. O. (1997). (CD-ROM) Primal Pictures Ltd, London.
   [Visual storm of images and explanations of the Dynamic, 3D structure of this limb: interactive CD, surgeon's-eye view illustration of our spatial challenge.]

3. Hay, F. D. (2010). EMT (epithelial-mesenchymal transition) Concept and Examples from the Vertebrate Embryo. In: Savagner, S. (ed.) *Rise and Fall of Epithelial Phenotype*, pp. 1–11. Landes Bioscience, Georgetown, TX.
   [Embryologist's view of epithelial sheets and their relationship with stroma – from the veteran leader of the field.]

4. Yamada, K. M., Pankov, R. & Cukierman, E. (2003). Dimensions and dynamics in integrin function (a review). *Brazilian Journal of Medical and Biological Research* **36**, 959–966.
   [Understanding what holds cells down to their matrix, integrins, from a prolific authority of the cell-matrix domain.]

5. Sands, R. W. & Mooney, D. J. (2007). Polymers to direct cell fate by controlling the microenvironment. *Current Opinion In Biotechnology* **18**, 448–453.

6. Yannas, I. V., Tzeranis, D. S., Harley, B. A. & So, P. T. (2010). Biologically active collagen-based scaffolds: advances in processing and characterization. *Philosophical Transactions, Series A, Mathematical, Physical, and Engineering Sciences* **368**, 2123–2139.
   [References 5 & 6: Reviewing trends in polymer scaffolds and natural protein sponges to mimic 3D culture for cells.]

7. Reddi, A. H. (2000). Morphogenesis and tissue engineering of bone and cartilage: inductive signals, stem cells, and biomimetic biomaterials. *Tissue Engineering* **6**, 351–359.
   [The case for deriving 3D signalling from natural embryo development: developmental biologist's strategies for engineering tissues.]

8. Novosel, E. C., Kleinhans, C. & Kluger, P. F. (2011). Vascularisation is the key challenge in tissue engineering. *Advanced Drug Delivery Reviews* **63**, 300–311.
   [Case for the importance of vascularisation: but is it tackling Mount Doom or a measured afternoon ascent?]

9. Hadjipanyi, F., Mudera, V., Deng, D., Liu, W., Cheema, U. & Brown, R. A. (2011). First implantable device for hypoxia-induced angiogenic engineering. *Journal of Controlled Release* **153**, 217–224.
   [Paper explaining the practical use of cell hypoxia in 3D culture to engineer accelerated blood vessel integration.]

Biodegradable scaffold – also known as 'Builder's Bamboo'. This scaffolding, erected in Shanghai for the repair of low-rise buildings, can be regarded as 'biodegradable' for civil engineering. We expect it to gradually degrade over time through environmental effects. Inset: bamboo scaffold, ready to go.

4 Making Support-Scaffolds Containing Living Cells

Bulk material compositions for holding cells naturally

4.1 Two in one: maintaining a synergy means keeping a good duet together

Those readers with previous experience of tissue engineering might notice something a little out of the ordinary with the scope of this chapter. There is a strong thread in the field which likes to deal with the 3D biomaterials (scaffolds) as one distinct issue. The acquisition, processing and expansion of the various possible cells which are likely to be used is then introduced as a quite separate subject. At first glance, this seems to be a pretty reasonable structure, based on a mix of logic, habit and expedience. After all, these are two very different disciplines. Also, when we are faced with major problems, it can be good to separate the component tasks, often into the main areas of expertise. However, this must still be balanced by the core tissue engineering need for joined-up, collaborative thinking.

As an example, we can be pretty sure that the value of joined-up collaboration is well known to the world's largest shipbuilders, Hyundai Heavy Industries of South Korea. When they start on a new mega-ship design, it is certain that the people

Extreme Tissue Engineering: Concepts and Strategies for Tissue Fabrication, First Edition. Robert A. Brown.
© 2013 John Wiley & Sons, Ltd. Published 2013 by John Wiley & Sons, Ltd.

running the flow simulation tanks, complex as their task is, are in close contact with their colleagues who design the internal space and accommodation. Furthermore, neither would feel comfortable about spending much time on a design without consulting the marine propulsion (engine) section. After all, no one gets paid until the ship (i.e. *the functional unit*) can carry what the customer wants, where and how he wants it carried. For example, cargos as diverse as 0.5 million boxed Barbie dolls, 3,000 paying passengers, two roads worth of course gravel or a kilometre-long tank of Saudi Arabian crude oil all need to be carried where and when they are required.

Indeed, shipping and tissue engineering could easily share a motto: 'functional carrying devices should be fully adapted to protect and carry their specialist cargo wherever that cargo needs to go – or there will be tears'.

In tissue engineering, two critical groups who commonly collaborate come from very different (disciplinary) tribes. These are the cell biologists and polymer chemists/material scientists. We can just about imagine (Figure 4.1) the level of commercial disaster that would result from the appearance of a gleaming new bulk oil tanker in place of a passenger liner for the return leg of a Caribbean cruise. No matter how carefully the stripy deck-chairs are positioned between the oil discharge pipes and gas vents, 3,000 hungry New York tourists would *not* be amused by the new accommodation as they slip out of the harbour at Antigua.

It is likely that tissue engineering cell-scaffolds can suffer at least the same disastrous level of functional mismatch. Furthermore, the consequences are likely to be equally dire, though we may still have a lot to learn about recognizing and measuring such functional clangers when they occur in tissue engineering. It is not a bad basic lesson, however, based on the ship-design analogy, to keep the key specialist tribes working closely together. As we saw in Chapter 1, maintaining close contact between tissue engineering tribes is critical – so what better way than to merge their most important *shared* contributions into a shared chapter?

4.2 Choosing cells and support-scaffolds is like matching carriers with cargo

After examining how the diversity of disciplines contributing to tissue engineering have shaped the subject (Chapter 1), it may not be a surprise that the tangible effects of that diversity are still emerging. As we have seen, one of the earliest concepts was that the building blocks for the fabricating new tissues would be:

(i) one or more 'suitable' cell types (ideally with stable, tame and reproducible characteristics, matched to the target tissue);
(ii) a 3D porous, cell-support scaffold, 'suitable' in its physical characteristics (mechanical properties, pore size, connectivity and so diffusivity) for cell attachment and synthetic activity.

(a)   (b)

Figure 4.1 A blissful outbound voyage though the Caribbean on the newest cruise liner would soon lose its magic if passengers were met by an oil tanker for the return journey. Photo (b) © iStockphoto.com/phlegma.

Instantly it is clear, though, that these two central pillars of the subject are buried deep in very different disciplines – one being cell biology, the other biomaterials science and polymer (including natural protein/polysaccharide) chemistry. In fact, task delegation between the two (cell and materials) tribes could not have been easier. In effect, a common operating pattern of work has developed in much of tissue engineering based on:

- 'We can make some scaffolds, can you find some cells that will live in them?' or
- 'We can generate some really promising cell types. What have you got for us to deliver them in?'

The reader might get the impression that this resembles more of an after-the-event collaboration than an example of ground-up biomimetics.

The luxury of hindsight allows us to question why trial-and-error experimental cycles of different cells in different matrices kept going for as long as it did without a more critical look at its basic logic. There has always been a chance that, one day, it would generate a serendipitous discovery of some special technique for really good 'tissue' growth. In the main, however, we have not been outstandingly lucky in this. It turns out, unfortunately, that it was not such a benign circular scrabble as it seemed; there was a hidden downside.

In fact, keeping the two sides of the 'how do we make a 3D tissue' question separate has encouraged a completely artificial and unhelpful separation into two approaches and philosophies. Part of the tissue engineering community has spent its time asking:

(i) What cells shall we use for 3D culture and how do we get hold of them?

The rest have been concerned with:

(ii) What is a good scaffold material to carry someone's cells and how big do the pores need to be?

As with most imperfect logics, the flaw soon floats to the surface as an even more difficult problem. In this case it immediately creates a third, and unavoidable question:

(iii) How do we get the cells selected by one group of workers *into* the material (fixed in place and living) developed by the other?

In effect, by segregating the two core tasks, we may have made things harder, by expanding the need to *merge* the other two, i.e. creating **the cell-seeding question** (Figure 4.2).

As we shall see later, this is no shrimp of a problem – it is a great white shark of an issue which

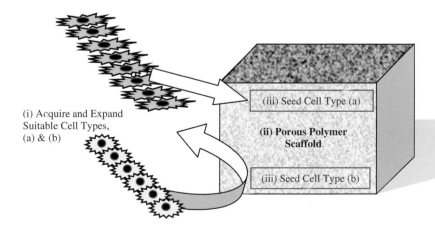

Figure 4.2 The idea could not be simpler or more basic: (i) Collect and expand cells which we hope will synthesize and assemble the tissue fabric. (ii) Fabricate a porous polymer which will temporarily hold these cells in appropriate 3D conformation. While this makes it easy for the two tribes to work on the individual components – it also means we have created a third stage. (iii) Cell seeding.

Text Box 4.1 Exercise 1

Imagine you are trying to find the most bizarre and incongruent mismatch of cells and substrates you can possibly think of, in order to generate models of cancer or diseases.

- Design a research team by linking together three different specialist groups into a collaboration, and justify your selection.
- Then design how you would instruct them to go about hunting for ways to generate the most

non-physiological, pathogenic cell and matrix cluster they can manage.
- Now give four examples of how you would measure the construct properties after two weeks in culture, to demonstrate how 'distressed', disrupted and diseased your 'tissue' had become.

Tip: Imagining you are a member of one of these groups will help you explain your plan.
[Guideline: approx. 2 pages: 60–90 min.]

---

surfaces again and again to rip lumps off our best efforts.

Aside from generating an extra question, this artificial division makes it much easier for experts to pre-design the 'solution' around either a promising cell preparation or good-looking scaffolds. Unfortunately, each of these approaches on its own is only likely to generate poor tissue engineering solutions (Exercise 1 in Text Box 4.1). This is the metaphorical equivalent of a bulk gravel carrying ship waiting patiently at Rotterdam's busiest refinery wharf to load up with liquid gas!

What once seemed a prudent and conservative approach can now appear, in retrospect, to be just too low a target to have been realistic. In our shipping analogy, we could imagine trying to salvage the situation by welding and sealing the hatches, bulkheads and drain-channels in our gravel carrier so as to hold in a few cubic meters of liquefied gas. Really, though, it should be clear that we are labouring to fix a problem of our own making. Now hold on to that thought for a while – we shall return to it shortly. This issue forms the core of our first glimpse of extreme tissue engineering for this chapter: the case of '**aiming low and still missing**'.

4.3 How like the 'real thing' must a scaffold be to fool its resident cells?

The 'real thing' here, of course, is the natural mature, healthy tissue in question, with all the complexity implied by that phrase (see Chapter 2). The

assumption in this chapter will be that it is, at least in the early stages, both logical and pragmatic to develop cells and their support material as a single, integrated unit. To do this, it is first necessary to understand the basic mechanisms by which our chosen cells will use their 3D surroundings (Chapter 3). The idea is that this will give a good indication of the *essential elements* we need to mimic for this or that cell-tissue-injury system. We might reasonably expect such intellectual tools, or basic knowledge, to enable us to design genuinely functional 3D cell-support systems.

The more astute (or wizened and cynical) readers may detect the faint odour of 'shallow logic' in the last paragraph. Although it is apple-pie-obvious that we should get enough background knowledge in the first place, in biomimetics we have the greatest difficulty predicting when we have 'enough'! A slightly more honest question might be to ask, What are the five or six top features that a 3D support material really *must* have if it is to fool our (*insert name of our cell type*) cells to doing what we need of them? This really is a crucial question, but it is so full of caveats that it merits a full in-depth analysis.

Firstly, the choice of five or six features is pretty arbitrary, though it is, empirically, a useful starting point. The history of the field suggests that, where it is known, the identity (and significance) of a reasonable spread of features can make it is possible to make a half decent stab at a design. Five to six generally does give a reasonable spread! The selection criteria are important, though. These are *must-have* features, necessary to 'fool cells' – the classic core of biomimetics – not features to satisfy government regulators or university accountants, but must-haves from the viewpoint of our cells. In fact, we are looking for the least we can get away with in terms of bio-compromise. After all, if we do not find simpler compromises but aim *only* to be more and more 'like nature', our solutions become incredibly complex and impractical to produce – in fact they *become* natural systems rather than mimics.

There are informative examples of the 'simplest effective signal' principle in nature, which are both illustrative and useful for calibrating our estimates of how simple 'simple' can really be. Figure 4.3 shows one: the red dot on the lower beak of a humble herring gull. Possibly you have noticed this while you were at the coast; or more probably not. But it has not escaped the bird behaviourists, who concluded that this dot is the target at which gull chicks aim when pecking at the parent's mouth. When the chick pecks at the beak-spot, the parent is stimulated to regurgitate partly digested fish, which is, oddly, appetizing food for the young gulls. For the gull this is a simple but effective information loop. The point is how surprisingly simple this signal is. Indeed, the chick will perform the full pecking ritual at any red spot painted on a piece of wood. The rest of the parent bird is unnecessary, except of course to supply the half-eaten sardine.

Figure 4.3 The humble herring gull has a red dot on its lower bill (arrowed). This demonstrates what we are aiming at in biomimetics, as the dot is all that is needed to stimulate the chicks to feed from the parent's beak. So strong is this key message that chicks will peck at a white stick if it has a red spot painted at one end! © iStockphoto.com/faith donmez.

This example illustrates how we can sometimes fool bio-control systems by input of really easy, simple pieces of information, *once we know what that minimum is*. If we were into gull engineering, we might generate energy from dot-pecking chicks. For tissue engineering, we need to know the minimal cell-cues needed for our 3D material supports.

The problem is the huge breadth and detail that these cues *could* take. We can start with the surface chemistry of our material. This modifies how different proteins from the body fluids bind to the surfaces when they are implanted, and so how cells attach (or don't), because, of course, different cell-types use different attachment proteins to go about their business. Then there is the substrate stiffness, or lack of it. There is much more to this than high and low 'squishy-coefficients'. Cells in natural tissues are adapted to (and so respond to) physiological tissue material surfaces with complex 3D zones and layers. Sometimes these can have continuous stiffness gradients. You can find examples of these for yourself; pinch your skin in a series of close places, running around from the back of your neck to the front of your throat. Alternatively, gently stroke the skin the back of your hand up and down and watch what moves. Now keep stroking and looking, but move around to the palm and see how much less moves!

The possibilities get more daunting still when we realize that these material properties are normally asymmetric and, worse still, they can be dynamic. 'Dynamic' means that properties change with time and even with rate of motion, as the materials move and water is displaced (visco-elastic behaviour). This happens during simple, everyday bending, stretching or compression. Such movements generate many secondary effects on cell physiology, for example enhancing tissue fluid movement/perfusion parallel with (rather than across) anisotropic fibre materials.

You can get an idea of the rate-dependent and dynamic tissue properties by doing the stretching-your-elbow-skin exercise. Figure 4.4 shows how the loose skin over your elbow will stretch and bounce back. This is both age- and direction-dependent. Snap-back becomes faster and greater where we test along a father, son, grandson series. Interestingly, the recoil tends to be complete when you bend

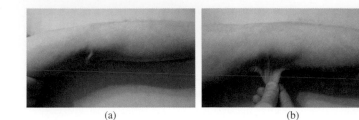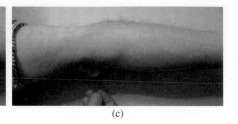

| (a) | (b) | (c) |

Figure 4.4 The elbow skin snap-back test. Try it.

and then unbend the elbow, but incomplete if you pull down and release (as shown in Figure 4.4c). This suggests that the direction in which the skin is extended matters, i.e. it is anisotropic.

We can see that factors which mimic *special* features in the cell environment not only offer a rich vein of control mechanisms but also represent small parts of very complex systems. Our extreme tissue engineering task is to identify the critical but minimal components of these complex spatial-mechanical cues which cells can recognize as 'key signals'. In other words, what are the 'herring gull red dot' factors?

In contrast to this biomimetic hunt amongst complex natural signals, cell-support materials (scaffolds) are presently selected and designed on biomaterials grounds. These are, at best, rather simple, identified largely in the search for prosthetic biomaterials. Indeed, the assumption that ultra-simple, even non-mimetic factors can determine cell outcomes seems to be at the core of the tissue engineering dream. The dream suggests that we will persuade cells to recapitulate biological fabrication with simple biomaterials, plus a handful of other cues, including protein growth factors. When it comes to designing the materials, we hope these will provide the special, key cue (gull-factor) deceptions.

As a result, a modest number of options are regularly revisited, occupying a great deal of literature with increments of new cell types on slightly altered surfaces. Some are simple tricks, learned from cells in monolayer culture, such as surface patterning. These can include producing grooves, channels, pits and humps in the cell-support material surface, or occasionally between layers, where we are dealing with the cues deep in the bulk of materials. Indeed,

these can be potent cues to guide cell motion and cell shape. They can indirectly alter behaviours such as proliferation, differentiation/gene expression or protein synthesis and export. However, they are manly derived from phenomena in 2D monolayer cultures. In effect, they are highly simplified parodies of natural control systems, based on little knowledge of their operation. Worryingly, these may be our best examples.

Other popular design features concentrate on factors such as material porosity and the degree of pore interconnectivity. This is the inevitable legacy of the need to get cells to migrate deep into many conventional support materials. They are the consequence of the cell-lethal conditions used in polymer and/or 3D scaffold manufacture. *Porosity* also gets far higher up the 'must have' list than it deserves in response to the largely faulty dogma (see below and Chapter 3: Mount Doom) that cells will die in a hideous, hypoxic agony if not adjacent to medium.

Worse still, it is still possible to see key design criteria quoted which include the words '*cheap*', '*generally biocompatible*' or '*approved by government regulatory offices*' (such as the FDA). To have these in the 'top 10 of biomimetics' seems to represent a grand misunderstanding of the high aims of biomimetic engineering. In particular 'cheap' is as appropriate as souring economy aluminium alley for Rolls Royce turbo fan blades.

Based on such analyses, the newcomer could be forgiven for thinking that tissue engineering is an ambitious dream, populated with extremely low targets. For example, when we make a successful off-the-shelf skin graft to treat the most high-burden problems of aging Western societies, its high market value will definitely *not* require that it is made from cheap, recycled waste products. Similarly, if we are intent on making *absolutely nothing* that could be new, progressive or adventurous, then sticking firmly with previously FDA-approved starting materials is clearly the way to go.

Indeed, as pore sizes evolve to greater than a few cell diameters (i.e. towards being *less* mimetic), we now see that they actually undermine (albeit accidentally) a central requirement – namely, 3D cell growth (see Chapter 3). Other factors become poorly mimetic as a result of such dubious, ingrained priorities, including:

- inappropriate cell clustering;
- sharp and distorted nutrient and gas perfusion gradients; and
- non-physiological access to, and export of, bioactive proteins, especially in terms of *rate and direction*.

More of these later.

In contrast, systematic identification of truly defensible 'simple' cell-cues is still in its infancy. Examples include:

- Fine grooves which stimulate many cells to elongate and align.
- The use of very soft or much stiffer materials to elicit fibroblast quiescence and low motility, as opposed to cell division and migration.

So, in answer to this section's question, even at this early stage we can be encouraged that there are some minimalist cues that will elicit useful and complex cell responses. The tough thing is to understand the language.

4.4 Tissue prosthetics and cell prosthetics – what does it matter?

It is possible to argue that the development of 'cell-scaffolds' marks a movement out of the era of tissue prosthetics and into that of cell prosthetics. As we discussed in Chapter 1, we can now produce

excellent artificial hip joints made of metal and plastic. They carry out the function that was once performed by the natural tissue. However, they never work better than when they were first fitted, as they cannot repair, renew or regenerate.

It is possible to view the TE scaffold as a *prosthetic repair tissue* – in other words, a structure that supports and enhances the natural repair process. As such, its role is temporary, lasting only during repair. Whatever the finer philosophical points, the key to the prosthetic versus engineered implant difference lies in the idea that a good TE 'scaffold' disappears as a natural tissue appears, whereas the ideal prosthesis lasts as long as possible. Such a big difference demands maximal attention.

Fortunately, we have a BIG clue to keep us on track. Where the central tenet, of '**essentially temporary**' is allowed to move down the *must-have* priority list, then we return rapidly (with the help of familiar surgical imperatives) into the world of prosthetics. This is especially true in those areas where early tissue mechanical strength is the surgical demand, such as in bones, joints and large blood vessels. Repairing or (when it becomes possible) regenerating tissues are soft, weak and vulnerable; weak blood vessels burst under high blood pressure, and immature tendons will snap when they are on the end of mature muscles. Both the patient and clinician would like mechanical function to return in hours, or at least days, rather than weeks. In the case of the vascular surgeon, however, it might not be so much what they *would like* as that their patients tend not to survive the rupture of a major blood vessel.

Figure 4.5 shows the distinction between basic prosthetics and tissue engineering logics, in the form of their intended *function* versus *time*. Although these are notional, rather than real, data plots, the point is clear as their respective dynamics are distinctly opposite. The rate of change of function over time will depend on the specifics of each example and the anatomical sites they occupy.

Nevertheless, prosthetic function inexorably declines from an early stage high point. This high point is their great surgical advantage – the patients leave and are happy. Engineered implants, by

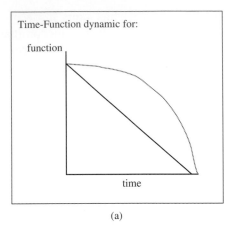
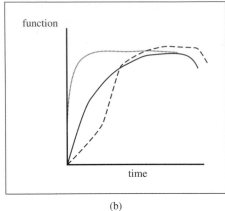

(a)   (b)

Figure 4.5 Notional (or otherwise aspirational) function-versus-time plots for (a) prosthetic implants and (b) engineered tissue implants.

contrast, are unlikely to meet local tissue demands immediately. Even perfectly matched tissue grafts would need to integrate, forming mechanical, vascular and neural links with the host tissue margins. But once that bio-integration is effective, the great advantage of tissue engineering becomes a reality – their function just goes on renewing, like other tissues.

The result of this conflict of possibilities has its effect on the respective support materials. Here there is a tendency to drift the function of *cell*-support scaffolds, for tissue engineering, such that they incorporate as much *tissue* support as possible, at time zero. This, to use the analogy of a box of chocolates, is to want all the hard centres and none of the soft, which always comes with unintended consequences.

Most commonly, we are tempted to design and fabricate very stiff, strong and long-lived scaffold materials as we mimic the function of the *finished* tissue rather than immature or repair tissue. Examples of this would be the seeding of fibroblasts onto high-tensile strength rope-like materials and calling it load-bearing tendon, or putting vascular cells on non-degradable polymer mesh tubes. The truth is, of course, that these *are* close to or actual prosthetic implants – just with added cells. Indeed, evidence now suggests that they do not ever really make the transition to the real, functional tissue that

we wanted, and biomimetic engineers suspect that cells may even be getting the *opposite* signals from those we intended.

One of the best established examples of this is the use of stiff, strong cell-seeded materials in ways which stress shield the resident cells. Stress shielding is an effect where one very stiff part of the system (in this case, the support material) carries most of the overall load, shielding another less stiff component (the cells). In tissues, the result of stress shielding by stiff materials is that cells are not exposed to the physiological strains which regulate normal tissue remodelling. In stress shielded bone and fibrous tissues, cells appear to switch off tissue fabrication, repair and remodelling behaviours. This is well recognized, for example, where stress shielding of osteoblasts leads to bone loss around stiff orthopaedic implants. Certainly, many tissue cells need to be exposed to at least basal maintenance levels of regular mechanical strains (deformation) if they are to maintain or rebuild their tissue materials.

The first lesson of this example is that over-reliance on strong, stiff cell supports is perilous to good tissue engineering, even when it looks like a good idea in the short term. Viewed more positively, we can conclude that transmission of external mechanical strains to attached resident cells is an important, basic cue for connective tissue growth. Consequently, it is certainly important to be careful when we blur the distinction between 'scaffold' materials designed to support tissues (i.e. prosthesis-like), as opposed to those which support cells. Materials providing whole-tissue support can potentially produce exactly the *opposite* effects from those we are aiming to achieve with cell supports. For this reason, we shall, from now on, remind ourselves by referring to 'cell support' materials instead of just 'scaffolds'.

4.5 Types of cell support material for tissue engineering – composition or architecture?

When we come to design our *extreme* cell support material, is it better to focus on the composition of the material or the 3D architecture we fashion it into? There was certainly a tendency in early tissue engineering to start from a point of intense faith in our favourite polymer material and work out from there. To return to our shipping analogy, this is equivalent to deciding first to build in high grade steel, aluminium, cast iron or wood, and only then to fit the cargo-type into our choice. Clearly, composition affects the eventual 3D space we can fabricate, but do we start with a promising material, or design the cell-carrying space and *then* look for the materials which are most suited?

To tackle this question, we need some basic background. First, we should try to consider the distinction between composition and 3D architecture of support materials *from the viewpoint of the intended resident cells*. Is it possible to distinguish the dominant source of the cell cues we would most like to provide (accepting that such predictions will be fuzzy)? Molecular composition, or 'the substance' of the material, will dominate many of the final properties, from its surface chemistry to its gross mechanical strength.

In fact, these two points of contact, *surface* and *bulk*, form a helpful division (Figure 4.6). Material composition is a combination of both surface and bulk properties. The first – surface properties – are typically just nanometres or, at most, microns deep. But they profoundly affect

how cells bind and interact. Commonly, these act as short-term and dynamic cues. The second – bulk properties – tend to dominate the large-scale characteristics. These provide the longer lasting features, such as mechanical strength and overall gross-structure survival time.

So, for this section, we shall consider only simple 'surface-deep' support material features in relation to material composition. It is still complex, and the really fancy stuff can come later.

4.5.1 Surface or bulk – what does it mean to the cells?

Any exposed surfaces can directly affect how cells use and interact with the construct (cell binding, motility, contraction, surface degradation, etc.). Such surface chemistry presents a range of possible charged groups or hydrophobic/hydrophilic areas, or it can alter the nano-stiffness. However, it can equally act indirectly, given that almost all implanted surfaces immediately absorb a cocktail of proteins from adjacent body fluids. Some of these proteins are natural mediators of cell-binding, while others just coat the surfaces and so *reduce* cell binding by occupying space but not attaching to cells. This means that the original material surface chemistry can be selected to take up desirable cell-binding proteins.

In addition, surface chemistry can influence substrate mechanical properties at the cell level (i.e. local to individual cells – a few micrometres into the bulk material). This surface material coating, which can represent the cell world, is often dynamic in its nature. Where the material is bio-degradable and formed of proteins that are digestible by cell enzymes, it will be rapidly removed and replaced. On the other hand, the underlying bulk material, comprising the stable, major part of the material, largely determines overall mechanical strength and survival time, being unavailable to the actions of surface attached cells.

4.5.2 Bulk material breakdown and the local 'cell economy'

Let us take a closer look, then, at some examples of the bulk material compositions we have mentioned.

Figure 4.6 Surface and bulk qualities of Christmas cake (left hand panel). Bulk fruity-nutty-cake material is surfaced by a mechanically stiff, micro-porous layer (arrows). A complex sub-surface interface layer is formed by compliant, visco-elastic marzipan (arrows: right hand enlargement).

It is worth clarifying that, in the biomaterials field, *biodegradable polymers* are often 'biomimetic' only to the level of being tolerated by tissues. Many of these, in addition, break down by simple hydrolysis in aqueous fluids. In contrast, *biological materials* are formed from bio-molecules (proteins, polysaccharides, etc.), though not necessarily in native form. Finally, *natural materials* are native and tend to be derived directly from parts of animals or plants. Materials which are *not* bio-degradable are excluded here, as they seem to fail the simplest, defining test of tissue engineering. They cannot be replaced by native tissue.

A useful way of dividing cell support materials is based on biomimetic function that has become apparent in modern tissue engineering, and this identifies three generic groups in terms of their functional similarity to native tissue materials. These are, broadly:

(i) predominantly synthetic materials (synthetic polymers, ceramics, soluble glasses, etc.);
(ii) predominantly natural polymer materials (*native* proteins, peptide sequences, polysaccharides, etc.);
(iii) hybrids (composites) between (i) and (ii) having synthetic parts linked to natural domains designed to mimic some natural functions.

The logic of this classification reflects (a) the ability of some support materials to work as an integral part of the natural cell-matrix remodelling process, versus (b) those which just break down or dissolve irrespective of cell activity. This is *cell-dependent* versus *cell-independent* breakdown.

Being part of or not part of the tissue remodelling process is a bit like being a small district in central Chicago trying to live, trade and bring up families – but using Euros instead of US dollars. They cannot work, play or trade together. They are functionally disconnected – the opposite of integrated. Suddenly, this now allows us to see another great paradox. One of the most common ideals quoted by tissue engineers (in this case the biomaterials tribe) is that the 'scaffold should degrade at a rate corresponding to production of new tissue'. In other words, it needs to be quantitatively 'replaced', in a **cell-controlled** system. But how can our ideal *ever* be realized when the material degradation process is 'dollar-priced' but the cells only have Euros in the bank?

It is easy, though, to see why natural, cell-integration materials might not be the first place to start. Natural biological materials may integrate with cell physiology, but they are also difficult to fabricate reproducibly and controllably because, by definition, they are biologically complex, with possible immune and infective problems. In contrast, despite minimal participation in the 'local cell economy', many synthetic biomaterials are relatively simple, predictable and reproducible to manufacture, with a safe clinical history. This, then, is the pragmatic balance which extreme tissue engineering must resolve.

4.6 Three generic types of bulk composition for support materials

In effect, this function-based classification comes down to the philosophy that lies behind how support materials are expected to interact with their resident cells. Like any good design process, it requires an

Text Box 4.2 'Bulk' and 'surface' material properties

It is important to be clear that for many polymer materials there is a major functional distinction between the vast majority, or '*bulk*' of the material and the thin *surface* coating, which is accessible to cells (Figure 4.7). Two spatial, cell-material relationships are possible, where:

1. Cell support surfaces are essentially dense and impermeable over areas much larger than cells (e.g. hundreds of μm). To compensate for the cell-impermeable nature of such surfaces, they are formed into interconnecting pores, at the sub-millimetre (100s of μm) scale. In this case, cells normally have little access to the supporting, sub-surface bulk polymer (e.g. synthetic μ-porous plastics and ceramics). As we will see later, their high bulk : surface ratio can be reduced by increasing this μ-porosity sometimes to more than 80 per cent of total volume, with the support of even stiffer and denser pore walls.
2. Cell support surfaces are much smaller than cells (e.g. nano-fibres of hydrogels or electro-spun substrates). Such *interstitial 3D cell seeding* within the material allows cells to move around and envelope the nano-fibre surfaces. This can dramatically reduce, though never completely eliminate, the bulk-material volume from which cells are excluded.

In either case, cells only touch the outer surfaces of the 3D support material, where the 'main load-bearing elements', are (1) the single gently curving μ-pore walls or (2) the many surrounding nano-fibres.

Cell signals from the support materials seem to be a combination of surface (bio-)chemical features plus substrate stiffness, measured against the internal cell-cytoskeleton. **Surface chemical** signals (e.g.

nm-deep protein layers) can interact with cell membrane receptors. However, **mechanical signals** can affect cells from much larger depths, many μm below that surface, such as soft/hard layers, gaps and splits. Curtis and co-workers[1] have compared this sensitivity to deep mechanical properties to the 'Princess and the Pea' fable, where a cell-princess feels a stiff pea through many softer layers.

For biodegradable or cell degradable support materials, of course, this relationship is, by definition, *dynamic*.

Breakdown and release of the *surface layer* potentially alters both cell-receptor and mechanically mediated cell signalling, e.g. time-dependent surface changes due to:

(a) *loss* of surface layers (e.g. uncovering deeper substance) by cell or chemical action; or
(b) by surface *re-covering* due to deposition of exogenous material, often extracellular matrix.

Hence, surface composition and properties of material supports are often different from those of the bulk material, but are disproportionately important to cell behaviour. This simple distinction has supported a large body of research into control of cells by surface modifications. Examples include synthetic polymer surface modifications, such as:

- alteration of surface charge or hydrophobicity to encourage protein binding;
- direct modification, typically coating with a tightly bound layer of active protein/peptides.

However the cell dynamic is important to remember here. Even covalently attached proteins will be broken down rapidly.

Reference

1. Curtis, A. & Wilkinson, C. (2001). Nantotechniques and approaches in biotechnology. *Trends in Biotechnology* **19**, 97–101.

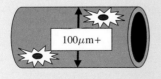
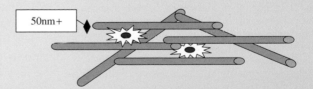

Figure 4.7 The left-hand diagram illustrates how surfaces might look/feel to cells in a type (1) relationship. They are on the walls of a deep solid material with a 'core bulk' (dark) zone with little or no role in cell signalling. In contrast, in the right-hand diagram, by changing the scale and architecture of the material to give a type (2) relationship, cells are enmeshed by the (pink) surface material with a much smaller proportion of 'excluded' bulk volume.

understanding of the practical trade-offs that each one is making. Once we understand this, we can make informed cost-benefit choices for our specific application.

Without doubt, none of these options are wrong to use but, equally, none will be appropriate for *all* applications. The days of listing the 'ideal properties of support materials for tissue engineering' in a single opening slide in a talk should now be gone. We are – belatedly, perhaps – in the era of learning the rules. Ideally, this should allow ETE to evolve into a strictly human form of **intelligent design**(sub-division 'cell-support'). As a result, the ability (or not) of our materials to participate in the natural cell-matrix 'economy' can become a useful defining feature of the cell-substrate combinations we investigate. These options are illustrated in Figure 4.8.

4.6.1 Synthetic materials for cell supports

The distinction between materials that cells can degrade naturally, by enzyme breakdown, and those that they cannot degrade (only dissolve) is

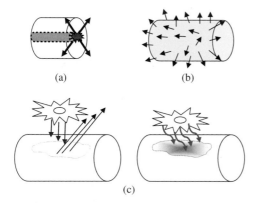

(a)   (b)

(c)

Figure 4.8 (a) Polymer core dissolution (e.g. PLGA: arrows indicate core-first loss of material). (b) Surface dissolution (e.g. phosphate glass: arrows indicate surface to core loss of material). (c) Natural polymer material (e.g. collagen protein). Attached cells secrete digestive enzymes onto the support materials (left hand diagram – red arrows), breaking down material into protein/peptide fragments which are released from the surface and lost. This leaves denuded areas around the cell footprint (dotted shape). Right hand diagram: same cells deposit new protein material (i.e. extracellular collagen-matrix – green arrows) to replace the digested material.

fundamental, at least to cells. It is relatively simple to imagine the viewpoint of a cell attached to a benign, but slowly dissolving, support material. A human analogy might be your fourth day trapped in a factory making coloured-ice food novelties. You can suck on an ice-strawberry, revel in a choc ice or nibble an ice-hamburger – they are harmless, but tasteless. After four days, though, they would certainly not be giving you many of the signals of satisfying food and treats that you are expecting. Figure 4.8(a,b) summarizes the two basic modes of degradation seen in synthetic materials. These are inside-out (i.e. core first) and outside-inwards. In either case, they provide physical support for cells which could then produce new tissues.

The first mode of degradation of synthetic bioresorbable polymers (Figure 4.8a) represents their curious habit of degrading faster at their core than at the surface, even though this is a hydrolysis process. We are more familiar with water degradation, where materials are lost from the outer surfaces, which are in close contact with the water, e.g. where river torrents wear into soft bedrock, salt crystals dissolve and steel rusts and flakes away.

A common example of core-first degradation is the copolymer poly(lactic-*co*-glycolic acid) (PLGA). This is formed from graded mixtures of poly-lactic acid and poly-glycolic acid, both of which hydrolyse in water but at different rates and patterns. Blending of the two in a range of proportions produces bulk materials with a range of compromise properties. The poly-lactic and poly-glycolic acid backbones break into their organic acid constituents (lactic and glycolic acids) on reaction with water. This reaction is dependent on water access, but the hydrolysis also accelerates at low pH. Consequently, as water seeps into the core and generates an acidic core pH, this leads to faster local breakdown far below the surface.

For high load-bearing structures such as PLGA sutures or bone-pins, the effects of this can best be seen in their strain-time profile. Strain under a sub-fracture load remains constant for many weeks or months as the core degrades but the outer shell carries the load (Figure 4.9). Then, quite suddenly, the structures will fail as the outer crust fractures,

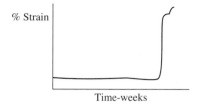

Figure 4.9 Diagrammatic plot of strain over time for PLGA (PLA or PGA) under a sub-fracture stress load. This shows the low deformation (percentage strain) of the material, constant for some months, followed by sudden and complete (catastrophic) failure.

incidentally releasing the inner degraded 'goo' of organic acids. At the multi-micron level, cells are growing on a stable, high-stiffness substrate for long periods, proliferating and laying down matrix. Then, suddenly, at some time point depending on the type and thickness of the PLGA fibre or strut, this constantly stiff cell support undergoes mechanical failure, instantly transferring major loads onto previously shielded cells and matrix. There have been concerns about the sudden release of acidic core degradation products (goo) onto cells, but in practice this, and any associated inflammation, seems to be quantitatively modest.

Figure 4.8b illustrates a more familiar pattern of aqueous dissolution of materials. In this case, the example is soluble phosphate glass. Phosphate glass is literally a glass-like material in which the silica component has been replaced by phosphate. The resulting materials react with water in complex manners to form soluble phosphate products, but in a strict outside-in direction. In other words, there is a constant loss of surface materials to the aqueous media as surface molecules of the glass dissolve and wash away. These can have rapid dissolution rates which can be reduced in a controlled manner by addition of trace contaminants to the bulk composition, such as iron or manganese. 'Bioglass' materials developed for hard tissue replacement are similar, but retain a substantial silica content and so are very slow to dissolve.

In both these cases, material is lost from the outer surfaces which are exposed to an aqueous environment. This, incidentally, means that release of these inorganic ions is proportional to the surface

area-to-volume ratio (hence simple architecture). This outside-to-in pattern of dissolution has advantages, not least the gradual release of dissolution products, as opposed to the 'all at once' sudden release for PLGA and others. However, a molecular layer over the surface of such substrates is constantly being removed and released, exposing deeper levels. Where this action is rapid, it can play havoc with cell or protein adhesion.

This, then, is an example of a major affect of bulk composition on surface properties, where proteins and cells are constantly re-attaching at rates proportional to the material dissolution. Attachment of engineered cell layers, such as epithelium, to the surface of such materials is likely to be temporary (see Text Box 4.2).

Both of the outside-in and inside-out degradation patterns are, however, entirely dependent on the presence of an aqueous medium. They are bio-degradable only by virtue of the fact that they lie in an aqueous environment when placed into living cell cultures. In this case, the term 'bio-degradable' does not imply any linkage or deeper dependence on the proximity of the living part of the system; they merely share the same water-based environment as the cells.

A key factor here is that dissolution is relatively simple to predict. In any definable system, the rate of loss of mass from the synthetic material will be dependent on standard physico-chemical factors, including the chemistry of the hydrolysis process, local pH and temperature and reactant equilibria (dependent on diffusion and mixing rates through, out of and away from the material), etc. However, the term '*relatively*' should have rung alarm bells, particularly for biologists. The basic physico-chemical drivers are, indeed, calculable, but not necessarily with any accuracy for dynamic and spatially complex bio-systems. So, while it is true that dissolution and failure rates for synthetic supports can be modelled and predicted for cell-free, and some simple cell, culture systems, predicting the fate of this group of materials after implantation *in vivo* is quite another fish market.

In contrast, biological materials (Text Box 4.3 and Figure 4.8) work in a completely different manner. Figure 4.8c illustrates the starting point for this contrast. Specifically, native biological materials can be degraded by embedded cells and, in some cases, cells will also add new matrix to fill in the holes made by this degradation. Because it is now becoming clear that this difference is so important, our summary of these material types will be unusually weighted – short on synthetic, long on native materials. This is the reverse of many tissue engineering uses past and present, but it may better explain where it will go in the future.

4.6.2 Natural, native polymer materials for cell supports

We have made a great deal of the ability of cells to 'use' and 'remodel' the bulk material of our synthetic or native support materials. In our own lives, the slow, inevitable degradation of the fabric of our house might leave us feeling anything from mildly concerned to panic-stricken. Where we are on this spectrum is determined by our repair skills and access to suitable tools and materials. For those of us who are handy with a hammer and a screwdriver and live close to a DIY store, it is an opportunity to improve and customize our living space. Steering clear of the dangerous practice of humanizing cells, we can see from this analogy that **the (chemical) nature of the bulk materials *can* have a greater impact on later 3D structure than expected.**

This is important enough to merit a second – short time-base – analogy. We have all probably eaten ice cream cones. Some cones have the cold stuff (plus chocolate and sweet bits) tightly wrapped in a colourful cardboard-and-foil cone. These are jolly, entertaining, insulating, easy to grip – but inedible. The other type is squashed down into a biscuit cone, which you can eat. Most likely, your relationship with the first, card-cone-support system is simple, consisting of: (i) a brief pleasure-support role, (ii) tearaway moment (optional), (iii) splat flat, then bin. The biscuit cone elicits a completely different set of interactive behaviours. You eat the biscuit as the ice cream is consumed, in parallel, partly nibbling *as you go*, snapping off portions to scoop up ice cream, reshaping and remodelling *as you go*, mixing and repositioning the ice cream flavours, choc-chips and sugar sprinkles *as you go*.

Critically, different types of people (just like cells) have their own behaviour patterns for this coupled spatial remodelling process. But, at its core, the behaviour is transformed by *your relationship* with the material; in particular, your ability to consume the biscuit (or cell-support material) dominates when, where and what you do. The point here is that we should expect the composition of native materials to have a greater impact on final tissue structure than that of synthetic materials.

Where natural material composition is concerned, the need for physical strength and bulk volumes largely excludes all but two natural polymer groups – proteins and polysaccharides. Generic examples of fibre-forming, aggregating protein materials are relatively small in number. Most work has focused on collagens, fibrin, silks and fibronectin – but that is about all the choice that is available. It is worth noting the obvious, that all silks differ from the other three types in *not* being vertebrate proteins (Text Box 4.3).

Many **polysaccharides** are available, but even fewer of these are mammalian. Polysaccharide aggregates can be given significant mechanical strength, either naturally or by chemical processing, and so are common as cell substrates. Examples include starch, agarose, chitin (chitosan) and hyaluronan (or hyaluronic acid). Starch is a glucose polymer whose properties, from dough to pasta, are completely familiar to us, courtesy of bakeries, patisseries and pizza suppliers. Agarose is a galactose-based refined seaweed product (algae, hence 'alginate') which is commonly used in biology to form water-rich (hydro) gels. These handily undergo sol-gel transitions at temperatures which can allow cell incorporation (on cooling). Chitosan, a chemically de-acetylated, poly-D-glucosamine, a derivative of chitin, is made from crustacean shells (i.e. a by-product of the shrimp industry, chiefly *Pandalus borealis*). Again, it can be made into tough polymer materials with controllable 3D structure.

Text Box 4.3   Biological: but is it native, non-native or biomimetic?

For non-biologists, the apparently mild tautology of 'native biological materials' may have caused gentle irritation. Actually, though, it is not self-reinforcingly obvious, and the reason why is worth explaining. It is arguable that almost all living cell products are 'biological', as they were formed by living systems. Their origin dominates. However, it is equally clear that some have been processed and modified so far from their origins that they take on quite different properties, and so deserve a different name.

For example, wood may be dried and seasoned, but remains much the same, such that it can undergo rotting by microorganisms, much as it would on the forest floor. However, when pressure-injected with creosote fixative or shredded into an embedding resin (e.g. chipboard), it has radically altered stabilities. Similarly, we are keen to distinguish between cell-free sheets of native collagen material (e.g. dermis, fascia, small intestinal sub-mucosa – SIS) and leather, which is strongly cross-linked by tanning. Such processing is designed to divorce the biological material from natural bio-degradation processes such as bacteria or fungal action or digestive enzymes. This makes them much more stable for everyday use.

We would say, then, that such processed biological structures were no longer 'native'. This is a particularly important distinction in our subject as, by definition, native materials *will* be available for cell-mediated degradation and remodelling via cell enzymes, but non-native biologicals will not. Chemical cross-linking (e.g. with glutaraldehyde or carbodiimide) of collagen materials, like leather production, renders them both resistant to remodelling and non-native. They are, then, fairly questionable candidates as tissue engineering supports.

An interesting parallel distinction lies in the term 'biomimetic', since, like 'biological' this is a term which is far more useful once qualified. Designing a material as a copy of *any* bio-system or structure *can* be biomimetic. But in biomedicine, we are normally interested in mimicking at least mammalian – and preferably human – systems. We are less bothered about bacterial slimes, crustacean limbs or plant cell walls. Hence it is helpful to specify what, in general, is being copied. This rigour makes it easier to ask what part of mammalian biology is mimicked, for example, by alginate seaweed gels or chitin? (See also Chapter 2.)

Hyaluronan, formally known as hyaluronic acid, occurs widely (though rarely at high levels) in many mammalian tissues, and so is bio-medically mimetic. It is a strongly anionic polysaccharide, composed of repeating disaccharide units of glucuronic acid and N-acetyl glucosamine, making it a member of the glycosaminoglycans family. In its native form, it is strongly hydrophilic, forming very long but unbranched, randomly folded chains. Its composition and charge means that it binds many times its own mass of water to form gels or highly viscous fluids. It has also been processed at an industrial scale into more stable and physically strong materials by progressive levels of cross-linking. These are formed into sheets and fibres, and some are in current clinical use, notably as perforated sheets for the support of dermal repair.

In fact, hyaluronan materials are particularly useful for illustrating the rather typical compromise spectrum common in biological materials – that of increasing modification versus loss of native biomimetic properties. The problem here is that, like many polysaccharides, hyaluronan chain structures do not really self-associate in any organized manner to form a solid aggregated material (in contrast to proteins, below). As a result, the 'solid' material structures frequently have indifferent or really poor material properties.

Unmodified hyaluronan, for example, falls somewhere between a weak gel and a viscous fluid, depending on its water content. This, of course is the native format, where it is most biomimetic. To 'improve' its physical properties, hyaluronan is commonly cross-linked by chemical treatments to produce useful solid materials. Unfortunately, as the extent of cross-linking increases, the ability of cells to use, digest and remodel it is gradually lost. This is similar to how the tanning process turns biological skin collagen into leather, which is impervious to cell enzymes as well as to bacterial decay. As a result, hyaluronan materials, with low level cross-linking, remain biomimetic but degrade faster and are less strong than highly cross-linked forms.

A small number of **natural proteins** can usefully be fabricated into practical biological support materials: the collagens, fibrinogen/fibrin, fibronectin and the silks. On the whole, these have mechanical and support functions in nature, either at the cell-support or the gross-tissue levels. An interesting common feature of all these is that they are able to self-assemble from relatively low molecular or monomeric forms. All of them form fibrillar materials by side-to-side aggregation and the axial accretion of many thousands or millions of monomers. Since the component monomers are in the nanometre scale and the gross materials can be metres in length, overall aggregation is not limited, though the component fibres can be, lying in the nm–µm range of diameters.

Again, a common feature is that fibre elongation can be driven by application of fluid shear forces (producing shear-aggregation or pseudo-liquid crystal behaviour). In other words, the directional shear of moving fluids tends to extend and elongate the protein material as it aggregates, potentially elongating and aligning fibres, like spinning candy-floss though air. The value of shear-driven aggregation in water-rich systems is not surprising as it simultaneously expels water and aligns the long, thin molecules, physically packing them close together. In many cases, fibre formation can be defined and driven by this processes of **aligning dehydration**.

Flow alignment also supplies us with a useful example of how natural material fabrication might occur Figure 4.10. Development of **silks** springs partly from their remarkable material properties and long track record as suture materials. Much of the modern focus, especially on their use for tissue engineering, has grown from work of Vinney (2000) and Volrath & Knight (2001). Their dream is to understand how spider or silk-worm spinnerets achieve shear-aggregate fibres from the fibroin protein monomer.

Biomaterials and tissue engineering companies have formed around technologies for using silk-like materials towards applications as diverse as nerve

Figure 4.10 Silk fibres being extruded from spider spinnerets. Reproduced with permission © 2004, Dennis Kunkel Microscopy, Inc.

regeneration conduits, cartilage and bone replacements. These are elegant developments, using subtle biomaterial modifications which initially seem strongly biomimetic. However, the counterbalance to their safe history is that silks are proteins from a *very* different animal phylum – the Arthropoda. So should we ask how much the silks mimic **mammalian cell systems** and, before lavishing cost and effort on engineering silk materials, determine if there are more appropriate mammalian proteins we can use?

Two mammalian blood-plasma protein candidates are fibrinogen/fibrin and fibronectin. Each has been as a 3D material or components of materials in significant numbers of applications. Both are human in origin (even potentially autologous, from the patients' own blood), and available in industrial quantities through the plasma fractionation industry. This means that they are non-immunogenic and, even though they can carry human pathogens, these are already assessed and cleared by the plasma fractionators, so are as safe as other human plasma products.

Fibronectin can be (shear-) aggregated from bulk solutions of the native protein to give dense fibrous materials. This process may be similar to that used

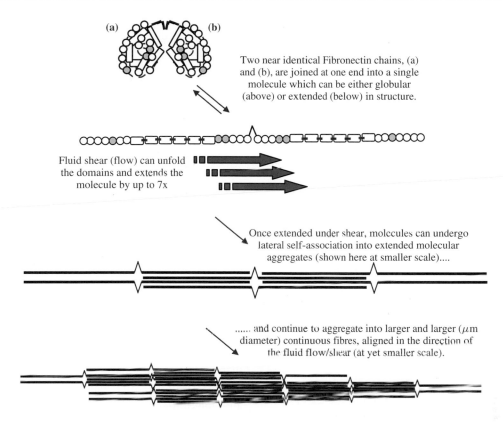

(a) **(b)**

Two near identical Fibronectin chains, (a) and (b), are joined at one end into a single molecule which can be either globular (above) or extended (below) in structure.

Fluid shear (flow) can unfold the domains and extends the molecule by up to 7x

Once extended under shear, molecules can undergo lateral self-association into extended molecular aggregates (shown here at smaller scale)....

...... and continue to aggregate into larger and larger (μm diameter) continuous fibres, aligned in the direction of the fluid flow/shear (at yet smaller scale).

Figure 4.11 Diagram illustrating how some proteins (e.g. fibronectin) may extend and aggregate under directional fluid shear – shear aggregation – to produce fibres and fibre alignment. This mechanism predicts that aggregation will be enhanced where molecules and fibres become anchored to fixed points and where fluid viscosity increases.

by cells in nature to aggregate fibronectin fibres during tissue repair, growth and cell migration. Fibronectin shear-aggregation involves stretching out of the native molecule by application of fluid shear (Figure 4.11). This changes it from a globular to a rod-like molecular structure, and so promotes lateral packing into fibres. Once aggregated, the bulk *and* surface properties are highly biomimetic and ideal for cell adhesion and guidance. Indeed, fibronectin is exactly the substrate deployed naturally in the early stages of tissue repair for exactly these functions. However, as always, excellence in cell guidance comes with a cost and fibronectin materials have limited gross physical strength. Consequently, applications so far have been restricted to promotion of guided nerve and spinal repair through guidance conduits. Many other applications could benefit from this bulk material as

a integral component, just as fibronectin monomer is widely used to surface-coat synthetic materials.

Fibrinogen/fibrin is already available as a self-assembling clinical material, sometimes known as 'fibrin glue' or 'fibrin sealant'. Fibrinogen is the soluble monomer precursor fibrin matrix in every-day blood clots. When our blood (or the cell-free protein liquid part, called plasma – Text Box 4.4) clots, the *coagulation cascade* is activated, (long enzyme pathway – not relevant here) This leads to formation of active thrombin, a protease which rapidly clips off specific tail-sections of the soluble fibrinogen protein, producing fibrin. Fibrin immediately self-aggregates into a fibrillar gel, which then progressively cross-links and shrinks to form a dense gel-material.

Surgical fibrin glues have been commercially available for many years, with separate fibrinogen

---

Text Box 4.4 (Vampire) trivia for non-biologists: blood, plasma and serum

Oddly enough, our most dramatic of body fluids is handily colour-coded. Whole blood (cells plus liquid) is, of course, red. However, this colour comes from the red blood corpuscles and, if we remove the cells, the remaining thick protein solution is *yellow* (straw-yellow, rather than canary-yellow, for the more discerning). However, it matters *how* you get rid of the cells. When we centrifuge cells away in an anti-coagulant (i.e. no clotting), we get *plasma*, which includes its original fibrinogen and coagulation factors. If we are cheap and just let this clot (allowing the fibrin and cells to form a solid clot), we get the yellow liquid comprising the blood proteins *except* for the fibrin, etc., used to promote cell growth in culture. Just to be confusing, fibrin clots alone are pure *white*!

---

and thrombin solutions 'ready to go' in two conjoined syringes. The two liquids are mixed and expelled when the plungers are pressed, so that the fibrin gel sets at the point of application. Tissue engineers have found this trick useful, as have surgeons repairing blood vessels, urethra and intestines, where it is also important that the joints do not to leak.

Such fibrin glues clearly also represent a handy means to produce 'instant' cell supports which can be assembled as and where needed. Importantly, the fibrin gel forms around the desired cell population and may also be injected directly into tissue spaces. Fibrin glues are sometimes loaded with drugs or protein growth factors to assist cell function, with the obvious advantages of a rapid, convenient and off-the-shelf system. In addition, traces of coagulation factor XIII (also known as plasma transglutaminase) cross-link the fibrin fibres, increasing the gel stability, while fibronectin (also a contaminant) is incorporated and assists cell attachment and migration. Less obviously, polymerization by a physiological enzyme means that cell seeding is at the time of fibrin-fibre formation. Cells are trapped, right from the start, within a meshwork of fibrin fibres, just as they would be in the body.

This is major. Put another way, the time needed for cell-seeding and infiltration is *zero*: there is *no* cell-seeding stage.

Unfortunately, free lunches are rare events, even in this branch of tissue engineering, and fibrin-based support materials have their own limitations. Being gel materials – albeit dense gels – they are inherently poor in load bearing. In particular, evolutionary pressures do not seem to have driven fibrin clots to develop substantial tensile or shear properties, or to hold sutures well. In addition, most repair sites have the cell and enzyme machinery to digest away fibrin, leaving it with survival times of only days or even hours.

Typically, then, fibrin-based cell supports are either excellent in directing (certain) cell functions and timed events, or alternatively, they elicit wholly inappropriate biological responses. Examples include the distinctive patterns of cell adhesion and migration, supported by fibrin. Specifically, keratinocytes attach poorly to fibrin and tend to grow *beneath* a fibrin layer. On the other hand, thrombin and fibrin degradation products have significant downstream biological effects on fibroblasts. These are powerful and biologically pre-programmed effects which may or may not be welcome as we engineer tissues.

The final family member, **collagen**, is one of the most widely used tissue engineering materials of all. It may also be the most misunderstood and confusing. It, too, self-associates into fibrils, sometimes with help from shear or enzyme action. However, its main route to fibril aggregation is more 'crystal packing' than 'enzyme drive' and it has more in common with silk than with fibrin. Essentially, at physiological pH, temperature and ionic strength, the distribution pattern of surface bonding along collagen monomer molecules matches *so accurately* that molecules stack together side-to-side. However this only happens when each touching partner monomer lies one quarter staggered to its neighbour. As a result, collagen self-aggregates naturally to give *very* long, strong and tightly packed fibrils. The basis of its strength is the highly regimented molecular packing implied by the strict quarter-stagger. Each new monomer can bind into a growing fibril in just one position, relative to its nearest neighbours. This positional dictatorship comes from the exposed amino acid sequence of each molecule and their limited ability to bend and wiggle (hence, 'no wiggle room'). Such a semi-crystalline molecular packing is the characteristic of collagens and silks, producing strong fibrils, distinct from those of fibrin and fibronectin.

Collagen materials and cell-supports have been fabricated in many physical forms, using a range of processes, and this generates a good deal of confusion. It would be more accurate to consider the term 'collagen materials' as the name of an entire club rather than that of one of its members. As a rule of thumb, collagen sponges are very different to collagen gels, and in both cases there are native, cross-linked, soluble, insoluble and aggregated forms. For most of these examples, though, the differences are mainly of structure rather than composition – in other words, beyond the subject of this chapter (but see Chapter 5). To stick to the topic of composition, we can divide collagen materials into four generic starting materials (predominantly animal-derived type I collagens). These are:

(a) highly cross-linked, insoluble collagen (e.g. reconstituted from shredded suspensions);
(b) tropocollagen (native monomer);
(c) atelocollagen (enzyme-extracted, monomeric with small, key end-sequences cut off);
(d) gelatine or heat-denatured collagen.

This series excludes two much quoted natural 'materials' *containing* collagen:

(i) decellularized whole animal tissue preparations (e.g. small intestinal sub-mucosa-SIS); and
(ii) matrigel, a basement membrane analogue, rich in type IV collagen, from cultured tumour cells.

Both are left out here as they are not fabricated from extracted collagen (bottom-up) but by simplification and processing of whole animal or cell products (top-down cultivation). We will return to this later, but essentially it is the difference between starting complex and working to make something simpler, versus building up complexity from simple starting components. The bottom-up-top-down fabrication distinction (Figure 4.12) is essential here, partly as it defines a central spit-line in biomaterials fabrication. Also, including top-down natural materials could bring all connective tissues and cell cultures, however heterogeneous and poorly defined, to the same discussion.

To elaborate further on the four types of collagen-derived materials:

(a) *Insoluble collagen materials* have a long history as haemostatic sponges and tissue supports. They can be as crude as homogenized tissues, reconstituted and freeze dried, but even these processes remove many original tissue components, such as cell debris and non-collagen proteins. In these materials, the structural elements are made up of shredded fibre bundles, mostly in the sub-millimetre size range.

(b) *Tropocollagen* is the intact, mostly monomeric, acid-soluble form. It is a promising starting point for bottom-up fabrication and readily aggregates into nanometre-scale fibrils and gels. Unfortunately, the extracted yield of this collagen from most mature tissues is low, due to cross-linking (so it is better from juvenile tissues).

(c) *Atelocollagen* is also a soluble, monomer-rich collagen like tropocollagen, but it solves the yield problem by the use of a protein-digesting enzyme (a protease, e.g. pepsin) to break down much of the tissue except for the collagen triple-helix. This cuts the cross-links by removing the short non-helical end extension or 'telopeptides' of the collagen, where most cross-links are located (hence 'A-telo-collagen'). Unfortunately, telopeptides are also important for normal fibrilogenesis, so atelocollagen is rather poor at forming gels – a bit of a death-blow for

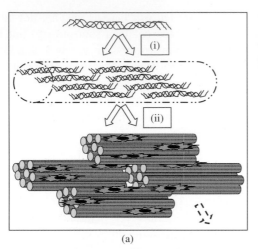
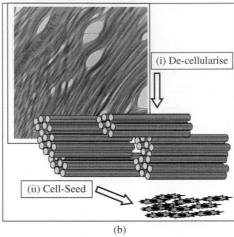

(a)　　　　　　　　　　　　　　　　(b)

Figure 4.12 Scheme illustrating the two routes which have evolved for preparation of native collagen-based materials and cellular constructs for engineering of tissues. (a) shows the so called 'bottom-up' approach of assembling the smallest available building units which it is feasible to use. In this case, 'smallest' is collagen monomers or fibrils, with cells into the required 3D architecture. (i) Purified (normally type I) collagen monomer, acid-soluble collagen, comprises a three-chain elongate helix, 300 nm long by 1.5 nm diameter. At neutral pH and 37 °C, these spontaneously aggregate into quarter-staggered semi-crystalline structures which produce long cylindrical fibrils ($\approx$30–100 nm diameter), trapping any living cells (or non-living particles) within their mesh. (ii) Physical expulsion of fluid from this gel produces a collagen-cell mesh construct dense enough to fabricate a simple living tissue. The 3D complexity of such tissues can be increased (as discussed in Chapter 6), again by incremental addition of components to each layer and by adding many more layers. (b) Shows the reverse strategy, 'top-down', in which a native tissue is first harvested from a suitable animal source; these can include a number of internal fascias (e.g. intestinal and bladder) or tendon and dermis. The dense collagen network (i) is then stripped of its cell content, using a range of disruptive and extraction approaches, to leave a decellularized but otherwise intact collagen material. Any immunological challenges or infective agents are also, hopefully, removed with the cells. (ii) By definition, cell seeding can then only be done after this stage. This means they are met by a dense-packed fibrillar material, where only the surfaces are available (i.e. even though cell debris can get out, whole cells are too large to get in). In effect, animals cells were embedded physiologically in the mesh (they made it around themselves), but these $\approx$15 μm nominal diameter cells were removed by fragmentation down to an easily extractable molecular scale. Replacement $\approx$15 μm human cells cannot possibly get back into the nano-fibre mesh simply by seeding. Some can, and they will, over extended periods of culture or *in vivo*, but only because they enzymatically disrupt and remodel that very special native collagen architecture which was the great advantage of this approach. This strategy has a long and successful history in *prosthetic biomaterials*. For example, replacements for defective human heart valves were developed using pig valves cross-linked, sterilized and cleared of immunogenic epitopes by glutaraldehyde treatment. While these remain effective prosthetic implants (life $\approx$15 years plus), their permanency is clearly not appropriate for tissue engineering.

bottom-up materials fabrication based on gel formation.

(d) Finally, *gelatine* is readily available, cheap and forms gels. However, it is made by boiling collagen until it breaks down and loses its native triple-helix structure. All gone are the triple helix, the tensile properties and quarter-stagger molecular packing of native collagen gels. In the case of gelatine gels, most of the collagen biomimetic properties are also consigned to the junk heap. Perhaps the greatest loss is that of tensile strength and resistance to enzyme degradation. Even the material aggregation process is different, as they gel in a manner completely opposite to collagen – on *cooling* rather than *warming*. In effect, gelatine gels are poorly biomimetic, weak and bio-unstable.

To summarize the above, the insoluble collagen materials family (a) are made by aggregating ready-polymerized clumps (μm scale) of collagen fibrils as their bottom-up building blocks. Soluble (monomeric) collagens of families (b) to (d) start smaller and aim to aggregate single molecules into (nm diameter) fibrils and so gels. Of these, acid soluble tropocollagen (b) is the most biomimetic. Whilst atelocollagen and gelatine (c) and (d) get around some inherent drawbacks of tropocollagen, they trade this off against biomimesis.

Dermal repair offers us a great exemplar sequence for these types of natural material applications. Surgeons have used living skin grafts to fill up holes in their patients for many centuries but, while this has improved steadily, other top-down technologies have also developed to make:

- non-living alternatives from preserved cadaver dermis, and further from:
- animal dermis (de-cellularized and cross-linked to reduce immune reaction and infection).

An example of this is found in Permacol™, a sheet of porcine dermis, treated to remove cells, chemically cross linked and then washed and dried. Cadaver and **top-down** processed tissues such as these have helped address the problems of the shortages of, and inability to store, graft tissue, and have also helped to avoid inflicting further wounds on already sick people at the donor site. In parallel to this, though, we have seen the emergence of **bottom-up** approaches based on soluble and insoluble collagen starting compositions (i.e. low and not-so-low bottom-up).

Integra™ (or collagen-GAG sponges) forms a good example of not-so-low bottom-up dermis fabrication. Shredded insoluble collagen is bonded with traces of glycosaminoglycans (GAG: sulphated polysaccharide) by extreme dehydration. This gives a stable, native porous collagen sponge which can be implanted to recruit local cells and support dermis replacement following burns or trauma injury, just like a skin graft. However, while Integra™ mimics dermal collagen and stays around to fill in the patients' gaps for long periods, it does not contain or deliver cells (the drying/bonding step guarantees this, as it is cell-lethal).

Enter soluble collagen (tropocollagen) gels, containing human cells which promote local healing.

Apligraf® is a native, acid-soluble collagen gel, already interstitially seeded with human dermal fibroblasts to improve and speed up dermal repair. As part of its production, the initial cell-collagen gel is cultured for some days to allow its fibroblasts to contract and strengthen what was initially an extremely weak gel, producing one which can at least be handled. So we have here an example of both insoluble collagen *and* soluble tropocollagen gel implants. Interestingly, where each is biomimetic, they still straddle the same compromises. Integra™ is a strong and durable support for endogenous repair, but without its own cells, whilst Apligraf® is mechanically weak and remains only a few days, yet delivers repair-enhancing cells and growth factors in that time.

Bottom-up collagen processing may have received a technology boost in recent years, to break this compromise loop. The big problem with processes which use resident cells to increase collagen fibril density of gels is that they depend on cell forces to expel water. This is slow and costly but, most of all, the forces are too small to produce the material density we need. Native collagen gels start at <0.5 per cent collagen (in fact >99.5 per cent water). After cell contraction, we can only get this as high as 1–2 per cent collagen (still 98–99 per cent water), which remains extremely weak compared with dermis (17–20 per cent collagen).

This huge strength gap has now been bridged by a technology called collagen plastic compression. In this process, controlled amounts of the excess water are rapidly forced out of the hyper-hydrated cell gels under external load. The whole process takes minutes, rather than days, and it leaves the cells unharmed (indeed, with nothing to do). It produces a tissue-like construct of any required density up to 30 per cent collagen, and makes it feasible to fabricate strong collagen-based 'tissues' *around* the cells (Figure 4.12a), without asking them to do anything.

The previous biomimetic compromise is gone and, incidentally, the process is an order of magnitude faster. Viewed in this way, direct fabrication technologies could eventually impact on tissue engineering in ways similar to the way transistors once changed electronic devices.

4.6.3 Hybrids: composite cell support materials having synthetic and natural components

Over the last decade in tissue engineering, there has been a gradual dilution of the early view that synthetic cell-supports represent the route of choice for reasons of reproducibility or regulatory approval. This has been accompanied by an increasing willingness to consider the benefits of biological or natural materials, and so by implication the need to work to solve their disadvantages. Central to this is the growing understanding that cells need to recognize and fully utilize the support substrate if we reasonably hope to achieve good biomimesis. Part of this movement has found its output in the development of novel families of hybrid support materials with both synthetic and biological – or at least biomimetic – components.

The initial logic of hybrid cell support materials springs from a large body of experimental effort to improve the cell attachment and utilization properties of synthetic polymer supports, most commonly PLGA. This started with coating the polymer material surfaces with recognition peptides for the cell attachment receptors known as integrins, most commonly RGD(S) peptides (Text Box 4.5).

At some point, a research movement of 'Scientists for Cell Attractive Substrates (SCAS)' was formed when a group of biomaterial scientists got together with a breakaway group of the developmental biologist tribe of tissue engineering (possibly in one of the better bars in Zurich). They got to discussing the mimicking of processes from early embryonic developmental as an approach to engineering tissues.

The theory they produced runs something like this: 'We now know a whole mess about growth factors, signalling cascades and cell phases/movements which are involved in 3D embryo growth, so let's recapitulate *bits of it* for tissue fabrication.' We can now only guess at the conversation that led these two groups to hatch their vision. The rather hazy legend has it that it started like this:

Biomaterial scientists: *It's been a rubbish month in the lab. We coated all our scaffolds with every kind of RGD peptide. Some were hanging off on long spacer-arms, while some were tight bound to the polymer. All we got was 5 per cent cell seeding, and **they** dropped off the surface and died in a week. To make it worse, the 95 per cent that fell off onto the support dish self-organized into a neural network and have started planning chess moves.*

Developmentalists: *We know what you mean. We got just the same neural net to form in monolayer, but it flatly refused to suggest any more than three moves ahead until we put it into a 3D apartment block with options to remodel the upper floors.*

So, the two groups formed a **joint research committee** and set to work making polymer scaffolds which cells could attach to and move through by degrading the matrix substance. At the same time, the moving/remodelling cells would release factors useful for controlling the arrival of other cells (perhaps from the next building). In effect, the

---

Text Box 4.5 Engineering with peptides to hold cells down

In the main, cell-matrix binding acts through a family of cell membrane receptors (the integrin family). While cell membrane integrin expression can be quite distinct between different cell types, many recognize and bind to very similar short amino acid sequences on the most common matrix proteins. The functional part of these peptides frequently has the sequence arginine-glycine-aspartic acid (+/− serine), hence they are known as RGD sequences. Another integrin recognition sequence, YIGSR, is commonly associated with basement membrane attachment sites. It has long been a dream of tissue engineers to use these peptides, attached to biomaterial surfaces, to improve or control how cells attach.

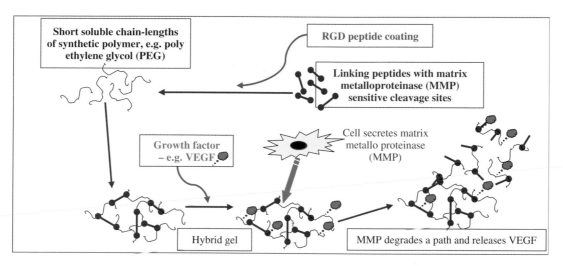

Figure 4.13 One example of how a hybrid gel of increasing complexity might be built up. Soluble PEG chains, potentially with RGD peptide coating, are linked together by peptide sequences such that the complex forms a hydrogel matrix. The peptide link sequences contain protease sensitive sequences which allow cleavage by cell-derived MMPs. With incorporation of the growth factor VEGF to stimulate angiogenesis, this gel becomes a designer substrate for local cells to move through, degrade and attract blood vessel in-growth (Seliktar *et al.* (2004) and Lutolf & Hubbell (2005)).

design of this is a simplified version of how we understand natural cell-matrix systems might work (Figure 4.13).

Unlikely (but memorable) as this fiction might be, it illustrates the core philosophy of this emerging group of *hybrid* cell support materials. The concept of how to go about the hybridization is based on linking ever greater varieties of modular parts (biomolecules, such as peptide recognition and cleavage sequences, growth factors) onto a structural polymer backbone. Cell-specific matrix cleavage is achieved by linking short support-polymer sequences together through synthetic peptides with a cell-enzyme cleavage site (e.g. in Figure 4.13, the ubiquitous matrix metalloproteinase, MMP-2). This linking is designed to aggregate the complex into a 3D cell support (normally soft), which may also carry RDG attachment sequences. Cells moving over or into such matrices naturally secrete proteases and so also degrade the support as they move.

A second thread of developmental biomimetics has evolved in this system involving cell-signalling growth factors. These speed up (or slow down) other cell processes which are needed to assist the new tissue formation. A prime example of this is to encourage angiogenesis or blood capillary (micro-vascular) in-growth for bio-integration and nutrient supply. Favourite amongst the factors used are vascular endothelial cell growth factor (VEGF) and fibroblast growth factor (FGF). Alternatively, platelet-derived growth factor (PDGF) and/or transforming growth factor-β (TGF-β) are candidates to promote connective tissue formation, for example in dermal tissue engineering. Such systems, though, can need a great deal more understanding of bio-control than we possess, in terms of the combinations, quantities and sequences of factors needed to achieve any given response.

Since many growth factors need to be free to leave the material and enter the target cells, physiological growth factor binding molecules have been coupled directly to the support polymers. These binding molecules then hold growth factors onto the support material until it is removed by an adjacent cell. Other cells can take up growth factor signals from the surface of the material, as they do in nature. In one example of this (Figure 4.13), heparin, a natural binding molecule for a family of growth factors, is chemically immobilized to the polymer matrix. Simple mixing of the material with heparin-binding growth factors, such as VEGF or FGF, loads the support matrix with control factors, which are now accessible for uptake by cells. Such direct cell-uptake would not be possible if the factors were directly bound to the matrix.

Such biomimetic surface and bulk modifications now present the possibility for full blown hybrid biomaterials where synthetic polymer supports are given complex cell-responsive functions. This tackles the difference between synthetic and natural-biological cell-supports as described in the previous sections (4.6.1 and 4.6.2). By making the synthetic polymer and peptide sequences in a manner which forms gels on mixing, it is possible to trap cells within the matrix as it forms (interstitial seeding). This also creates a matrix which it is possible for cells to degrade directly with their enzymes. These represent (semi-)synthetic analogues of native matrices such as collagen and fibrin as described above. The example illustrated in Figure 4.13, from the Hubbell research labs, is based on soluble polyethylene glycol as the synthetic backbone, but others are possible. Many of the component ideas are based on earlier work using fibrin materials or conventional surface modifications of PLGA materials.

In theory this leads us to a 'Lego' type of system for building up and tailoring biologically mimetic support materials to match any given local cell/tissue application. Sadly, no one has really reported on chess-savvy cell networks, so we cannot test the original myth against a rated chess master.

Excitement generated by hybrid materials is real and justified, though it is important to ask where they lead in practice. They clearly represent new tools to understand in detail the key cell events and molecular sequences which are so important to tissue formation, especially in 3D. However, as tissue engineers, it is critical for us to know if their utility is in *understanding how* tissues are formed or as practical cell-support materials *ready to apply* to implant fabrication.

One such analysis is to compare the advantage/disadvantage profile of the hybrid itself with the two hybridized elements. **Synthetics** are reproducible, cheap, easy and safe to produce, but are poor biologically. **Natural** materials are in many ways as good as we can get biologically, but they can be difficult to produce and their very naturalness raises questions of safety and predictability. The hybrid systems can be judged on the balance of how much they bring of the *best* or the *worst* of their components.

It seems likely that hybrid matrices are presently more research tools than imminent, practical implants. This is partly because of production cost complexity and partly as our understanding of the workings of cell-matrix systems is still rather too simple for us to assemble the best 'Lego' polymer parts. In addition, sharper readers may have noticed that the discussion of hybrid materials so far has focused exclusively on composition. Spatially when, where and how things happen, and the control of material-mechanical properties, μ-shape, asymmetry and direction, are hardly yet on the agenda. Thus, our analysis might suggest that hybrid materials have lots of practical potential for tissue engineering but are longer-term prospects for clinical applications. There is a strong probability, however, that they will inform how we design and fabricate the compositions of ever more biomimetic implants.

4.7 Conclusions

Tissue engineering relies heavily on the idea that substantial control can be exerted on cell processes through the surface and bulk (composition) properties of its support materials. In terms of options, there are three clear general concepts on composition: synthetic, natural and a hybrid of the two. Despite the enormous spectrum of combinations, this division is helpful as it allows a rational analysis of the *functional* cost benefits for cells in any given application. The task of the tissue engineer remains to develop supports which have *biomimetic* and *dynamic* cell control properties for the cells and site (or bioreactor) where they are to be used. Meanwhile, next-generation cell support materials will reduce the major compromises we presently have to make, while hopefully avoiding the prosthetic logic. In particular, this means identifying the *key* biological processes that we need to promote or to shut down.

In recent years, the pendulum has probably swung away from the dominance of traditional, unmodified synthetic polymer support materials. Natural (usually protein) materials have attracted greater interest, but a great deal of basic work is needed to understand how to fabricate these to the level needed. In this, we are learning more from how natural protein fibres are produced in natural systems, particularly the relative perfection of their packing, and how they influence cells at so many levels. Hybrid matrices, may provide major clues as to what is needed for better natural and synthetic support materials by showing what controls are *really* important (and when) for any given tissue engineering objective. Meanwhile, there is now a growing argument to engineer natural materials systematically in the same way we have done for synthetics.

Further reading

1. Pachence, J. M., Bohrer, M. P. & Kohn, J. (2007). Biodegradable Polymers. In: Lanza, R., Langer, R. & Vacanti J. P. (eds.) *Principles of Tissue Engineering*, pp. 323–339. Academic Press, Burlington, MA.
2. Abou-Neel, EA. & Knowles, J. C. (2009). Biocompatibility and other properties of phosphate-based glasses for medical applications. In DiSilvio, L. (ed.) *Cellular Response to Biomaterials*, pp. 156–182. Woodhead Publishing, Cambridge, UK.
3. van Dijkhuizen, R., Moroni, L., van Apeldoorn, A., Zhang, Z. & Grijpma, D. (2008). Degradable polymers for tissue engineering. In: van Blitterswijk, C. A. (ed.) *Tissue Engineering*, pp. 193–221. Academic Press, London.
4. Brown, R. A. & Phillips, J. B. (2008). Cell responses to biomimetic protein scaffolds in tissue repair and engineering. *International Review of Cytology* **262**, 75–150.
   [References 1 & 4: Good starting points to taste the diversity of composition and behaviour which are available in materials.]
5. Tognana, E., Borrione, A., De Luca, C. & Pavesio, A. (2007). Hyalograft C: hyaluronan-based scaffolds in tissue-engineered cartilage (review). *Cells, Tissues, Organs* **186**, 97–103.
   [Focused view of some forms and uses of hyaluronan as support materials.]
6. Volrath, F. & Knight, D. P. (2001). Liquid crystalline spinning of spider silk. *Nature* **410**, 541–548.
   [Inspiring review of 'where to go and why' with silk, from two true experts and major players in advancing the use of silk.]
7. Fratzl, P. (2008). *Collagen Biology and Mechanics*, pp. 506. Springer, New York.
   [Full throttle, modern review of collagen in all its aspects. Try Chapters 1 & 3. This pins down collagen, but you need to be strong.]
8. Seliktar, D., Zisch, A. H., Lutolf, M. P., Wrana, J. L. & Hubbell, J. A. (2004). MMP-2 sensitive, VEGF bearing bioactive hydrogels for promotion of vascular healing. *Journal of Biomedical Materials Research* **68**, 704–716.
9. Lutolf, M. P. & Hubbell, J. A. (2005). Synthetic biomaterials as instructive extracellular microenvironments for morphogenesis in tissue engineering. *Nature Biotechnology* **23**, 47–55.
   [References 8 & 9: Pair of Influential paper describing examples of engineering natural functions into non-natural support materials.]
10. Vinney, C. (2000). Silk: Fibres: origins, nature and consequences of structure. In: Elices, M. (ed.) *Structural Biological Materials: Design and Structure-Property Relationships*, pp. 295–333. Pergamon Press, Oxford.

Understanding the scale where we are operating is absolutely critical when we consider 3D shape and spatial organisation. This pyramid of cannonballs forms a splendid 3D shape example. Held together by the square brass 'monkey' (arrowed), it has generated a legend and common English saying. The trouble is that when we actually *measure* at the scale that matters, the legend seems not to stack up.

5 Making the Shapes for Cells in Support-Scaffolds

Constructing tiny Galapagos for cells

In basing so many of his illustrations of the evolution of niches on the very special Galapagos Islands, Darwin made a particularly intelligent selection. The island location makes the local niches excitingly special, capturing and engaging the imagination. At the same time, the special effect of being an isolated island group has tended to greatly simplify the picture we see. In other, contiguous locations in a large land mass, the same process, repeated over the millennia, has left an unthinkable complexity which we still pick over.

Multi-dimensional complexity is the essence of niche evolution in bio-systems, but looking full-face into that complexity is too daunting to understand. Like studying the sun during an eclipse, Darwin's genius was to illustrate the evolution of niche complexity through simplified versions – on remote ancient islands.

Extreme Tissue Engineering: Concepts and Strategies for Tissue Fabrication, First Edition. Robert A. Brown.
© 2013 John Wiley & Sons, Ltd. Published 2013 by John Wiley & Sons, Ltd.

In our small way, extreme tissue engineers are trying to understand and, for short periods to recreate, *grossly simplified* cell niches* in order to achieve

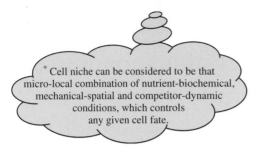

* Cell niche can be considered to be that micro-local combination of nutrient-biochemical, mechanical-spatial and competitor-dynamic conditions, which controls any given cell fate.

our bio-fabrication goals. We are trying to build small, isolated but biomimetic 'islands of cell-space' with our cells in situ and working for us. In other words, we are aiming to build our equivalent of a Galapagos – simple enough for us to cope with, given that the full niche is so mind-meltingly complex. Here, we take a look at ways in which we can and are making Galapagos at the cell-scale.

5.1 3D shape and the size hierarchy of support materials

There is a saying in English when the weather gets really cold that it 'could freeze the balls on (or off) a brass monkey' – commonly shortened to *'brass monkey weather'*. Legend has it that this saying comes from the old naval days where cannon balls were stacked in pyramid shapes by the side of their guns. Triangular trays with holes machined to be close-fitting and grip the cannon balls would hold each layer and so keep the pyramid of balls together (for an illustration, see any good Hollywood pirate film). These were known as 'monkeys' and made from brass (front piece illustration). The story was that at very low temperatures, the difference in thermal contraction between the steel balls and brass monkey would either force off the balls or make them stick (in both cases rendering the warship useless).

This is a rather quaint analogy for our cells in their neatly fitting niches within an engineered 3D structure. Providing the physical shape and dimensions of the supporting material are made appropriate to the cells or cannonballs, they will

be sustained in a stable, 'comfort' position and will thus be ready to perform their function when needed. The importance of 3D space and shape is equally highlighted in this analogy by the *apparently* subtle temperature-sensitivity of the (brass monkey) support-system.

The trouble here is that this story is now regarded as an urban myth (or Hollywood pirate-story), for many reasons. First, in reality cannonballs were stored safely below decks in long *wooden* trays ('shot garlands' – i.e. no metal-to-metal-anomalies). Also, a fatal flaw in the story is that the difference in brass-iron expansion coefficients is small, producing only ≈ 1 mm change over a 1 metre long monkey if the temperature changed by $100\,^{\circ}$C!

The moral of this is that if we want to produce control systems with shape and 3D structure, it is essential to do the measurements and find out *exactly* which part of the length-scale hierarchy our cells (or cannonballs) are sensitive to. Hollywood can prosper from arm-waving ideas, but tissue engineers will not.

Figure 5.1 shows a scanning electron micrograph of a collagen gel (very biomimetic) embossed with a pattern, derived from a fine fibre mesh. At the gross, millimetre-scale (top), the 3D pattern resembles that of the original embossing mesh. It is an orthogonal pattern, with opposing fibre strands arranged at right angles. As we look at higher magnification, in the in $1-100\,\mu$m scale (middle panel), the pattern becomes strongly parallel. In the lower panel it is clear that, at the sub-μm scale, the 30–50 nm diameter collagen fibrils which make up each ridge are completely random in orientation. Not visible here (but drawn), there is a further level where collagen molecules which make up the fibrils (1–2 nm diameter) again have a parallel alignment for significant distances, as this is how the molecules aggregate into fibrils.

In other words, the **same structure** has at least four distinct patterns of 3D orientation, depending on the scale-hierarchy that is sampled. There is, then, a **'reading-window'** effect for any given bio-structural hierarchy, which makes it important to consider at which scale these structures are being read (i.e. the scale at which they act). A single cell would mainly sample and utilize structures

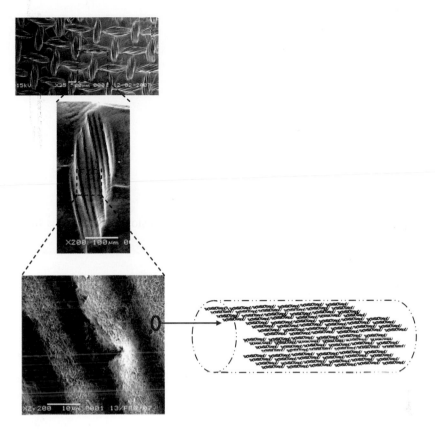

Figure 5.1 A pattern embossed into a single, dense-compressed native collagen gel. Scanning electron micrographs of the same site (embossed with nylon mesh), but at three increasing enlargements. Top, bar = 500 μm; middle, bar = 100 μm; bottom, bar = 10 μm). The diagram (bottom right) shows the 'parallel' pattern of collagen molecular packing within each of the fibrils (i.e. the fine strands just visible in bottom micrograph). From: Kureshi, A., Cheema, U., Alekseeva, T., Cambrey, A. & Brown, R. (2010). Alignment hierarchies: engineering architecture from the nanometre to the micrometre scale. *Journal of the Royal Society Interface* **7**, S707–S716.

in the middle range of this hierarchy. In contrast, the integrin receptors embedded within the cell membrane would bind to 3D structures in the mid-nm range (lower panel). Only where cells form larger clusters, or syncytia, could they utilize mm-scale patterns.

Finally, at the other end of the spectrum, parallel macro-molecular structures in the low nanometre range of the hierarchy might influence protein-protein interactions and glucose mass transport. The hierarchy paradox means that, like expansion and cannonballs, 3D structure does not matter simply because we can see it. This is a function of our microscopy. 3D structure of support materials *must* be measurable to the cells that we hope will use it. If the scale of the structural cues is too large or too small, they cannot affect cells directly. Like the sound of a dog-whistle to the human ear, it might as well not exist.

The take-home message here, like so many before, is pretty obvious once stated, but no less important to actually implement. Our task here is (simplified) 3D niche design for cells but, to achieve this successfully, it is critical to get our minds down to the cell and molecular level in the scale hierarchy. Not least it is necessary to measure features and events we have engineered around our cells in order to make sure that our designs really do work as we expect. This is, after all, just good hypothesis testing. The surprise perhaps, is that we still have so much basic biology to learn after so many centuries of study.

Then again, that is what followed from Darwin's beautiful simplification.

5.2 What do we *think* 'substrate shape' might control?

Cell-support architecture is basically the way that the bulk composition is structured in 3D to hang together in space. This also governs how the surfaces are made available for our cells to hold onto and grow over, either alone or attached to their neighbours. That same 3D organisation and pattern of surfaces also dictates the rate and direction that soluble molecules (nutrients, oxygen, proteins) move, enter and leave. If that were not enough, support-material architecture also controls mechanical properties of our constructs, particularly at the cell/micro-scale.

To use a human scale analogy, it is easy to understand why the material properties of glass lead to its use in particular functions. Think of glass and we think 'hard, brittle, optical clarity'. But it is still striking to look at the wide variety of detailed functional properties which can be achieved by fabricating different μ-structures. Figure 5.2 suggests some of these, from the transparent table top to opaque etched decoration and the sharp microparticle abrasive paper.

Clearly, then, chemical composition matters, particularly in the basic and the gross functions, but 3D spatial organisation of that material across the length-hierarchies dominates the more subtle and dynamic properties. So, if we agree that the extreme challenge is to reproduce *some* of the subtlety and dynamics of native biology, where would you, the reader, choose to concentrate your efforts?

To list and organize the 'tissue-engineering-critical' properties which are controlled by architecture, we can start at the sub-nano/molecular level and float up the size hierarchies. At this level, small nutrient/metabolite molecules such as sugars, amino acids, phosphates and O_2/CO_2 pass rather easily through different nano-structures. However, the direction and rate of movement is dictated by the packing and alignment of the bulk material.

Figure 5.2 The same composition of glass is shown here, used in a number of different 3D architectures. This makes a reflective glass mirror and transparent table surface (i) smooth, down to the nm scale; (ii) opaque, with μm-scale etched surfaces (decoration on the decanter) and (iii) abrasive glass paper comprising mm-scale random glass fragment particles, for scouring painted surfaces (i.e. the 'green' glass-paper, rear). Although their compositions are the same, they each have contradictory properties and completely different functions, based on the scale of their surface structure (ultra-smooth versus abrasive; transparent versus opaque).

The effects of bulk structure on small molecule movement at this level can also be dramatically affected by distortions of the structure during dynamic mechanical loading (Figure 5.3).

The same factors affect mass transport of control macromolecules, such as proteins, though at a different scale – in the nano/sub-micron range. Such protein movements dominate export, import, remodelling processes and repair. This, in turn, means that support-material structure controls cell function (cell-matrix attachment, cytoskeleton, shape, division, differentiation) plus communication and cooperative behaviours. In turn it will influence cell-cell attachment & dialogue, migration, sheet/layer formation, contraction, fusion, synapsis).

So, aside from that little lot, how else *could* 3D μ-structure affect the things we are most concerned about in tissue engineering?

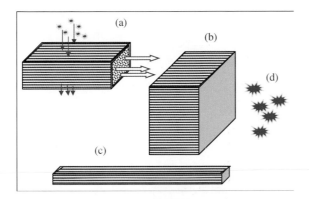

Figure 5.3 Where small molecules are passing through a fine, aligned fibre structure they may pass freely across the structure or be deflected. In (a), small molecules entering from the upper surface are shown being 'deflected' orthogonally, moving more parallel to than between the fibres (i.e. path of least resistance where size shape is a constraint). Smaller molecules might be expected to diffuse more equally in all directions. What, then, would you expect to happen to molecular movement when the same material is deformed (compressed) to be shorter and fatter (b) or stretched to be longer and thinner (c) – or where it was cycled, slowing from one extreme to the other? Alternatively, what would happen if the diffusing molecules had an average radius 10 or 50 fold greater (d)?

5.3 How we fabricate tissue structures affects what we get out in the end: bottom up or top down?

How we fabricate tissues is the subject of later chapters, but at this stage it is worth outlining, at the most general level, which overall strategies are available to us. This highest level of strategy refers to the '*direction*' in which the chosen fabrication route occurs: **top-down** or **bottom-up**.

Production of petroleum products from crude oil may be considered as *top-down* processing, whereby highly complex starting material is refined down and sometimes modified into its useful components. Starting materials vary from 'Brent crude' and 'West Texas Intermediate' to heavy crude, bitumen and tar. Depending on their viscosity, complexity and sulphur content, they can be used to produce different products. If, or when, it becomes economic

to manufacture synthetic fuels by catalytic assembly of natural methane, or even from elemental carbon and hydrogen, this would be a *bottom-up* process.

In short, we can either *refine down* complex (often natural) products or *build up* from simple purified building blocks. Importantly, in both directions, there can be many starting points, depending on what we choose as the 'top' or the 'bottom'. The top can be unbelievably complex and messy, while the bottom can be uneconomically and atomically basic and pure, but at each extreme the costs and technical difficulties soar. It is not accidental that when petroleum was cheap, our ancestors did not bother refining the prolific tar-shale deposits that we now use.

The crude oil analogy here, however, is incomplete as it only covers chemical composition. We are considering here assembly of 3D materials with stable shape and substance. In the case of fabricating shape, we might consider making a boat by hollowing out tree trunks as a top-down approach. The bottom up alternative would then be to make and shape a clinker-built boat with nails and smooth planks. Similarly, digging passages, doorways and rooms into soft rock is top-down house building, while manufacturing rectangular bricks, tiles and cement is how we generate a bottom-up housing market (Figure 5.4).

These analogies teach us that there is a utility-tension between these two approaches. Top-down tends to be simple and economic, leaning heavily on what is easily available, but using crude starting shapes and materials. The relative simplicity of this fabrication has to be balanced against the restrictions on what you can do (i.e. the starting material dictates what is possible). By contrast, bottom-up shifts the effort and expenditure to making and assembling the material building-block. This is initially more complex and expensive, but this investment opens up the possibility of making *exactly* what is needed, where it is needed.

Cell support materials which are made predominantly from synthetic polymers (e.g. poly-lactides or poly-glycolic acids) are, almost by definition, made by bottom-up processing. In this case, we go from simple chemical monomers assembled into

(a)

(b)

(c)

Figure 5.4 The building of **cave houses**, as opposed to free-standing brick or wood houses, can be viewed as two contrasting construction methods. The first (cave house building) represents a top-down form of construction, where an existing complex 3D structure (the rock cave) is cut out, selectively dug away and simplified down to give the desired shape and function. On the other hand, a free-standing house is fabricated bottom-up, from nothing, using basic building materials of bricks, planks and tiles. The difference is clear in these cave houses in Southern Spain, where (a) is a home built into a rocky outcrop; its chimney (**) and garden (b) show the idea. Larger dwellings are seen in larger rock faces (c), where the entrance (red arrow) and patio (open arrow) might look like a standard house, but the upper windows and balcony (double arrow) give it away.

polymer chains and then aggregated to fibres or sheets for assembly into 3D structures. Mineral materials, such as hydroxyapatite, can be made top-down by grinding up extracted bone mineral into powders, but they are more commonly built up from their inorganic chemical constituents.

Natural protein materials, however, present a far greater variety of options, illustrated here by extending the collagen example which we started in the last chapter under 'composition'. Possibly the best example of the two-directional, bottom up/top-down approach to engineering materials structure can be seen in collagen-based cell support materials. There are two counter-current logics. The first relies on **refinement down** of whole (collagen-rich) tissue structure, as against **assembly up** of simple collagen building blocks (Text Box 5.1).

The top-down approach uses native animal tissues, with all their advantages of natural architecture, strength etc. Top-down processing involves the removal of as much as possible of the animal cells, cell debris and other antigens, minimizing risks of rejection or infection. Although some versions are in clinical use, this remains 'work in progress' to improve clearance and prevent unwanted inflammatory responses. At its best, this tissue (top-) down approach generates excellent biomimetic structure with exceptionally strong and biocompatible properties, ready for cell seeding. This is exemplified by the processing of porcine small intestinal submucosa – SIS – to provide implantable platforms, some of which are currently in use clinically.

For top-down fabrication it is essential that the animal cells of the original tissue are broken down into small enough fragments to allow these to escape from the dense tissue mesh of matrix which is required. However, as Shakespeare might have put it, 'there's the rub![7]' Given that the natural pores, tracks and channels in these dense native connective tissues are mainly nano-scale to a few μm, there is little chance to force multi-micron scale

---

[7]'*To die to sleep, to sleep, perchance to dream; ay, there's the rub . . .*' Part of the 'To be or not to be . . . .' soliloquy in Shakespeare's *Hamlet*. 'Rub' was a sporting term for an obstacle in the game of bowls which diverts a ball from its true course.

Text Box 5.1 Differences of opinion about what the 'bottom' is in bottom-up

'Bottom-up' and 'top-down' are terms more common to fabrication technologies and engineering than the biosciences (i.e. the biology-tribe members). Perhaps predictably, this means that the complex-top can be much more complex than we are accustomed to in engineering. For example, we probably need to recalibrate the concept of 'complex' versus 'simple' where biologists are involved. Their natural reaction is that, despite its complexity and variability, a cell can be considered as a 'basic building unit'. The justification here is that to qualify as 'living', we are pretty well stuck with cells, however imperfect, they are as *the* bottom unit.

This recalibration is just as necessary for our definition of what we choose to class as 'bottom' for protein materials. For example, in the case of collagen, it can be taken as reasonable that the native collagen monomer is the basic building block for the material. However, to go up a scale in 'basics', pre-aggregated collagen fibrils could be used as base units to build our materials with a new bottom-up. On the other hand, to go down a scale, we *could* assemble strings of collagen-like amino acid sequences into designer proteins with some of the properties of collagen materials.

The question is, since amino acid sequences are smaller units, are they then the true bottom, from which we go up? The answer is, 'not really'. These are crude distinctions based on process 'direction', and there is no moral high ground in having a more basic or smaller bottom – just the normal rewards of pragmatic function. In fact, if this were true, the peptide guys would be undercut by organic chemists synthesizing amino acids from carbon and nitrogen.

If any reasonably definable basic building blocks allow us to assemble what we need, economically and reproducibly, that is OK. In the example here, synthesizing complex collagen protein 3D materials from isolated amino acids is not practical, economic or functional at present. At the other extreme, large fibril aggregates, in the form of shredded collagen tissue fragments, are certainly cost-effective and simple, but they can be functionally limited. This presently favours the use of intermediate level starting units, namely collagen protein monomers aggregated into fibrils around cells. However, in reality, the base point changes with time as it is sensitive to technology and society's expectations.

cells into that mesh when it comes to re-seeding without substantive disruption. The trouble is that 'substantive destruction' of the native architecture reduces the great advantage of the top-down processing. Current approaches to achieve this interstitial seeding (into the collagen mesh) is either to wait until incoming cells break open their own multitude of pathways into the bulk collagen (often *in vivo*), or to cause substantial artificial disruption to the mesh using detergents or ultra-sound.

In contrast, using the bottom-up route, collagen is assembled from small, monomer or oligomer units (Text Box 5.1). The collagen nano-fibrils aggregated in this way are far smaller than the cells that they enmesh, just as they are in living tissues (Figure 5.5b). This is made easy where collagen fibril assembly, or gelling in general, occurs under physiological (i.e. cell-friendly) conditions, as happens around any resident cells. The result is that cells are seeded **interstitially** (i.e. into the fabric of the fibril material) from time zero.

In terms of our previous 'composition' distinction between synthetic biodegradable polymers versus cell-degradable natural polymers, there are clear consequences for the success of our fabrication of μ-structure. Not least, the dynamic of controlling the μ-structure of the natural polymers is immediately given over to the resident cells to develop, change and remodel, as they are programmed to do. In contrast, where resident cells have no ability to remove synthetic polymer structure, their ability to perform remodelling processes is restricted or even removed. This means that the bottom-up effect of being able to fabricate nano-structured support matrices, with cells enmeshed at t_0, is complemented (using protein supports) by enabling cells to maintain the matrix biomimetic structure as time progresses (t_{0+n}).

So, in short, synthetic, bio-soluble polymers block natural cell-tissue dynamics, whereas natural polymer cell-supports give up the dynamic of process-control to the cells. Note: the risk of

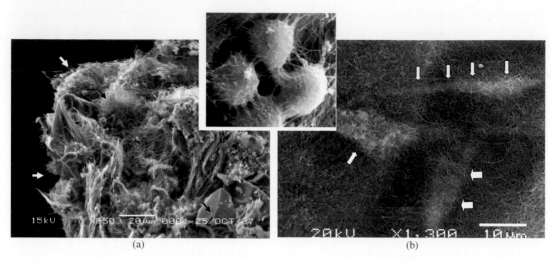

Figure 5.5 (a) Soluble phosphate glass particles, <20 μm (black arrows), embedded in a nano-fibrillar collagen gel (white arrows), compressed to a tissue-like density. (b) Cells interstitially seeded into the fabric of the collagen fibril mesh-work, and again compressed to a tissue-like density. Main picture shows three enmeshed fibroblasts just below the surface (arrows); inset shows a group of four cells just retained by a few surface collagen fibrils. Reproduced by kind permission of Dr. Tijna Alekseeva, UCL.

giving up control is that cells can (and do) get the programme 'wrong', especially *ex vivo*.

5.4 What shall we seed into our cell-support materials?

We now come to a major tissue engineering-scaffold distinction which we have seen emerging more and more regularly in the previous sections. This is the way in which some materials can be assembled and formed into 3D shapes *around* resident cells, while others must be *seeded* with cells after fabrication. The former can be described as 'interstitial-' or even t_0 (time-zero) seeding, as opposed to the latter, 'surface-' or post-fabrication seeding.

Some readers may already be starting to see why the stage and manner of cell seeding is, in fact, one of the most important factors in determining the level and type of 3D structuring we can achieve. Once again, it comes down to the scale hierarchy question. Most mammalian cells we would like to use range in size from 12 to 25 μm in diameter (when spherical). Thus, for cells seeded post-fabrication, some form of access 'pores' of 100 μm or more will be needed to allow even small cell groups to penetrate to deeper zones. 10 μm apertures might allow the squeeze-through of single flattened cells and so painfully slow seeding. The result is that post-fabrication seeding materials have 3D μ-structures *forced on them* in a manner which has nothing to do with biomimetics, bio-control or cell function. In contrast, interstitially t_0-seeded materials are *not* forced to grow in this way.

The cargo-carrying ship analogy of the last chapter is one we should continue to develop. Interestingly, that cargo does not necessarily have to be cells alone. Depending on the nature of the cell support and its intended application, there are many ways of loading with nano-micron scale particles, macromolecules (hormones, growth factors and enzymes) or even small molecule cargos (e.g. drugs or metabolites). Fibrous cell-supports, formed by gelling under physiological conditions, permit the widest range of interstitially seeded cargo-carrying options. More robust cargos, such as hydroxyapatite particles, mimicking hard tissue substrates such as bone, can be added to delicate, natural protein materials as well as those made under cell-lethal temperature, pH or solvents.

Figure 5.5a shows the appearance of soluble glass particles (<20 μm diameter) cast into a collagen gel after compression to near-tissue matrix density. Larger or heavily charged bio-molecules (typically >100 k molecular weight) can be incorporated directly into the fabric of many support materials – for example, those with a nano-fibrous mesh or pore structure with slow or cell-dependent breakdown. However, in many cases, delicate or smaller macromolecules, drugs and metabolites can be carried in artificial vesicles or nano-particles, which are themselves trapped within the support material. Such delicate cargos commonly need to be loaded under physiological conditions, either to preserve the cargo itself or its vesicle wrapping. Again, this can be simpler for gelling and nano-porous support structures.

The use of support materials which carry alternative bio-control cargos in addition to their complement of cells has grown in importance with the need for complex, agent-delivery systems at the implant site. This is increasingly seen as a major point of contact with conventional pharmacological and bio-molecular therapies, and a rich opportunity for collaboration with the 'pharmacology' tribes.

Continuing our extended example analysis of collagen-based cell-support materials, it is clear that cells can be seeded onto the surfaces of top-down fabricated, whole tissue-derived materials such as SIS (see Section 5.3). Bottom-up fabrication of cross-linked collagen sponges, involving freeze drying or glutaraldehyde treatments, must still be seeded post-fabrication. However, supports made using native fibril self-assembly (i.e. native collagen or other self-assembling hydro-gels) aggregate at the nano-scale, around cells, to give tissue-like 3D cell-matrix structure. This is interstitial cell distribution within the material in a nano-fibrous mesh.

5.4.1 Cell loading: guiding the willing, bribing the reluctant or trapping the unwary?

We are about to analyze shipping analogies involving human cargos. With the aim of side-stepping any political correctness sensitivities which might raise nasty historic or geo-political connotations of 'mass transport of people', we shall specify 'troop ship' or 'worker's ferry'.

Clearly, once we reach the point of loading living cells as part of our cargo, we are well inside the zone requiring delicate biological treatment conditions. This is much the same as for passenger transportation, where cabin space and leisure conditions are critical. Paradoxically, loading of cells *can* be easier than non-living cargos, because cells will, under the right conditions, move into the deeper parts, much as passengers will distribute between decks, given incentives, time and a map. Without good instructions, though, both cells and passengers are as likely to go anywhere they fancy – or nowhere.

Much of the earliest work in engineering tissues, using synthetic polymer cell-supports such as PLA and PLGA, was designed to encourage cells to migrate, as far and as fast as possible, from the surfaces where they were seeded, to the deeper zones. Once cells had attached to the surface, the aim was to make it as easy as possible for them to populate all levels of the construct. This affected how 'pores' were viewed and designed. The architecture of these cell-supports was fabricated with this as the primary aim, giving rise to porous foams, non-woven, woven and knitted meshes. Fibre-based materials tended to produce structures with complex, sometimes highly asymmetric patterns, comprising interconnected pores or channels with a huge range of sizes and shapes.

The variability and pattern of these depends on whether the fibres are woven, non-woven, knitted or completely random-enmeshed (like cotton wool) and on the variability of fibre diameter. Non-fibre-based materials can have more regular pore structures, described numerically as mean pore diameter, percentage porosity or interconnectivity. These terms are found commonly in descriptions of sponges or films used to support cells, where pore distribution can be:

(i) homogeneous throughout;
(ii) larger in superficial areas; or, in some instances,
(iii) laser-drilled from top to bottom through otherwise dense films (e.g. Hyaluronan: 'Laserskin').

For these materials, pore density, diameter and connectivity are important features, though notably, these range widely across the scale of several 100 μm, i.e. many cell diameters.

In terms of our troop-transport analogy, this would be a little like designing a vessel with huge open spaces, so that the troops can get on board as fast as possible (Figure 5.7). However, large, open living spaces are not so great for the occupants during long, rough sea crossings, unless it is assumed that the troops will reshape the space once they are aboard. In a way, this was the logic behind early cell-supports: get the cells to fill all the 3D holding spaces *as fast as possible* and assume that they will construct what they need to 'be comfortable' for the longer term. Unfortunately, this assumption *is wrong*. It might happen where the troops have ambiguous, vague instructions but no detailed operation manual (in the case of cells, read 'ambiguous control cues'). Alternatively, the problem might be that building the necessary 3D comfort structures, cell remodelling, takes the troops or cells too long and the mission goes 'wonky'. The troops are forced to roll about pointlessly in unsuitable spaces for weeks, arriving exhausted, sick or mutinous—just like our cells.

Clearly, there is a place for wide open spaces to transport close-packed troops, but this would be a 'landing craft' (Figure 5.7), which has a completely different philosophy and purpose. The name in this case is the clue: landing crafts are designed for short, quick, on-off *landings* (maximum duration, hours). Critically, where this is the *primary* design aim, there can be no assumption that the troops will do anything other than 'be transported'. So, just as with troop transportation, it is important to work out exactly what we want our cells to achieve *before* they are pushed aboard our cell-support constructs. This includes pre-launch understanding of exactly what we can *realistically* expect our cells to contribute to the process.

To wind up the cargo ship analogy, it is clear that to achieve simple, maximal *carrying capacity* (whether cells, troops, bananas, oil or gravel) the best designs will have the least internal 'dead space' created by walls, floors, corridors and partitions. In effect, the empty cargo space *can* be arranged, either as one large void or as thousands of tiny sections. But the overall ship volume will carry much more if the 'cargo' is in one or two huge compartments, as it is in the case of oil or gravel.

On the other hand, as we saw earlier, **how the cargo arrives** matters. Bananas do not travel so well in five metre high piles. People and cells are even less tolerant of rush-hour transport conditions for more than a few minutes. This, then, represents a very real tension. People or cell-cargos can travel long periods in single, well supplied and tailored compartments (Figure 5.6) or crushed together *en mass* for very short periods (Figure 5.7). Critically, this factor must take priority over simple carrying efficiency, otherwise nasty (surprising) things happen to the cargo. After all, both human and cellular cargoes have a 'choice'. So let us have a closer look at why this is the case.

The special thing about soft, living cargos (i.e. not gravel/oil) is that they *must* be supplied with

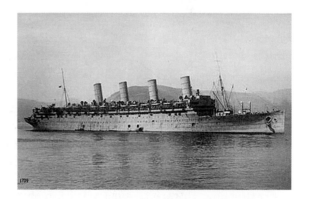

Figure 5.6 The Cunard liner *Aquitania* was used for long-distance troop transport in World War II. Men travelled in cabins designed for passengers as well as occupying cramped space between the crew and the engines. Provided the men were tightly controlled (i.e. obeyed orders), they kept to their windowless cabins deep in the ship and away from the decks. However, mutinous, overcrowded troops in hot weather will push themselves from the bowels of the ship onto the deck or to the best cabins, unless enticed with some creature comforts. And so it is with cells within the interior of a deep material. © Cunard Line.

Figure 5.7 This Canadian troop landing ship uses Plan 'B' for carriage of people, namely putting them all together in a single space. The mutiny problem is solved, everyone is equally uncomfortable and anyway there is no choice. Unfortunately, this is not really what is needed in scale, function and durability (i.e. in our terms, not biomimetic of a tissue), e.g. there is short survival time and the space is unlikely to be made larger. Note, there are two groups aboard: crew (red ring) and cargo-troops (white).

two key inputs if we want to keep them usefully functional for any length of time:

(i) feeding; and
(ii) information.

What is more, if these requirements are not supplied, and in good time, the 'cargo' may not just fail to do what is wanted, but may find surprisingly unpleasant things to do instead! In the case of troops, this might be anything from imaginative and unruly behaviour to outright murder and loyalty realignment. Cells can move to the wrong places, build up junk tissues, break down good structures or just die. And of course, we are much better at commanding troops than we are at controlling inherently mutinous cells! In fact, even after we have succeeded in forcing or tempting them to climb aboard our carefully designed support materials, cells have little to encourage them from moving almost anywhere or congregating wherever they want to. In the case of troops on a long, hot voyage, this might be on

deck or in the higher, breezy cabins. For cells, it is over the outer surfaces or in the uppermost pores of our constructs.

It is an important similarity that, as with large groups of people, the greater the crush of cells, the harder it is to supply nutrients and oxygen to those in the middle of the crush. In the same way, it is increasingly difficult to get different groups of people in the centre of crowds to respond to complex commands. Again, this is easy to understand for the troops in the landing craft (Figure 5.7). If we give lunch boxes or tea to the men at the edges, it will be a long wait and thin rations for those in the middle of the deck (consumables get consumed!). Equally, if our orders are shouted across the deck, it is increasingly difficult to direct these orders at any particular group of troopers, other than those at the edges. The ones on the middle may not hear, or, worse still, may hear incorrectly above the background.

In contrast, troops on the long-range troop transport ship can have food and drink delivered as it is needed, down corridors and between decks (rather like the way vascular conduits deliver blood). Similarly, information and commands can be targeted in time and space to just those groups of troops who need them. For example, the bridge engineers and sappers might be sent maps of rivers and fences they need to prepare to cross; catering platoons could be supplied with daily cooking or nutrition plans; transport specialists might be given the latest vehicle repair manuals.

Viewed in this way, basic design priorities for 3D construct structures can set as simple delivery vehicles or complex transport systems, depending on the task they are needed for. Quite simply, we need to tailor our cell-landing craft or cell-troop transport vessels to match the application we have in mind.

The first question we need to ask of ourselves, then, is 'what do we expect or hope that our seeded cells will do?' One common answer is that they should leave the support material to migrate out into the tissues and work with the locals (i.e. the patient's cells). This underlies much current stem cell thinking – for example supplying neural stem cells to treat Parkinson's disease. Alternatively, the idea might be that these cells release growth control factors which will activate or direct local host cells to do a better job, for example cells providing growth factors to promote wound healing. These can be regarded as cell therapies for which the support material is a (short term) delivery vehicle – **a landing craft**.

However, the aim might be to seed in cells which will work away on the support material to form a native 3D graft tissue. These may typically be expected to form organized barrier layers and zones or to lay down a load-bearing bulk of extracellular matrix or form cell-lined channels for blood or axons from the host. In this case we are asking for major cell activity, IN the support material, over extended periods of time. We are now looking at serious internal 3D structure to allow prolonged cell supply with nutrients, materials and individualized detailed instructions. This is the **troop transport ship** concept.

5.4.2 Getting cells onto/into pre-fabricated constructs (the willing and the reluctant)

Getting cells into position on, or into, the support material of choice has occupied a major proportion of the average tissue engineer's time and published output. The **engineer's** approach is to *place* them where we want them to be, as part of the controlled fabrication process. **Biologists**, in contrast, have tended to look for *persuasive* ways to use the cell's own guidance and movement mechanisms. One example approach has been to provide micro-nano topography (grooves, ridges, channels) of a size and shape that cells recognize and react to. Another is to generate diffusible growth factor gradients.

The theoretical aspirations of such bio-inspired approaches have started out high, aiming to place multiple small groups of cells here and there in the 3D space. In practice, however, these have often given way to increasingly modest goals (Text Box 5.2). It has, in fact, become increasingly clear just how much more detail of the basic biology we must understand before these become practical,

robust control mechanisms. In truth, these bio-routes have now been reduced to targets which are either depressingly modest or self-defeating in how long they may take to develop.

It is now clear that a great deal of practical effort has been expended in past strategies just to get the cells to attach at all to some of the synthetic polymer substrates. As a result, a 'lowest common denominator' philosophy has come to dominate, which simply aims to grow epithelial cell sheets *over the construct surfaces* or to tempt stromal cells, such as fibroblasts and osteoblasts, to migrate down to deeper parts of the construct. Forming surface epithelial cell sheets can be achieved:

(i) by *pre-growing* a sheet (in 2D culture) over a period of some days, until it is at least a confluent monolayer and, ideally, is differentiating into the required multi-layer specialist structure and can then be laid onto the construct; or, alternatively,

(ii) cells in suspension can be dropped onto, or flowed slowly across, the upper surfaces of our construct, such that as many as possible attach and grow, again eventually forming a continuous sheet.

Getting cells (e.g. stromal, connective tissue cells) into the **deeper, core** or **interstitial** zones of the 3D construct has either been very easy or difficult, depending on the type of scaffold support material used. As we have seen in the last chapter, cell support materials fabricated using harsh, cell-lethal conditions must, consequently, have their cells added at a later stage of assembly. These materials are predominantly the synthetic polymer, ceramic and glass-like materials.

Where we must have such a *separate* cell seeding stage, a range of ingenious methods have been developed. These include simple surface application of cell suspensions, with some cells falling into or being drawn down into the pores. This can be assisted by centrifugation, blotting or controlled flow. Gradual movement of cells down into deeper zones is now commonly promoted by gentle fluid flow (circulating perfusion) around the construct during culture. This flow can help to reduce the

Text Box 5.2 Examples of aspiration versus reality in biological cues to 'engineer' 3D cell distribution

If we expect to direct and energize, for example, growth/migration of axons to re-connect a facial nerve (perhaps as a result of a road traffic or dental surgery accident), then what is needed? First, axons must attach to the available substrate and elongate along the most direct (straightest?) track possible from one end to the other of our implanted construct to bridge the injury. At its best, this might aspire to get axon regeneration to reflect the density and fascicle distribution or branch structure of the original nerve. In reality, topographical and substrate-material guidance is some way away from being this specific or robust, especially *in vivo*. Our present efforts have produced re-growth which looks more like (b) than (a) in Figure 5.8. This is disappointing, as early non-degradable tube conduits achieved much the same results by just confining outgrowth using narrow silicone tubes.

In another example, we aspired to promote rapid micro-vessel in-growth (angiogenesis) to 'feed' our

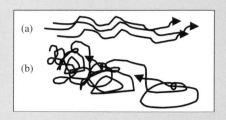

Figure 5.8

implants by providing an artificial gradient of the angiogenic growth factor VEGF (vascular endothelial cell growth factor). Unfortunately, it turned out that in isolation and at practical dose-levels, VEGF alone produces a very imperfect, leaky, tortuous vessel bed, resembling tumour vessels. As a result, many workers now just seed constructs with free endothelial cells or 'stem-progenitor' cells that may develop into micro-vessel cells. The hope here is that these cells 'know' how to organize themselves into 3D tubes, which *might* then go on to form vessels joined to the host circulation. This is truly modest and not really an engineering-type target.

sharp diffusion or cell-consumption gradients of nutrients which often form close to the construct surface in culture. However, even modest fluid shear can damage surface cells. The design of such 'pre-fabricated-seed-later' methods tends to be strongly influenced by this driver, sometimes to the exclusion of good 3D structural biomimesis. After all, the presence of 3D deep cells is a defining feature.

In contrast, cell seeding and in-growth need not be major problems for those cell support materials (commonly gels) which can aggregated under physiological conditions. As we have seen already, such hydro-gels are often – though not always – natural materials, and not all of these can be interstitially seeded at the time of fabrication.

Put simply, then, pre-fabricated supports require a separate step for deep cell seeding, while self-assembling materials can come ready-seeded, by interstitial cell enmeshing. The full importance of this difference is another defining feature of extreme tissue engineering (ETE).

5.4.3 Trapping the unwary: Seeding cells into self-assembling, gel-forming materials

As we have seen, once we leave the domain of pre-formed cell supports, our story changes dramatically. Certainly, we must now leave our shipping analogy behind as there are no clear parallels. The trapping of cells into a 3D gel-support material would be a bit like loading a ship by designing it to self-assemble around its cargo – interesting, but *extremely* impractical. In effect, when any one of the gel-forming materials starts to form (e.g. collagen, agarose or fibrin) a physico-chemical change is triggered to initiate monomer aggregation and fluid segregation at a nano-scale (Text Box 5.3). This aggregation almost always involves lateral and very close packing of many hundreds of thousands of molecules. Very close molecular packing is generally essential for the process, as physical proximity promotes the formation of either *many weak* or a *few strong* bonds, all of which are short-range.

Text Box 5.3 Fibre self-assembly into a 3D cell-support gel and the enmeshing of cells

Prior to gel aggregation, we must normally prepare a solution of the monomer molecule. This is a homogeneous dispersion of monomer molecules (collagen, fibrinogen, agarose, etc.), evenly and randomly distributed throughout the water phase (Figure 5.9a: pink represents the aqueous 'shell' surrounding the monomer). In order to understand this better, it is useful if we focus more on the water and less on the monomer.

Each monomer molecule is surrounded by a similar (average) water shell, mainly governed by the monomer : water ratio, i.e. the starting monomer concentration. In the example of collagen fibril aggregation to a gel, this 'ratio' is commonly around 0.2–0.5 per cent monomer, or 99.8–99.5 per cent water – 2 mg/ml protein in water. In this example, each monomer is surrounded by roughly 500 times its own mass of water.

While the term 'hydrogel' should already have tipped us the clue about this water dominance, the *extent* of its huge excess might be a surprise. When gelling is triggered, monomer molecules rapidly pack together side to side into dense 'fibre' aggregates (Figure 5.9(b,c)). Naturally, as monomer molecules pack closer and come into contact, much of the 500-fold

water excess must be redistributed. In effect, as the monomers get closer together they push out the water. 'Out' in this case is outside the fibre volume, or into the inter-fibre spaces (pink rings in Figure 5.9d).

The fibres become relatively dehydrated. While the overall ratio of protein to water is the same before and after gelling (still 500:1 – Figure 5.9b to d), its distribution is now non-homogeneous, starkly segregated into solid fibres containing little water, separated by watery zones containing little protein/monomer. The aqueous inter-fibre spaces can be considered as the pores. The gel forms when fibres are sufficiently entangled, or randomly enmeshed, to support transfer of mechanical loads across the material (i.e. to act as a 'solid'). Despite this, the **average ratio of protein to water has not changed** (still 99.8 per cent water).

When cells are present in the pre-gel stage (Figure 5.9b), their membranes prevent protein or water redistribution to or from the cytoplasm, so they are largely passengers in the segregation/gelling process. As a result, cells (or other particles) become passively enmeshed (Figure 5.9e) between load-carrying fibres. This is not to say that embedded cells are 'mechanically independent' of the load-carrying fibre elements. Even non-attached cells will be compressed by deformation of the mesh, and those attached to fibres will be exposed to complex tensional loads, just as they are *in vivo*.

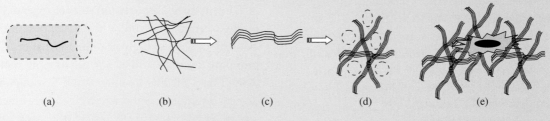

(a)   (b)   (c)   (d)   (e)

Figure 5.9

As the cells are suspended in a gelling monomer solution and become segregated away from the fibres and into the water-rich fluid phase, they become trapped by the tangle of the newly forming fibres. Scale is key here and, at this stage at least, the fibres are almost always in the nano- or small micron diameter scale (the so-called meso-scale).

In the case of our collagen example, fibrils would be ≈30–100 nm diameter, or 150–500 times smaller than living cells (nominal spherical diameter ≈15 μm). We can regard such systems as cells held evenly throughout nano-fibrous networks, surrounded effectively by fluid-filled nano-micro (meso-scale) pores. In this case, cells are physically trapped and there is no requirement for complex bio-attachment; the cells have no 'choice' (note: to repeat – 'cell-choice' is a human shorthand, not a real cell option). Importantly, this is the environment in which cells live in natural tissues, 'interstitially' distributed throughout the extracellular matrix material. Despite being trapped, they can easily move, either by physically pushing and squeezing between fibres (at this stage, these are very soft gels) or by degrading the protein fibres to form discrete channels, as they do *in vivo*.

Clearly, the gels we are discussing here often have *very* high fluid contents and are, correspondingly, weak. Recent technologies for gel compression and controlled fluid removal have changed this, and offer the real possibility of extending our control of where and how much volume the water occupies. In other words, we no longer have to accept the arbitrary (and high) fluid : cell : fibre ratios which simple gelling leaves us with. We are back in control. These compressed, partly dehydrated gels have all of the biomimetic and cell-entrapment properties of the best extracellular matrix gels, but with the capability to provide usable properties around the cells.

The plastic compression process involves controlled expulsion of excess fluid from between the fibrils by suction or mechanical force (for detail, see Chapter 6, Figure 6.10). In this, the hydrogel, with its interstitial cell population, is subjected to directional fluid expulsion under combinations of compressive load and blotting, typically through a single fluid-leaving surface (FLS). Where fluid extraction is upwards, into a porous plunger, the FLS is at the top, allowing the process to be carried out in conventional multi-well culture dishes. It also means that many layers can be compressed in sequence, on top of the first.

All major parameters of the compressed sheets are then controllable, from the dimensions of the final gel to its collagen and cell density. Each compressed layer is produced in minutes, typically at around 50 to 100 μm in thickness, so multi-layering can be useful. Hybrid layers are simple to form, for example with direct addition of mineral particles, channel formation, incorporation of additional proteins or even synthetic polymer meshes. This

strategy provides a new route for accurate engineering of natural materials, comparable with the way that synthetic polymers are made. A core difference, however, is that it is achieved without harsh conditions that can kill resident cells or denature the proteins.

Compression-fabrication of cells into their support material, as described here, is a process that has more in common with biological synthesis than with human industry. It is a bit like building a normal sized house with inflated hollow rubber bricks, then deflating all the bricks so that the house shrinks to doll's house proportions. The advantage is that it is much easier to fabricate the detailed internal 3D structure at the larger scale (with the bricks inflated). Similarly, it is also much simpler to position the desired groups of inhabitants, especially in the deeper rooms, *before* the house is shrunk (deflated). It gets around the compound problems of our need to make precision structures at the **cell scale*** when we are forced to work at the **human**

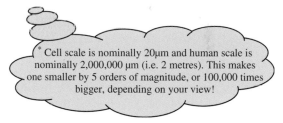

* Cell scale is nominally 20μm and human scale is nominally 2,000,000 μm (i.e. 2 metres). This makes one smaller by 5 orders of magnitude, or 100,000 times bigger, depending on your view!

scale, and being at the same time restricted to using difficult, complex building materials under assembly physiological conditions. Taken together, the material and physical restrictions alone make this a tall order. Having the ability to dodge the 'scale problem', with this shrink-compression trick is an essential enabling factor.

To conclude, then, for pre-formed cell support materials (e.g. synthetic polymers or ceramics, requiring cell seeding) there is a cost benefit tension to be analyzed for each tissue engineering application. We must balance the poor biomimesis inherent in large, shallow cell accumulations and surface-deposition of cells against their advantages of simplicity of production and use (i.e. the basic landing craft analogy). On the other hand, do the requirements of our application demand that cells are placed into specific locations and patterns, often deep in the 3D structure, where the local environment supplies instructions (bio-cues), nutrients, etc? In the latter case, good biomimesis comes at a cost, namely the expense, complexity and uncertainty of producing natural materials. In some ways, these resemble the trans-oceanic troop transport ships, like SS Aquitania, but their production technologies are not as well understood as those of synthetic materials (Text Box 5.4).

This will need a significant research effort. The first ('landing-craft') option can suffice for quick, minimal-culture applications, where delivery of cells alone is enough for repair. But to engineer functional, 3D *graft* tissues, with prolonged culture, such minimally mimetic supports may not be sufficient. In this case, the 3D structure of natural, interstitially cell-seeded support materials is better, and the traditional problem of high water content/poor mechanics may have been resolved.

5.5 Acquiring our cells: recruiting the enthusiastic or press-ganging the resistant

Where do we get our cells from, once we have decided how and when to cell-seed? This subject is huge. Happily for this chapter, though, much of it is otherwise known as **cell and developmental biology**. Consequently, like so many other core components of tissue engineering, it is knowledge and technology which is accessible from *elsewhere*, as and when needed. The trick for tissue engineers is not necessarily to have an encyclopaedic knowledge of cell biology (that is for cell biologists). Rather, those who need to use this part of the tissue engineering landscape need to understand the location of the solutions. These are the river crossings and mountain passes which can be vital to any successful journey to a tissue engineering application. But most importantly, we first need to ask ourselves *why* we need to cross these hazards at all.

The twin kingdoms of cell and developmental biology are home to some of the most vigorous and dynamic of the tissue engineering tribes. There are innumerable possible combinations of cell types and phenotypic shifts, and such shifts in cell behaviour function are the essence of how embryos develop, wounds repair and, more darkly, how tumours form. These cell shifts comprise the 'mountain ranges' which lie across many tracks of tissue engineering, and down from these comes the torrent of risk and opportunity which we might call *stem-progenitor cell biology*.

In the early stages of 3D culture (see Chapter 3), the concept was to isolate and grow cells directly from the tissue which we needed to regenerate. The main question at that point was whether to use cells from the patient (autologous cells) or from a safe donor (allogeneic cells). Hospital-based initiatives and service industry models have tended to concentrate on autologous cell sourcing. On the other hand, manufacturing industry models aim towards producing reproducible 'packages of bits', off the shelf, for use *now*. These tend to favour using allogeneic cells, but examples of both models are common. More recently, the aspiration has been to prepare early adult stem or progenitor cells from suitable sources and force them down the required cell lineage.

The tension between allogeneic and autologous sources has not disappeared – it has just taken a back seat. The stress has now shifted to trying to supplying *suitable* combinations of biochemical, spatial and mechanical signals, in sequences which fool uncommitted cells to become the cells we want. In common with many targets set by the bio-tribes, we must take careful note of the 'suitable' caveat. It is now very clear that this caveat is shorthand for the need for much more basic knowledge.

We certainly do need to understand stem-progenitor cell controls much better if we aim to use them in an engineering sense. It is also true to say that this topic is not really the *new research question on the block* which it might seem to be. Indeed, there have been determined efforts to tempt bone marrow stromal 'stem' cells to make bone or cartilage since the pioneering work of Friedenstein, Owen, Howlett and others, some decades ago (Text Box 5.5).

Other sources of adult stem/progenitor cells for engineering include adipose tissues (fat – a convenient by-product of liposuction), skeletal muscle,

Text Box 5.4 Micro-porous versus nano-porous cell support scaffolds

Getting cells into their support materials: a story where success is a problem and less can be more.

With a generous slice of hindsight, let us analyze the consequences of having to populate either (i) self-assembly or (ii) pre-fabricated cell support materials with cells. This time the task in hand is to analyze the *primary* consequences, when we must have a separate, specifically designed *cell-seeding stage*. The 'logic-box' below summarizes why pre-fabricated scaffolds can suffer such poor deep perfusion of nutrients and oxygen. The obvious starting point is that pre-fabricated scaffolds need seeding, but self-assemblers do not.

2. submicron material pores are around *all* cells as spaces between the matrix nano-fibres, *but:*
3. such nano-pores present minimal diffusion barrier to oxygen, and small nutrients = rapid nutrient mass transport between cells;
4. clustering and 'consumption barriers' only occur at very high cell densities.

Paradoxically, then, nano-fibre self-assembled materials *minimizes* 'barrier forming' tendencies, whereas macro-porous materials promote cell-clustering, path blocking and, thus, poor mass transport to deep cells. In this case, then, *less really is more!*

Tip: The biomimesis of natural nano-fibrous materials may give us clues about how natural

Logic Box 1. Forced seeding of ≈15 μm cells needs 100–200 μm access. **Logic Box 2.** So, high Porosity + strength = very dense construct wall material.	**Logic Box 3.** Success = (i) Cells divide to fill the pores; (ii) High density cell colonies deep within dense impervious walls; (iii) Colonies form 'tissue' in deep pores.	**'BUT' Consequence Box.** Cell clusters block the pores and consume nutrients en route to deep zones: consumption + diffusion. + Minimal exchange across pore walls = major sub-surface nutrient/O_2 depletion.	**Outcome Box.** Acellular-dead core = shallow tissue constructs, needing perfusion/mixing and external media flow.

Figure 5.10 below summarizes 'before and after' cell seeding and growth; the more successful this system is in surface cell growth, the greater the problem of poor nutrient access (mass transport) to deep cells. In other words, success brings a fatal problem.

This analysis is completely upturned where we consider self-assembling, cell-enmeshing materials because:

1. a separate, forced cell-seeding stage is not needed (cells are interstitially located at time zero);

bio-systems work. If you have followed this analysis closely, you will realize that such nano-porous materials do not *only* enable mass transport of small nutrients to deep cells. This structure will restrict the movement of macromolecule *products* of cells (e.g. proteins, polysaccharides, etc), depending on molecular radius and inter-fibril spacing. This becomes a cell 'valve system' where access of raw materials to cells is free and non-directional, but export is restricted in rate and direction (by fibre material anisotropy), as it is in most natural tissues.

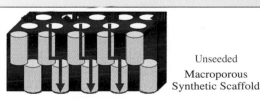

Unseeded Macroporous Synthetic Scaffold

Nutrient diffusion limitation: Via Pores = low. Walls = high

Successful Growth: Surface Cells block pores AND Consume nutrients

Nutrient diffusion: Pores Blocked-Walls Minimal.

Figure 5.10

Text Box 5.5　Bone marrow stromal stem cells: hot off the press?

Well, not really. Despite the current excitement about the use of adult progenitor cells, especially from bone marrow, for new-tissue-generation, it is actually not such a new trick. Friedenstein *et al.* of Moscow University (1966) were quietly working with others (Owen & Friedenstein, 1998) to understand how to use these same technologies back in the early 1960s – 50 years ago. To give this timescale some context, most family TVs were black and white (monochrome) and powered by heated glass valves! Modestly, these

researchers quoted the cell precursor ideas from still further back in time, from Burwell and others.

So the answer to our question is 'no' – generating tissues using marrow stromal stem cells is definitely *not* new.

References

1. Friedenstein, A. J., Piatetzky-Shapiro, I. I. & Petrakova, K. V. (1966). Osteogenesis in transplants of bone marrow cells. *Journal of Embryology & Experimental Morphology* **16**, 381–390.
2. Owen M. & Friedenstein A. J. (1988) Stromal stem cells: marrow-derived osteogenic precursors. *Ciba Foundation Symposium* **136**, 42–60.

cornea and blood. However, in many cases, these cells are vanishingly rare in the overall population or already partially committed to one fate. This can make them difficult in practice to find and expand, or less amenable to multiple uses. The problem is that the signalling systems involved seem to be immensely complex, rather like the patterns of noise, face and hand movements which we humans use for communication. Unfortunately, short of a few crude grunts and shouts, we do not really understand the cell language. While it is clear that an understanding of this biological language *would* have huge value to tissue engineering and regenerative medicine, it also seems premature to consider these attempts as 'engineering'. As an analogy, imagine trying to back-pack across the more remote parts of China knowing only 'xia xia' (thank you).

So, getting hold of the '*cells you want (i)*' (i.e. cell acquisition) is the first target. Then we come to the question of how best to processes these into a '*useful cell preparation (ii)*' and further, what is necessary to push these into exactly '*the type of cells we need*' *(iii)*? The current difficult concept areas (as opposed to technical hurdles) lie in the italics, *(i), (ii) and (iii)* above. Not all of the cell types that we would like to use *want* to be acquired (i), expanded (ii) or differentiated (iii). Sometimes we have volunteers (e.g. bone cells from bone marrow, corneal epithelium from the corneal limbus), but

all too often we are dealing with conscripts or, at worst, press-gang material.

For example, we might need cells which produce 'cartilage' – but which type of cartilage? There are different forms of cartilage (e.g. ear, meniscal, articular, growth), some of which grow, some of which do not, and different types will work or not work in different body sites. What makes these different cells different and how likely it is to alter their habits is not well known. Strategies range from taking cells:

(i) from the tissue we want *to make* (hoping they do not change);
(ii) from a similar but more convenient tissue (hoping they change a little); or
(iii) from un-programmed cell populations, hoping they will know exactly what to change into, and do it.

These are the 'same tissue' (volunteers), 'similar tissue' (conscripts) and 'stem cell' (press-ganged) options. It should be alarming, however, to note just how much 'hoping' is going on, and how some of these aspirations are *opposites*.

All too often, of course, the key tension is between the advantages and the difficulties of the acquisition and conversion stages. For example, adult stem/progenitor cells, in principle, seem to have all the advantages in terms of acquisition. They can be harvested from fat, muscle or bone marrow aspirates with *relatively* ease from many patients, making autologous grafts a possibility. Unfortunately, once acquired, these potentially ideal recruits turn out to need serious pressing, cajoling and training in order to expand their small numbers and to persuade them that they *are* what we *think* they are! All too often, with stem/progenitor cell sources, we find a hugely complex, poorly understood set of problems in driving them down desired differentiation pathways.

This is exemplified by the oldest work in the area – that of getting marrow stromal stem cells to form either bone, cartilage or fibrous tissue cells. This is a problem tackled to date through a host of growth factor and mechanical loading routes. In contrast, mature, differentiated cells, acquired from the target tissue itself, tend to be more numerous and active in expansion. They often are far less ambiguous in their need to be (re)differentiated – tendon, dermal fibroblasts or vascular smooth muscle cells being pretty well what they claim to be on the label. However, these willing troopers-for-the-cause can be hard to acquire, requiring significant lumps of deep tissue to be hacked out, extracted and sifted, through long processes.

What is more, the practicalities of dealing with sick patients can, as always, get in the way here. For example, cells from older donors (a frequent feature) divide very slowly, so need long expansion times. Other significant groups of patients (e.g. cancer) need to take cytotoxic drugs, so their cell division is seriously inhibited. Also, de-differentiation is constantly possible. In the case of cartilage cells, there is a constant tendency for them to drift towards a fibroblast-like phenotype in culture, and so stop making cartilage.

The technical details by which researchers currently attempt these three stages now comprises a large proportion of the conventional TERM literature. It also has, at present, a strong focus on techniques and phenomena rather than concepts and mechanisms, and so is beyond the scope of this work.

5.5.1 From cell expansion to selection and differentiation

Cell expansion is a curious term. At no point is there any intention to produce giant cells, pushing off the culture lids, or to have them straining to open the incubator door. The missing word which makes sense of the phrase is 'population' (after 'cell'). Outside the world of cell technology, it is probably a good idea to make sure the 'population' word is kept firmly in place. This is because the other implied (i.e. unspoken) technical aim is to expand *only* the population we *want*!

Expansion of a cell population, then, requires a technically specialized stage to generate the desired density of the required cell type(s). At the same time, though, there should be the smallest possible increase in the number of contaminating, less desirable cell types. To this end, much equipment and tissue culture processing has been developed, forming a distinct branch of the discipline in itself. However, while much of this technical detail is beyond the scope of this book, one basic message is important. It is, presently, almost always necessary to have a *separate* cell expansion stage. This is because:

- cells tend to be bad at doing 'division' and 'other activities' at the same time; and similarly,
- they also tend to need different growth conditions for fast division than for other activities.

Wherever the cells are acquired from, it is critical for cell therapies to generate enough active cells to carry out the functions that we predict will be needed to make the new tissue. In other words, there needs to be a cell population expansion stage to give sufficient[8] cell numbers for their function. It is commonly assumed that this function would be to 'fabricate' the new tissue. Though there is often rather little mechanistic analysis, the term *sufficient* cell numbers frequently seems to mean as large as economically/ethically possible. Indeed, there have been examples of risky, self-feeding logic in this

---

[8]Again, the scary bio-caveat '*sufficient* or *suitable*'.

area, suggesting that where a favourite cell therapy does not work as expected, then it just needs more cells to be effective.

However, there is another of our TERM tensions here which it is important to recognize. The thing is that many tissue engineers (mainly the cell-biological tribe) are far more fixed on the importance of getting the 'right cell type' for the job. As a handy analogy, we shall call this the tailor's dilemma.

Legend has it that tailors of fine garments continuously agonize over the relative importance of the quality and the quantity/amount of fabric they use for any given suit ('never mind the quality, feel the width'). In the case of cells for tissue constructs, the argument goes:

- Cells of a *suitable* phenotype (basic behaviour patterns) for the repair/regeneration task in hand are present in tissues because they perform that task in nature.
- It makes sense to use only the best cells for the job, as irrelevant or less effective cell types will just clutter the space, consume nutrients and degrade the process.
- Finally and critically, don't worry about starting cell numbers, as processes in the body often start with a tiny population of key worker-cells which proliferate to give the required numbers naturally.
- After all, cells divide; and the best place for this is in the body!

This, of course, is the 'never mind the width, feel the quality' argument and, on the face of it, it is a potent case. It certainly helps to distinguish the present strategy for the use of stem/progenitor cells. After all, (the argument goes) stem/progenitor cells may be very low in density but they have, in fact, evolved to do just what we want – provide a few cells which locally 'become' (i.e. differentiate into) the cells we need, then proliferate rapidly.

In contrast, (i.e. the 'width-matters' argument) goes: mature cells derived from the target tissue (e.g. cartilage, dermis, tendon) are adapted to *maintain* the fabric of an existing, intact tissue, not to build it from scratch. Indeed, when one smashes up and

extracts cells from the simplest of mature tissues, its resident cells are still not a simple homogeneous population that we can 'expand'. For example, there is good cell physiological evidence for at least two metabolically different fibroblast types in skin dermis (and then only because there are two commonly named layers to correlate them with). In normal joint articular cartilage (only 1–2 mm deep and non-vascular), there are perhaps three or four identified chondrocyte types. This increases if we include the meniscal cartilage (a favourite footballer's injury) and aging/degenerating chondrocytes.

What do we get, then, when resident cells are extracted from mature tissues and grown up to expand their numbers in culture? The result of this operation can get complex when cells from associated structures find their way into the culture, from blood vessels (e.g. smooth muscle cells), nerves (perineural fibroblasts) or adjacent and attached tissue layers. Each cell population in these mixed extracts can and *will* proliferate at different rates under standard culture conditions, so after a while one or two types may overgrow the rest while others die out completely (Figure 5.11).

These dynamics need to be understood if we are to devise any rational strategy for the cells we use. For the cell specialist (analogous to the fine tailor, feeling the fabric quality) this requires knowledge of:

- the cells present;
- culture conditions which drive some cells down required 'lineages' (differentiation tracks);
- conditions needed to reduce or kill off those types which are not needed.

For the non-cell specialist tissue engineer, (analogous to the tailor who looks just for the thickest suit fabric), the argument is that this level of control is implausibly complex and difficult to achieve from our present knowledge base. So, pragmatically, (s)he adopts strategies to get as close as possible to the cell type needed in a mixed population. Ultimately this relies on high gross cell numbers to ensure that there are at least *some* useful cells in the final mix.

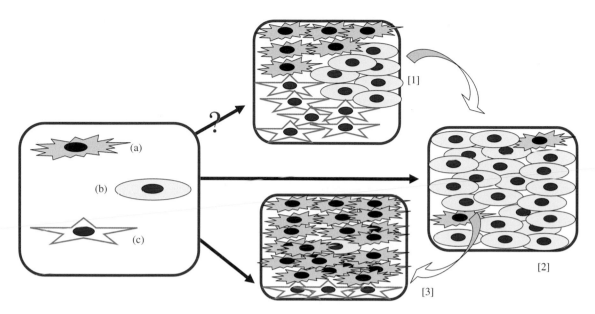

Figure 5.11 Basic strategies for 'cell expansion'. Outcomes range = (1) < (2) OK < (3) A pain. How many cells do we need for our particular application? In engineering systems, this could be a relatively formulaic problem. How much of the required work does each cell perform per hour; what is the total of this work that is needed across the whole tissue volume; what cell density is 'too low'? Unfortunately, we normally are pretty vague about *how* our seeded cells are going to achieve, or even contribute to, the regeneration we want. As if to amplify the problem, the distribution of cell (sub-)types might almost be different for every application. The diagram shows a typical set of unknowns, with (say) three cell types or cell phenotypes in the original population – (a), (b) and (c). [1] These might (at least for one or two passages) divide equally to leave the ratio of cell proportions the same (this is less common). [2] Alternatively, cell type (c) may not divide at all and die out, while (b) divides many times faster than (a), producing a culture with a very high (b) : (a) ratio. [3] Finally, it may be that one cell type (in this case (c)) only grows in one spatial zone of the culture vessel, for example forming an underlying layer. Of course, it is also possible that these culture types can change from one to the other with increasing passage number. It is a brave tissue engineer, then, who predicts precisely how these cultures make tissues in a 3D support material. So, the only safe answer to the question 'How many cells is enough?' (which we cannot know with much certainty) is 'As many as possible', i.e. the 'sufficient' caveat.

This tension between two logics, *refinement-selection* versus *pragmatic bulk*, sums up the present position for cell acquisition and processing in much of regenerative medicine and tissue engineering (Figure 5.11). However, as in the case of our analogy of the tailor's dilemma, it may be that there can, in reality, be no 'correct' or 'high moral ground' resolution. Rather, what we may have described is a necessary working tension.

It is, then, not actually a good idea to aim to 'remove' the tension, but rather to work to resolve it for any given cell type and application. After all, the tailor in the end can ask his customer what the suit is for and make an informed decision based on the *needs and requirements* of the fabric. This is a key phrase. The reader may also have noticed how often it has occurred in the preceding tissue engineering logic. Cell decisions depend almost entirely on what the tissue engineer considers is the *main function* of the seeded cells in constructing the new tissue. Unfortunately, as we have seen many times already, our basic knowledge of tissue repair, regeneration and remodelling is not always sufficiently robust to add this detail. Where such knowledge is absent, or collaboration with the right tribe is missing, we can drift to pseudo-engineering design.

Text Box 5.6 The hidden 'big question': how will we know the cell number that is 'sufficient'?

In engineering systems, this would be a relatively formulaic problem: how much of the required work does each cell perform per hour; what is the total of this work that is needed across the whole tissue volume; what is the net loss-rate of cells; what are the operating ranges of these values?

Sadly, since the biology does not yet allow us to be sure what these seeded cells are doing in any given tissue site, we find ourselves closer to guesses than to predictions. Currently, one of the most common and pragmatic rationales is just to increase the cell seeding density as high as practically possible, on the assumption that *more must be better*. Actually, this can prove to be a dangerous logic (i.e. without a mechanism of action, 'more must be better' cannot be assumed to be true).

As a result, where tissue generation, repair or regeneration fall short of expectations, there is a tendency to use the circular logic that the initial cell density was 'too low'. Consequently, the only safe answer to the question 'How many cells is enough?' is that we cannot know until we stipulate what they do in tissue formation.

The conclusion of this section, then, is that neither a simple 'cell expansion' nor the 'selective cell differentiation' approach can presently be considered a definitive one-size-fits-all answer. In some tissue/repair applications, crude separation and simple expansion of mixed cell populations will be sufficient. In others, much more cell selection and control of differentiation will be needed. The question to ask in order to progress, then, ceases to be which of the two tension-strategies is 'correct' (commonly it will be *neither*). Rather like the tailor's dilemma (see Chapter 7), we must be happy to work with the tension, understand what our cells actually *do* in each specific application and 'tailor' (sorry . . .) the strategy to that. At the same time, it is critical to research the underlying mechanisms of cell tissue repair and remodelling in order to improve our decision-making. In other words, we need to find out just what our cells *really do need to do* (Text Box 5.6).

This is another of our extreme tissue engineering moments where a concept shift emerges. It is inevitably complex, as biological understanding is still developing (sometimes out of engineered tissue models themselves – see Chapter 1). However, to get deeper into this *hunt-for-function* we should turn back to our earlier analogy, comparing cell support materials with troop carriers and landing craft. Look at Figures 5.6 and 5.7 (large troop ship and small landing craft). There are at least two basic functional groups of people being carried: the troops, who will carry out a role on arrival; and the crew, who operate the delivery system, i.e. the ship.

5.6 Cargo, crew or stowaway?

Basically, in the business of maritime transport of people with a job to do – be they troops. plumbers or wind-farm engineers – the ship will carry *at least* two distinct groups. These are the crew, needed to operate the ship/transport, and the workers themselves who do their various jobs once they disembark. In the same way, it is possible to consider two roles for the cells we put into our scaffolds and constructs.

5.6.1 Crew-type cells: helping with the journey

Cells which work on the scaffold during a culture stage or maintain the function of other seeded cells might be considered as analogous to the crew of our transport ship. In constructs that are cultured for substantial periods prior to implantation, this distinction between crew and troops can be an elastic concept. In effect one set of cells are needed now (vessel crew) the others are needed later, so one function can merge into another over time. For example, the fibroblasts which are seeded into the dermal equivalent collagen of Apligraf™ have the early stage function of contracting the loose collagen network down to a denser, tissue-like material before it can usefully implanted.

In some cases, mature, differentiated cells have been seeded into populations of stem/progenitor cells with the intention of enhancing stem cell differentiation towards that cell type. For example, bone cells grown together with endothelial precursor cells will push along the endothelial differentiation to produce spatially defined micro-vessels, as defined by the Kirkpatrick group (see Further Reading). However, chondrocytes or chondrogenic stem/progenitor cells, when seeded into synthetic polymer meshes (e.g. PLGA) can be cultured to lay down the initial rudiments of a cartilage matrix to support repair. In these cases, cells have a crew-like function, helping to prepare the tissue construct before it is implanted. In contrast, there is clearly the separate aspiration that some of the seeded cells will continue to make dermis and cartilage, or link up to host micro-vessels, after implantation. This function, beyond culture (the role of the troops who are carried in our analogy), is intended to help with tissue construction or integration at the implantation site. Cartilage is a special case here, as there is generally little expectation that surrounding, host-tissue cells will be recruited to help, and definitely no neural or vascular in-growth after implantation.

5.6.2 Cargo-type cells: building the bulk tissue

Bone marrow stromal stem cells are often seeded into constructs for bone implantation and regeneration (with or without a 3D pre-culture period). Since this is usually linked to pre-culture with osteo-inductive cues and/or selection to promote osteogenic behaviour of progenitor cells, they are clearly intended to have a bone-building function (and so are cargo-like cells). Experimental attempts to fabricate pulsating heart muscle, using cardiac myocyte-seeded 3D scaffolds, are clearly also intend to carry a cargo of cells with a function in building the new tissue. This is because mature heart muscle is composed of functional fibres, formed from many thousands of such myocytes all merged/fused together into 'syncitia'. Interestingly, most cell-seeding approaches have the default assumption that their cargo of cells

work mainly by rebuilding the bulk tissue after implantation. However, as we shall see, this remains an assumption.

Support, protection and integration functions

However, not all cells need to function as construction specialists in the host. Cell cargos can be designed to engineer good integration of new tissue margins with those of the existing surrounding tissues. For example, some researchers have added vascular endothelial cells to their constructs in the hope that these will speed up revascularization/angiogenesis or the in-growth of host vessels from the margins. Similarly, specialist cells called olfactory ensheathing cells have been used to promote and guide neurite regeneration across the margins of spinal cord injuries.

If we stay with our troop-ship parallel, these would be the inevitable platoon of sappers and logistical engineers. Large vessel constructs are frequently pre-seeded with vascular endothelial cells to line/coat the lumen. This is effectively a 'defensive' function, with the intention that such a lining sheet will prevent coagulation In this case, thrombus formation would rapidly block the vessel construct as host blood and blood platelets pass through and contact the thrombogenic wall components, such as collagen or polymer scaffolds.

Concepts surrounding the function of implantable support materials have expanded recently with the wider development of implantable slow- or controlled-release drug depots. Clearly, there is great practical benefit in the addition of relatively common drug agents to assist tissue engineered implants. Examples of this include the addition of antibiotics to skin implants, used in seriously infected wound sites, or anti-coagulants to prevent thrombosis around or in (peri-)vascular implants. Evolution of this branch of refinement research in one direction takes us into the control of release of the drug from the scaffold material so that it works over extended periods. A second branch has been in the binding or trapping of cell-regulating growth factors onto or into the scaffold.

Delivering commands and controls after docking

This track aims to control what the seeded cells do once they are implanted with their support material. Development of growth factor depots within constructs is directly analogous to giving troops on a ship their sealed orders to follow once they are deployed (or in the case of cells, after they leave the lab). Unfortunately, in the case of growth factor depots, it is rarely clear whether these orders are *just* for the troops or for the local resistance fighters (host tissue cells), who will also inevitably come into contact.*

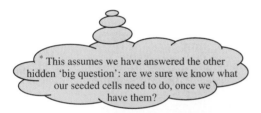

* This assumes we have answered the other hidden 'big question': are we sure we know what our seeded cells need to do, once we have them?

Rather like sealed orders, there can be at least *some* modest confidence that the seeded cells (one's own troops) will take some notice of them. However, what effect they will have on the locals (resident tissue cells) is pretty well anyone's guess. They may not even be in the local dialect. Growth factors on a scaffold can be made stable in the lab, but (like orders when they have been unsealed) they degrade and diffuse away in various unpredictable 'leaky' ways once they get into the tissues. The message-growth factors may conflict with local signals or be unintelligible to the locals when broken down or combined with other orders. So, when the ship lands, or the construct leaves our lab, the chances are that we lose control!

Perhaps you can identify with another analogy; imagine you are a baseball pitcher – in fact, the pitcher in Figure 5.12. The question is, what do you want the ball to do and how can you make best use of your control-window (that is, the period when you are holding the ball)?

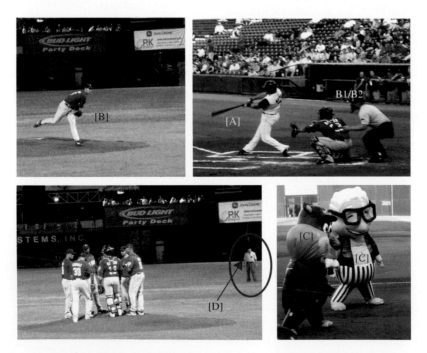

Figure 5.12 Window of control: control of the ball is what the pitcher does (top left) with help from his team (bottom left). But try as they may, once the ball leaves the pitcher's hand, control leaves the blue team. It may go to the batter [A], the catcher [B1] or the umpire [B2] – oops! Alternatively, the latex-wobbly guys might run off with it [C] or it may smack seven sorts of sense out of the sausage salesman [D]. Whatever happens, the pitcher must use all his skill/technology to control the fate of the ball as long as possible *after he lets go*. How would *you* impart post-partum control?

The best we can hope for is a 'window of time' where control can be applied. Unfortunately, this window is typically not just short, but its onset, duration and location can also be almost unknowable.

Simple examples of growth factors delivered to effect tissue responses include transforming growth factor beta 1 (TGFβ1) to speed up extracellular matrix deposition, or vascular endothelial cell growth factor (VEGF) to induce angiogenesis. Clearly, for either of these to be of any help to tissue repair (as opposed to actually doing harm – see Chapter 1), they need to work at specific times and locations. In fact, one of the few points we can be clear on is that when such growth factors go wrong, they can generate tissue scarring and disruption.

However, in another example of extreme tissue engineering, a new generation of approaches may be emerging. The idea here is to trigger specialized depots of seeded cells to generate an *appropriate* full physiological *cocktail* of growth factor signals to stimulate the desired response, at the chosen time and place. An advanced form of this strategy has been described for controlled angiogenesis, or the ingrowth of surrounding blood vessels into the construct or implant site. In this illustration, a very high density of selected (expendable, suicide-squad) cells are positioned as a depot, deep in the construct. Not surprisingly, this dense cell-depot generates its own local hypoxia, inducing local cell stress and even death, but it also elicits release of the full, physiological angiogenic growth factor cascade, just as a local hypoxia would *in vivo*. The result is that surrounding endothelial sprouts invade and form new blood vessels (Figure 5.13).

The key point of this example is that resident cells can be tricked into eliciting *perfectly normal* (i.e.

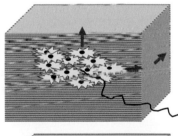

(a) Engineer a dense cells depot which will induce local hypoxia, cell stress or death. **Need-to-know:** diffusion path-lengths (blue arrows), matrix density diffusion coefficient for O_2 (for VERY dense substrates this *may* differ in different planes), cell activity and response to hypoxia (these are cell-type specific).

Real-Time O_2 monitoring probe

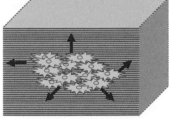

(b) Culture-incubate the engineered depot under normal or reduced oxygen conditions until hypoxic stress stimulates production of a full angiogenic growth factor cascade. **Need-to-know:** how long to culture any given cell type under the selected conditions for optimal release to the matrix (red arrows).

Implant directly as a living angiogenic depot, **OR......**

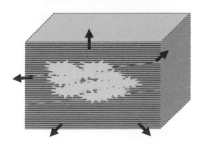

(c) Kill resident cells, for example by freeze-thawing. Implant whole construct or slices to allow angiogenic factors trapped in the matrix during culture to release. **Need-to-know:** Rate of release of angiogenic signalling proteins from the matrix; dependent on molecular radius of factors, material density in each plane and minimum diffusion path length, after slicing.

Figure 5.13 Scheme to show engineering of hypoxia-induced angiogenesis. Idea is to provoke a local cell-depot to make itself so hypoxic that it generates a burst of angiogenic factors. In practice there are some key need-to-knows which can only really be measured in a defined 3D cell-specific-model, ideally calibrated in terms of its core O_2.

physiological) cell-cell and tissue-tissue responses which involve whole cascades of growth control factors. This approach does not require detailed knowledge of the content and sequence of the cascade, just how and when it can be elicited. This is where the smart 'tissue engineering' comes in, since it is the definable properties of 3D engineered constructs which puts us in control. In this case, it allows us to dictate when the angiogenic cascade will be produced, and how long and where the growth factors will be released – in effect, where and when the response-window will open.

This particular example is known as engineered, or hypoxia-induced, angiogenesis. In fact, its success is based in availability of a biomimetic 3D model tissue, with its predictable diffusion properties, known diffusion path-lengths and culture periods. This allows us to control post-release cell responses, with only an incomplete understanding of the isolated factors.

It also forms an important illustration of how seeded cells can, intentionally or accidentally, produce key local tissue responses other than those of simple 'tissue building'.

5.6.3 Stowaway or ballast-type cells

Here our people-transporting analogy hits a divergence. In any case of mass transport (maritime or tissue), there is a larger or smaller issue of the stowaway (right down to fare-dodgers on the Underground). This concept cannot properly apply to cells, as they neither choose to ride nor pay a fare. However, there is an analogous problem, which we shall call ballast-cells. These are the cells we tend to ignore (or fail to acknowledge) in the design of 3D tissue constructs. They come along with those we wish to have, *on functional grounds*. We either cannot or choose not to eliminate them from our heterogeneous initial cultures (discussed above).

However, while it is often inconvenient or non-economic to eliminate ballast-cells, it is not good to ignore them when it comes to implant function. Depending on their density, division rate and metabolic habits, they will consume nutrients, oxygen and space, in competition with those cells

which do have a designed function. One clear example of this is the engineered angiogenesis described in the previous section. High densities of ballast-cells, for example, will add to hypoxia and unplanned angiogenic growth factor release. While this could be good for skin repair, it is less so in cornea or cartilage. In some cases, they will take on a non-designed function of their own (desirable or not).

Future tissue engineered constructs may, then, need to be pre-analyzed to quantify the effects, not just of cargo and crew cells, but also of any stowaway/ballast-cells. This analysis is increasingly easy using the quantifiable properties of the 3D construct itself as the test-bed model.

5.7 Chapter summary

To conclude, for pre-formed cell support materials (commonly synthetic materials) there is a cost-benefit tension which needs to be worked out for each tissue engineering application. On the one hand, we can tolerate the poor biomimesis inherent in the large, shallow cell accumulations which result from surface deposition of cells onto scaffolds and simple incubation (the basic landing craft). But on the other hand, it is possible to engage the strategy, involving the expense and complexity of systems which provide environmental conditions, suitable to control cell activity throughout a 3D structure (SS *Aquitania*).

There is a third strategy which may become increasingly attractive in this sector, and this is to minimize the culture period and implant the constructs at the earliest possible point. In the self-assembly sector, often using gel-forming biological materials, cells are enmeshed, biomimetically, throughout the fabric of the material (i.e. interstitial seeding, similar to tissues). The penalties come from the very weak mechanical properties of gels (water contents >95 per cent) and limited ability until recently to control either the gelling process or how the material properties are improved with time.

However, we can now conceive to start on the task of biomimetic 3D structures based on engineered native components, e.g. collagen, fibrin and silks.

Further reading

1. Jawad, H. & Brown, R. A. (2012). Meso-scale engineering of collagen as a functional biomaterial. In: Murray, M. (ed.) *Comprehensive Biotechnology* 2nd edition. Elsevier (in press).
[The idea of engineering materials out of natural proteins, as we make fabrics out of manufactured fibres: bottom-up.]

2. Curtis, A. & Dalby, M. (2009). Cell response to nano-features in biomaterials. In: DiSilvio, L. (ed.) *Cellular Response to Biomaterials*, pp. 429–461 Woodhead Publishing, Cambridge, UK.
[Review of how topography – surface structure – controls cells: by leaders of the field for three decades.]

3. Lennon, D. P. & Caplan, A. I. (2006). Mesenchymal stem cells for tissue engineering. In: Vunjak-Novakovic, G. & Freshney, R. I. (eds.) *Culture of Cells for Tissue Engineering*, pp. 23–60. John Wiley, New Jersey.

4. Johnstone, B., Yoo, J. & Stewart, M. (2006). Cell sources for cartilage tissue engineering. In: Vunjak-Novakovic, G. & Freshney, R. I. (eds.) *Culture of Cells for Tissue Engineering*, pp. 83–111. John Wiley, New Jersey.
[References 3 & 4: Two chapters from this practical guide on cell acquisition and growth, with practical examples of problems and solutions.]

5. Panetta, N. J., Gupta, D. M., Kwan, M. D., Wan, D. C., Commons, G. W. & Longaker, M. T. (2009). Tissue harvest by means of suction-assisted or third-generation ultrasound-assisted lipoaspiration has no effect on osteogenic potential of human adipose-derived stromal cells. *Plastic and Reconstructive Surgery* 124, 65–73.
[Modern example of cell acquisition – and such a mind-boggling title!]

6. Ghanaati, S., Unger, R. E., Webber, M. J., Barbeck, M., Orth, C., Kirkpatrick, J. A., Booms, P., Motta, A., Migliaresi, C., Sader, R. A. & Kirkpatrick, C. J. (2011). Scaffold vascularization in vivo driven by primary human osteoblasts in concert with host inflammatory cells. *Biomaterials* 201(32) 8150–8160.

7. Brochhausen, C., Lehmann, M., Halstenberg, S., Meurer, A., Klaus, G. & Kirkpatrick, C. J. (2009). Signalling molecules and growth factors for tissue engineering of cartilage – what can we learn from the growth plate? *Journal of Tissue Engineering and Regenerative Medicine* 3, 416–429.
[References 6 & 7: Taking hints from nature for combinations of cell types and growth factors in scaffolds to mimic development in 3D.]

8. Hutmacher, D., Woodfield, T., Dalton, P. & Lewis, J. (2008). Scaffold design and fabrication. In: van Blitterswijk, C. A. (ed.) *Tissue Engineering*, pp. 403–454. Academic Press, London.
[Review of conventional theory and practice in current use for biomaterials design in tissue engineering.]

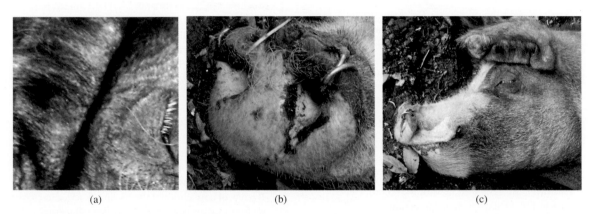

(a)   (b)   (c)

This is Daisy, the pig, viewed at different levels and orientations. Her eye and ear (a) in fine detail take on one form of *asymmetry* (top-to-bottom), whereas the nose (minus nose-ring-art) has a **bilateral symmetry**, but is *asymmetrical* into the plane of the photo (b). In (c) we can see her full face from one side, and she *seems* to be asymmetrical in all three planes – but this is because we are looking at a 2D picture. To analyze the 3D structure of any given part to be engineered, this illustrates the importance of reconstructing the *full* structure, based on a number of defined orientations and at representative scales (e.g. from mm and μm to nm).

6 Asymmetry: 3D Complexity and Layer Engineering – Worth the Hassle?

If cells built tissues the same way that men build bridges . . .

If cells built tissues the same way that men build bridges . . .

When analyzing asymmetry and hierarchical complexity, it is important to have a strong sense of scale. Analysis of scale-levels in biofabrication can provide some surprisingly obvious but useful observations. Figure 6.1 shows a simple composite illustration, comparing the proportions of human-scale and cell-scale fabrication. This helps us develop a semi-realistic picture of 'a day in the life' of a tissue construction cell, compared with human construction workers.

If we scale down the main building unit in the human world (i.e. man/woman) to that which we are expecting to build our tissue grafts (i.e.cells), we end up with a ratio in the region of 100,000:1 (10^{-5}). This assumes a nominal spherical cell of 20 μm and 2 m man, both of which are of course debatable but

Extreme Tissue Engineering: Concepts and Strategies for Tissue Fabrication, First Edition. Robert A. Brown.
© 2013 John Wiley & Sons, Ltd. Published 2013 by John Wiley & Sons, Ltd.

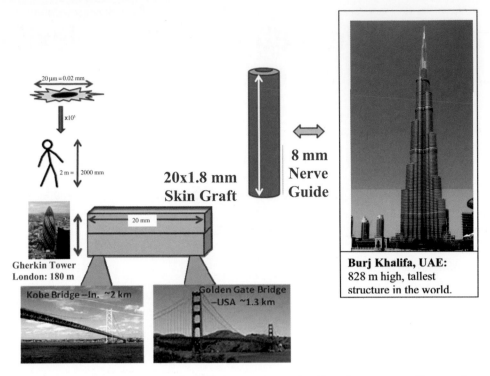

Figure 6.1 More issues of scale: comparing cell and human scales. Both cells and men must still move themselves, and the materials they assemble, across space proportional to their own size. The laws of motion, mass, energy conversion apply to both.

working approximations. This gives us a benchmark for some of the more modest target tissues, such as a small skin graft (say, 20 mm × 20 mm × 2 mm thick) which could at least have cosmetic value (0.0004 m^2 surface area).

If cells were men (in scale), this would be equivalent to them setting out to assemble a structure 2 km long. In length at least, this is the same as the Kobe bridge and 35 per cent longer than the Golden Gate. The width, of course, does not bear comparison in this example. A skin implant, for example, produced at 1.8 mm thick, would be modest for treating serious wounds, but it would still correspond, in human engineering terms, to the height of the Gherkin Tower in London. Again, though, this civil engineering example is only comparable in one plane – its height. The cells would really be aming to make a structure equivalent to the Kobe Bridge in

length, one Golden Gate bridge span in width *and* a GerkinTower thick. Quite a tall logistics order!

To add a little bio-medical context to this estimation, a large area of catastrophic skin burns (common targets for tissue engineering) might need graft surface areas in excess of 1 m^2, or 2,500 times larger than our skin graft example. A more common example would be short vascular implants or peripheral nerve graft conduits, which would have major clinical value for a host of post trauma reconstructive surgical procedures – if only we could make them! Even though we could make great use of modest 8 mm long conduits, this would *look*, to its resident cell poplulation, like the Burhj Dubai tower – the largest building humans have yet built.

Although these comparisons may at first glance look fatuous, it is how these structures 'look' to the resident cells that we need to consider. The logistics of mass (materials) transport and cell/worker mobility apply at both scales and, despite the many differences between the cell and human worlds, both operate under the same basic rules of physics.

6.1 Degrees of tissue asymmetry

Like so many parts of human self-perception and psychology, perception of our own (a)symmetry can be described either as a paradox (academic critique) or 'up the wall' (colloquialism).

The funny thing about our own bodies (*no*, not that!) is that they are, like most vertebrates, profoundly asymmetrical in two of our three planes, but surprisingly symmetrical in the third

(Figure 6.2a). If you slice us through the waist, we (happily) have a serious head-to-bum mismatch. Similarly, bisecting us top-to-bottom in a line between our ears and our hips leaves all facial features in one half (but none in the other); spine and buttocks in one (but belly and chin in the other); palms in one (but wrinkly backs-of-hands in the other). But slice us between the eyes and legs and the halves are remarkably similar.

We have a high uniaxial (left-right, lateral) symmetry with almost identical leg/arm length and facial sense organ symmetry – eyes, ears, nostrils*.

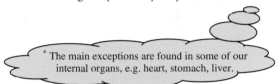

* The main exceptions are found in some of our internal organs, e.g. heart, stomach, liver.

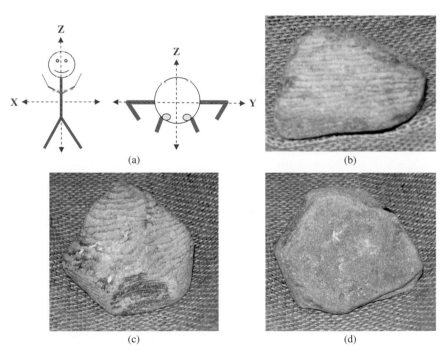

Figure 6.2 (a) Planes of human asymmetry can be seen if we imagine being sliced: front-to-back top ('Y'); left-to-right ('Z'); top-to-bottom ('X'). In general, we are 'mostly' asymmetrical: we are asymmetrical in planes X and Y but largely symmetrical in the Z plane. (b–d) The three pictures here are of the same fossil, viewed in three separate planes (arbitrarily, *x*, *y* and *z*, shown in (b), (c), (d)). It is a single fragment of a fossilized tree, with its growth rings running through. (b) is in the *x*-axis cross-section, (c) is the *y*-axis (showing growth rings eroded at an angle to the plane). But (c) shows the *z*-plane, parallel to a single growth ring (boring!).

Yet, despite our profound asymmetry in two of the three possible planes, we are completely obsessed with the *slightest* asymmetry in our symmetrical plane. We get seriously upset by the smallest difference in leg or arm length. Tiny differences between left and right ears or mouth shape are a personal (cosmetic) disaster, sometimes ending in fashion surgery. The small left-right differences in foot or breast size are common, but mean an instant visit to the airbrush studios in Hollywood. Yet we are profoundly asymmetrical beings, and this has considerable selective/functional benefits.

Asymmetry of our gross structure is not as irrelevant as it might seem to our considerations here, of tissue μ-structures. In fact, *gross* anatomical asymmetry makes it almost inevitable that the fine structures of the tissues in these body parts are also profoundly asymmetrical. After all, it means that they are made of essentially anisotropic material structures.

For example, arms, legs, toes and fingers inevitably have two quite distinct ends: one attached and one non-attached. This is complete mechanical polarity. Although we have a nice symmetrical pair of eyes, each has one end buried in the brain and the other end looks at the outside world (indeed, this page) through a transparent window – your cornea. The result is that all the building blocks and structural components of the tissues have a stark polarity (distal-proximal, in anatomical terms). This is a major consideration when we want to manufacture replacement tissue parts but, being so familiar/obvious to us, it is also easily overlooked. It is a good idea, then, to reproduce those native asymmetries.

In practical terms, and especially at the μm-scale, asymmetry in nature is the result of a series of layers, zones, hollows, channels and voids. Not surprisingly, then, the idea of reproducing and fabricating long-range asymmetries in structure by 'layering' has appeared as a fabrication strategy for both tissues and conventional goods[9]. Very crudely,

the way in which layering can be biomimetic is indicated in Figure 6.2(b–d) using fossil structure. Interestingly, this is grossly symmetrical in two planes and asymmetrical in one plane, just as we are.

6.2 Making simple anisotropic/asymmetrical structures

On the whole, even small solid-tissue elements are rarely random or isotropic in structure. In fact, our lab has a small competition running to identify exceptions, firstly in mammals, then in animals generally. This is not a trivial point as, in most cases, 3D structure and spatial complexity seems to be essential for basic tissue function. As a result, designs for engineered tissues which *start* with the aim of spatial uniformity or randomness, particularly at the nano-micro scale, need to be viewed with caution.

However, even with this seemingly sensible and basic proposition, we rapidly find ourselves coming up against one of the widest held tenets of tissue engineering – *homogeneous, random porosity is good*. This is particularly obvious in the design of biomaterial cell supports. These commonly claim to have random (sponge-like) interconnecting perforations and channels, with a close range of diameters, as a *major merit*. It appears, then, that we have stumbled on another of the paradoxes of tissue engineering. Perhaps we might learn something useful by looking beneath the supporting assumption, or rather the dilemma between these two basic and opposing assumptions.

The first assumption is likely to encounter very little discussion. This assumption is that it is good for us to aim at engineering *simplified* tissues, at least in the first instance. It comes with the further assumption that the resident cells will later have the capacity/information to provide all necessary spatial complexity. It is illuminating, though, that the ambiguity arises when we come to the assumption of what specifically we mean by 'simplified', and again we may see a hint of inter-disciplinary bias.

---

[9]To be specific, we can consider 'layering' as the linear or radial deposition of many thin sheets (similar or dissimilar), in sequence.

Clearly, our aim is essentially to assemble the 'bio' **components** into a 3D **structure** which **functions**, 'eventually'. Our simplification question reduces down to: 'In the early stages of assembly, is it better to simplify the composition or the spatial organisation of the construct, to achieve function?' Notice that the 'timing' caveats here are becoming ever more important to analyze (Text Box 6.1).

The big question now rationalizes down to: which one of these two biological complexities – composition or 3D structure – can we reduce down to its bare essentials *without* losing the ability to reach our goal of tissue function *in a reasonable time-scale*? In this chapter, we explore the possibilities of reducing composition complexity at time-zero to a minimum but *at the same time* building up spatial complexity, by layer engineering.

In considering the case for initial simplicity of composition, we shall return to analogy – in this case is the construction of a particularly beautiful building, the Blue Mosque in Istanbul (Figure 6.3a). We could start by assembling a very few key supporting components – stones, wood and concrete – to give the skeletal shape and basic template of this complex domed, fluted and colonnaded edifice. The strategy here is to produce a structure which resembles what we know we need in size, shape and proportions. Then we invite the sculptors, carpenters, goldsmiths, painters and plasterers (over period of time) to embellish and build on the initial skeletal structure, creating the beautiful mosque we expect.

Conversely, we might collect all of the hardwoods, fine carpets, paintings, gold leaf and ornate windows we would expect would contribute to the beauty of the mosque. But in this case, the multiplicity of components is stacked into simple piles or a single great mound, because there is little or nothing to support any spatial complexity. In this strategy, there is an attempt to assemble the compositional complexity *before* fabrication of any supporting structures.

We can see from this that *sequence* is an essential component of the construction, just as important as the choice of roof and floor tiles. Without the template of the 3D structure, there is nowhere for the wood, coloured glasses and gold to be worked in. In fact, the complexity they bring (out of sequence) might actually degrade the construction process through confusion. It is hard to imagine, in the latter case, how the exquisite form and function of the Blue Mosque could rise out of a design which brings all the necessary diversity of components but no structural template around which to hang them.

In fact, this analogy is a surprisingly close reflection of the development of structures in the vertebrate embryo. Tissues start as 'simple' templates, with a few basic components, and are sequentially remodelled to produce more and more complexity. With that increased spatial complexity, of course, comes subtlety of function. The vertebrate skeleton, for example, first condenses as a simple, cartilage-only replica of the final edifice. Gradually it is vascularized, calcified and articulated to form a fully functional, working skeleton, *though it still retains its original pattern*.

So, our first working assumption in this section is that design of early-stage tissue constructs needs to start with a short, simple component list. These

---

Text Box 6.1 Timing caveats

Timing caveats are important and become more so as we go on. We are particularly interested here in the simplifications that we *must* make at the start of the fabrication process, in terms of how they will impair function downstream. How long we can sustain the tension between making the 'simplification' and needing the 'function' depends on what we are trying to make, but the *time lag* is inevitably a component in our strategy. For example, it is common to assume that early-stage limitations of tissue function (e.g. in mechanical strength) can be improved by gradual *addition* of structural complexity, commonly through resident cell remodelling or growth. This time lag is key to the process design.

(a)

(b)

Figure 6.3 (a) The Blue Mosque – and the materials its walls, roof and floors are made from. (b) Inside the Blue Mosque, the apparently endless complexity turns out to be made of hierarchies of similar structures, repeated at the different length scales. Firstly, the roof mass is supported on huge columns and arches (left top, arrowheads). Between these, are rows of half domes, set within larger bays (ringed and arrowed in the lower view). Within each of these levels there are repeating half-rings of arches (curved arrow, top), each arch carrying its own repeated embellishment. It can be difficult to show the comparable 3D repeating structure of tissues in this way, except as diagrams, which can be found in any good histology text. This is because of how we slice and image solid structures (see Figure 6.2) and so have to reconstruct the third dimension. The inset diagram (right) represents a small repeating structural unit of bone. This comprises blood capillaries and nerves (red tube), running through Haversian canals (larger tube), surrounded by rings of bone matrix, bone cells and their own micro-scale connecting channels, known as lamellae (or layers – dotted red). Spot the similar repeating of layers across scales.

components should be carefully chosen as the basics needed for construction – even if the eventual structures are to be complex 3D shapes. This applies at all scale levels. It will form and then act as the template onto which further layers of complexity can be added, in sequence. In the case of biological structure mimicry, the term 'non-random complexity' should be translated as containing:

- asymmetrical layers (planar and radial);
- repeating zones;
- ridges and channels; or
- anisotropic fibrous sheets.

We shall shortly learn what these mean in more detail, but at this stage it is important to understand that they are useful terms and concepts for our task of producing reductionist structure designs. Importantly, these structures must cross many hierarchies, from the nano to the metre scale.

6.3 Thinking asymmetrically

The complex, repeating asymmetry of many tissue structures is so familiar to us (the owners) that we often do not notice it. Neither do we lay awake pondering *how* and *why* our eyelids, tongue, earlobes or fingertips are built up of so many repeating layers and zones. The layers are mainly asymmetrical, sometimes in more than one plane. They are interlaced with repeating dense/less dense zones and fibre-based anisotropies. These layers and asymmetries and structural repeats have a direction and go through a sequence of patterns.

At first glance, this complexity looks daunting, but then again so does the interior space of the Blue Mosque. Look at Figure 6.3b and notice how the (asymmetrical) arches in the roof are organized in strings through space, following the curved outline of the domes. Sometimes there are sub-bulges (with their arches) pressed out of the *natural arc* of that dome (white arrows in Figure 6.3b). In other aspects, different types of arch, with a higher aspect ratio and greater asymmetry, are arranged *in sequence* in the curved space (red circle in Figure 6.3b).

The arch is only one many of asymmetrical structures present when we look carefully enough at the mosque. Eventually, you should find that the 3D patterns from which the structure is made become more and more evident. Interestingly, as your eye starts to see this 'detail' of the structure, so the construction possibilities of how it might be put together also become clear. In effect, our eyes are starting to recognize the directions in which the various shapes and components have been built up. It ceases to look like a single, rather daunting mass of complex shapes, and it becomes a rational, repetitive sequence of layers and zones. This is when we recall that there is nothing supernatural in this architecture – it was fabricated by people (however inspired)! In other words, it is quite possible to do this again.

What is most helpful is to realize that it is made up of relatively simple building units (3D asymmetries) repeated in patterns and variants. With a little more practice, one can also identify those few basic, often *support* elements which are used in the first phases of construction to carry the weight and maintain the general shape of the building – in other words, the template structure.

In fact, these tend to be the foundations, walls, pillars (Figure 6.3b) and buttresses, which are often hidden or cut away in later stages. These *carry* or *support* the many smaller components to give the final complexity. The same can be true for biological tissues, as we shall see later. Interestingly enough, it is often those users and scholars of such structures – those closest to them – who are least likely to perceive them in this reductionist manner. But it is the role of anatomists and histologists on the one hand, or followers of religions who use such buildings on the other, to deduce meaning from the *overall* finished structure, not to reduce and simplify it again. In building tissues, that job falls to other tribes – and in this case, us.

In the interests of scale, clarity and indulgence (not to mention accessibility), we now switch analogies from the massive and drop-dead gorgeous, to confection! The familiarity of this example helps us to get right up and close to the questions of:

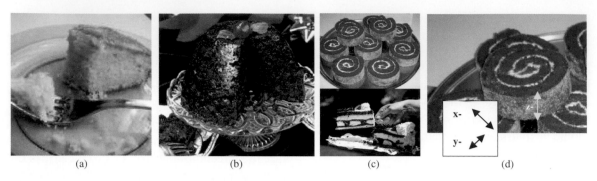

Figure 6.4 (a) Victoria sponge cake, at a gross scale, is compositionally *and* structurally simple. (b) Christmas pudding is compositionally complex but structurally simple (i.e. random). (c) Swiss roll (upper) and Black Forest gateau (lower panel) are both spatially complex, though only the Black Forest gateau is both spatially *and* compositionally complex (see text). (d) Swiss roll planes: *x, y* and *z*.

- How 'simple' can simple *really* get (and still be useful)? and
- Where does complexity start to take off as a problem?

Cake-world examples here start with the Victoria sponge, Swiss roll and Black Forest gateau, and lead on to the British Christmas, or plum, pudding (Figure 6.4).

At the scale we are familiar with, Figure 6.4a shows how the Victoria sponge is structurally homogeneous, especially inside. We shall come back to the 'scale-caveat' later (Text Box 6.2). In contrast, we find examples of 'non-homogeneous' cakes in 'layered desserts' such as the Swiss roll and Black Forest gateau in Figure 6.4c). These are inhomogeneous partly in terms of the very different substances which make up their mass (*compositional inhomogeneity*) and also in the way they are spatially organized is into distinct *zones* and *layers*, with separating *interfaces*. In other words, they are made up of different components, such as cake, cream, icing and fruit, arranged (asymmetrically) in *3D space* into layers, depots, tracks or zones.

Notice also the tendency for *structural repetition*, which is an important and extremely useful feature of non-random structures. This is useful for us to understand, as it is highly biomimetic but relatively simple to fabricate. Black Forest gateau (Figure 6.4c, lower panel), illustrates another important feature for biomimetic fabrication in its obvious top-bottom polarization. In other words, its layers

have a distinct *sequence* of shapes and compositions. Once we are aware of these layers and sequences, we also see their interfaces!

The '**interfaces**' are, in effect, lines of rapid change in composition, density or phase between layers or zones. Though easily overlooked, interfaces can also be key spatial factors:

- for producing biological function; and
- to assist fabrication and processing (more of this later).

It is often useful to consider an interface as just another, very thin, layer in the structure, which either fixes to or glides between its neighbours. It is important, though, that interfaces should not be seen as being *only* present between planar surfaces. Radial or non-linear interfaces are illustrated by the Swiss roll (cross-section in Figure 6.4c, top) and around the zones lying *within* the planar layers of the Black Forest gateau (lower panel).

Such radial interfaces are most commonly seen in natural tissues in the form of blood vessels, nerves and ducts, which can comprise 1–5 radial layers in cross-section (depending on the size scale). Each one represents an interface. Since nerves and blood vessels are effectively tube-like in structure, their interfaces are strictly 'radial-planar' in 3D. Those around spherical glands might best be described as 'simple radial' (as in the white cream zones in the Black Forest gateaux).

Our first conclusion, then, is that Repetitions, Interfaces and Polarities (RIP) are our most important tools for the design and fabrication of biomimetic structures.

Another useful distinction is illustrated by the Christmas pudding example (Figure 6.4b). This is the opposite extreme combination to the Swiss roll, being, perhaps, the most non-homogeneous material of the series in its composition. For those of you who have not experienced such a dessert, the visible lumps within its mass are composed of a huge variety of different edible materials. In fact, it has a very pronounced *compositional* heterogeneity: it is the essence of complexity in terms of what it contains. However, there is absolutely no structural organisation of these elements; they are randomly distributed in such a way that, when averaged over a significant volume, the 3D structure is homogeneous*.

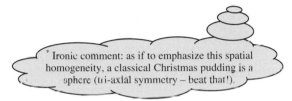

* Ironic comment: as if to emphasize this spatial homogeneity, a classical Christmas pudding is a sphere (tri-axial symmetry – beat that!).

Christmas pudding, then, is *spatially homogeneous* but, at the same time, compositionally complex. Perhaps the fact this is rarely a desirable template for natural tissues (i.e. poorly biomimetic) is a good lesson for us to take away as aspirant fabricators of tissues. Actually, it is a combination which is closer to that of adult mammalian repair tissues, i.e. scars (lots of good tissue components with poor 3D organisation – see Chapter 1). However, the body is all too good at making scar tissues, at all sites, without our help.

The second take-home-message, then, is that the far end of the 'success spectrum' is not the *absence* of a tissue, but the engineering of damaging scar tissue – Christmas pudding mimicry.

To delve deeper into our cake-related analysis of organizing spatial factors, it turns out that cake can teach us a great deal about different forms of symmetry, asymmetry and layering. Figure 6.4c illustrates two further levels of non-homogeneity which are easy to miss. At our comfortable visual scale ($\approx$1 mm and above), Swiss rolls are made of only two components – flavoured cream and chocolate cake – but these are rolled up into the characteristic

spiral. Swiss roll, then, is 'compositionally biphasic' (i.e. simple), but spatially non-homogeneous/complex. We can measure this either in terms of the rate of change of structure/composition or as the number of interfaces encountered across the cross-sectional diameter of a single slice (Figure 6.4d, x and y planes). This cake is both simple and complex.

In contrast, Black forest gateau contains numerous substances arranged into spatially distinct layers and zones, in a number of planes throughout the cake. This is the most tissue-like but the hardest to fabricate, particularly as it does not yet hint at the way biological complexity must extend down to the cell-critical micro- and nano-scales.

Could it be, then, that the Swiss roll embodies what we are looking for as the initial tissue template – a practical compromise between complexity and simplicity? Ideally, we are looking for the complexity that is essential for bio-mimicry, but it needs to be simple enough to fabricate. So if this is our spatial paradigm, it must be important to understand the secret of the Swiss roll and its clever compromise. **Why is its complexity so simple to generate?**

The clue lies in the number of interfaces crossed in the Swiss roll when tracking in the x–y plane, across its cross-section (Figure 6.4d shows x–y–z planes). There are approximately 12 interfaces in the x plane across the diameter this cake slice (excluding the cake/air edges (arrowed)). This would equal or even exceed the same measure in a Black Forest gateau. However, the key here lies in the z plane, where the Swiss roll scores a stark *zero* interfaces crossed (i.e. the tracking plane is parallel to the layers).

In conclusion, we can propose that the design and fabrication of spatially complex; hopefully, biomimetic tissue templates should be possible by repetitive assembly of many (compositionally) simple layers. **In effect, this is *layer engineering*.**

Perhaps, then, our detailed analysis of cake structure, in search of fabrication strategies, has not been such a self-indulgent exercise. It has provided a framework to enable us more easily to identify current mismatches between our targets and approaches, in particular where we again may be 'aiming low and still missing'. The high target which was originally set, to make tissues with complex structure and composition, has unwittingly forced us to rely on the 'relatively low aim' approach of persuading cells to make them for us. The modest success of this approach has tended to lower the aim still further, such that engineered tissues are considered successful if they achieve two or, exceptionally, three layers with distinct structures.

However, the message from the cake analysis is that spatial complexity *can* be achieved if standard engineering concepts are adopted to simplify composition. This is based on the use of repetition and polarity to generate many interfaces (spatial complexity) through layering. Repetitious assembly of simple components is a field that we humans excel in (at scales from mobile phones (nano-micro) to trucks and ships, and from microns to kilometres.

Furthermore, layer engineering carries the potential for evolving greater and greater complexity at larger and smaller size scales, as fabrication of the layers becomes more sophisticated. After all, unlike Swiss rolls, the layers **do not all have to be the same**. Increasing the options for the types of layer, layer sequence and changing layer thickness (i.e. more and thinner layers) will inevitably permit exponential increases in structural complexity, but with the same simple process.

Our next task must be to understand just how this layer engineering idea can be applied in practice to assembling *living* tissue. After all, making elegant, but non-living multi-layered structures, and then expecting to get cells into the right place, is not a good sequence. The first step is to take a much closer look under a stone which we have repeatedly skimmed past – **the scale caveat** (Text Box 6.2). In short, the scale caveat qualifies what we aim to make in terms of the size scale it must work at.

The size caveat, then, highlights something of a weak spot in our understanding of cell control needed for tissue engineering. Although it is well known that cells use and respond to substrate shape, structure and mechanical properties, it is much less certain how to *use* the language of these cues. In particular, how cells respond to different features on their top and bottom surfaces (e.g. non-2D, multi-polar) is not understood.

It is difficult, then, to imagine them being applied in the near future for the practical fabrication of tissue detail. In many ways, this is comparable to the problem of growth factors in tissue engineering. It is clear that they convey potent messages to cell systems, but we do not know the language. On the other hand, the opposite approach of *physically fabricating* a man-made, complex structural hierarchy (i.e. not reliant on cell synthesis and assembly) is seriously daunting, particularly in biomedical sciences.

We shall return to this later, as it may represent less of a real blockage and more a result of limited cross-disciplinary, between-tribe thinking. After all, the engineers and designers responsible for mobile phones can fabricate and assemble parts within the same size scales. However, unlike tissue engineers, mobile phone engineers cannot be tempted by even the most remote dream of finding a cell which will make a competitor to the iPhone – they have no choice! Paradoxically, there are no cells in a cell-phone.

6.4 How do we know which scale to engineer first?

The cake analogy teaches us that it could be a good idea to engineer small sections, or layers of the structure and *then* to create structural complexity through a subsequent assembly step. But which 'small-scale, simple component(s)' should we begin with?

Although it is possible to simplify our thinking on which target structures need to be assembled, it remains of huge practical importance to establish how we might assemble such 3D structures, with all that this implies for final shape, symmetry and anisotropy. First, many of the most common target tissues (and certainly the connective tissues) have key structural features which operate at all levels, from the gross (mm to metres) to the meso-scales (μm- to nm). In other words, it is now well understood that the non-cell (extracellular matrix) bulk of tissues is built up of hierarchies of repeating structures, aggregated all the way from the nano up

Text Box 6.2 The 'scale' caveat

The crucial thing about cakes is that we can only really do the analysis at the gross (visible) scale. Effectively, this is on the mm to cm range (or metres if you do party catering or gluttony). In clinical terms, this is the scale which most interests both the surgeon and the patient, as mostly this is where they measure success. However, in fact, it is really sub-millimetre and, indeed down at the μm and nm scales, where most of the critical cell-based business is done – and where success or failure lies.

The heart of this effect is easier to understand if we consider cells as tiny (20 μm) globular factories. Their business is to take incoming shipments of small, simple molecules, nutrients, oxygen, etc., sized in the (sub-) nano range, and assemble these into ever larger, more complex biological building blocks (proteins, fats, polysaccharides). Once synthesized, more and more of these building blocks undergo continuous, repetitious aggregation and assemble to form more cells, cell layers or cell masses on the one hand, and complex structural materials on the other (extracellular matrix: ECM).

Eventually, these aggregated structures come to our scale of attention as visible parts of our bodies, like pimples, fingernails or that welcome new lump of skin filling in the hole where the salad knife slipped. These personally important structures are 'simply' millions and millions of fibre-forming proteins (ECM building units at 10–100 nm scale) aggregated up to the mm-cm scale. The cell sits in the middle of this scale hierarchy like a pump motor, sucking up sub-nm matter and pushing it out into sub-metre body parts by endless repetition.

In effect, then, it is the size of cells and their repetitious assembly habits which generates the long-scale structural hierarchies we see in tissues. Each smaller level of structure is integrated with the next level of structure up, and so on. Thus it is often difficult to divide 'the function' into clear size scales. This is particularly clear in the load-carrying connective tissues, where structures such as the fibrils in tendons get larger by addition of more and more of the same small parts. In the case of tendon, collagen microfibrils (nm) aggregate into fibrils and then into fibres, eventually fibre bundles and mm-scale tendon fascicles. Each of these adds a little more to the functional (tensile) load-carrying capacity total as the fibre structures get thicker.

In this respect, a tissue would be much more like a high-rise hotel block than a holiday camp full of chalets; tissues tend to come as an 'integrated whole' rather than a loose federation of independent units (or huts). Remodelling the accommodation in the hotel is trickier than the chalet site, due to the interdependence of the hierarchy layers. It is the complete scale hierarchy which functions – either altogether, or nothing at all. So, to be truly biomimetic in our engineering, especially of mechanical tissues, we may need to build up by integrating small (cell-scale) units through continuous repetition.

Clues as to the 'important' scale, where we need to provide cell controls, come from research into the size and types of surface 'texture' to which cells respond. Features such as ridges, channels, pits and lumps (generically termed topography) change cell behaviour. At the lowest level, surface structures of less than 20 nm can trigger a few cell responses, while at the 'big' end, ridges or fibres of 10–100 μm diameter will cause many cells to align or move in specific directions.

This gives us a guide to the range of scales where the behaviour of cells can be influenced by their substrate. The lower end of the range seems to be governed by the dimensions at which cell membrane integrins can be clustered (at the points where cell membranes attach) to generate enzyme-based signals inside the cell. The upper end (100–200 μm fibre diameter) seems to be the level where the cell itself (20–60 μm when spread) is unable to tell the difference between a very gentle curve and a flat surface (just as we struggle to *perceive* the planet's curve).

References

1. Dunn, G.A. & Heath, J.P. (1976). A new hypothesis of contact guidance in tissue cells. *Experimental Cell Research* **101**, 1–14.
2. Clark, P., Connolly, P., Curtis, A.S., Dow, J.A. & Wilkinson, C.D. (1990). Topographical control of cell behaviour: II. Multiple grooved substrata. *Development* **108**, 635–644.
3. Loesberg, W.A., te Riet, J., van Delft, F.C., Schön, P., Figdor, C.G., Speller, S., van Loon, J.J., Walboomers, X.F. & Jansen, J.A. (2007). The threshold at which substrate nanogroove dimensions may influence fibroblast alignment and adhesion. *Biomaterials* **28**, 3944–3951.
4. Dalby, M.J., Riehle, M.O., Sutherland, D.S., Agheli, H. & Curtis, A.S. (2004). Use of nanotopography to study mechanotransduction in fibroblasts – methods and perspectives. *European Journal of Cell Biology* **83**, 159–169.

to the gross level, with almost no discernable lines of functional segregation. The most obvious, and in this case most relevant, example is the hierarchy of collagen organisation.

Collagen is a fibrous protein in which the 1.5 nm diameter, rope-like monomer forms the basic building block and aggregates in a semi-crystalline manner to form long fibrils. These range from 30–500 nm diameter, depending on age and tissue. The fibrils can be packed or woven into larger laterally packed aggregates called fibres and fibre bundles ($\approx$500 nm–500 μm) and, in a few tissues, to discrete, outlined fascicles ($\approx$mm: see Chapter 1, Figure 1.1). As in the case of the fibres of cotton or wool that make up the more familiar fibres of our clothes, collagen fibrils are of indeterminate length. This allows them to be packed, knitted or woven together in a regular manner, or randomly intertwined for long-range load transmission. The density and overall 3D architecture of this fibril network is what gives each connective tissue its characteristic mechanical properties.

Unlike man-made fibres, though, collagen fibril networks would typically enclose resident tissue cells. In contrast to the relatively fixed size and proportion of the collagen component, these cells are complex, constantly moving and shape-dynamic. Let us assume for now that they are roughly spherical (Text Box 6.3) and call them nominally 20 μm diameter.

As always, it is important here to get a clear understanding of the scale of this cell-fibril relationship. If we want to illustrate this relationship, it also needs to be based on the spatial relationship in native tissues, where **the collagen comes out of the cell**. For the purposes of this analogy, then, let us imagine we could reduce a sheep to be the size of a cell. This would make the strands of its fleece roughly proportional to the diameter of collagen fibrils as they aggregate around the cell. Mathematically speaking, fleece-hairs = collagen-fibrils: sheep = cell, as in Figure 6.5. This emphasizes the idea that hairs/fibrils are very much the product of, and so spatially associated with, the sheep (or the cell).

It also allows us to look at the next size scale up; what happens where lots of sheep get together, like the cells in a tissue? When considering a flock of sheep, it is reasonable to expect air spaces between the sheep in a flock. Clearly, this is not going to be true for cells in a connective tissue, which are separated only by collagen-rich ECM, with a few vascular and nerve channels. It is as if each sheep's fleece could grow to an enormous thickness. The woolly fibres become completely entangled, even bonded together in the inter-sheep spaces. This is exactly what happens to cells as their collagen 'fleece' becomes increasingly thick and cross-linked.

An interesting corollary of this sheep analogy starts to be very important to tissue function. If the fleeces of sheep in contact really did interlink and

---

Text Box 6.3 Cell shape and size

In systems such as this, where we are getting right under the skin of the '3D question', we can quickly get into difficulties over the changing shape of cells. In suspension, and with minimal attachment to a matrix or tissue culture plastic, most cells tend to become spherical, commonly in the order of 15–20 μm in diameter. A classic exception to this generality is the mammalian red blood cell, which is a bi-concave disc of 6–8 μm diameter and $\approx$2 μm thick, with absolutely no attachment and no nucleus.

Dramatically, things happened to cell shape when they attached down to a flat plastic culture dish. They spread out and move, generating complex and bizarre shapes. These resemble fried eggs in monolayer, as the soft cell cytoplasm is flattened and extended over the flat plastic, leaving the stiff nucleus standing high in the centre. In 3D culture, and in natural tissues, these flattened structures are rare. In culture, this spreading, sometimes to as much as 80 or 100 μm across, gives the wrong impression of large cell size and it makes discussion of 3D cell size and shape confusing. For simplicity here, we consider cells as non-suspended and spherical (at the time of seeding). How, when and why they change their shape from then on may be the target of future work.

(a)   (b)

Figure 6.5 The formula idea here is that where Cells = Sheep and Collagen = Wool: (a) sheep/cell producers will be embedded in the wool/collagen; (b) a sheep-wool unit would look like this in 'cross-section', much like a slice through a fibroblast or osteoblast plus its collagen. Photo (b) © Eric IsselShutterstock.com.

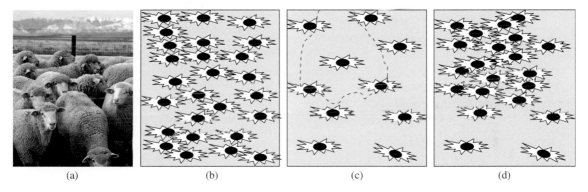

(a)   (b)   (c)   (d)

Figure 6.6 (a) Flock of sheep, © US Department of Agriculture. (b) Early, low-strength cell-rich matrix, before mass collagen deposition (Cells black, collagen grey). (c) Mature, strong collagen-rich functional matrix with load carried on the collagen. (d) Post-injury, the process repeats locally, with high-density cells making bulk collagen (red dotted line = injury).

join mechanically, the whole flock would move as one. This echoes the hotel analogy, above, illustrating how the building units in a tissue are functionally integrated. If the shepherd pulled hard on two or three sheep at one edge of the flock, the whole flock would feel the tensile load and be dragged in that direction. No stragglers, and no sheepdog needed!

Consequently, we can picture a densely packed flock of sheep (Figure 6.6a). As long as we stipulate that their fleeces are not just in contact but 'super-glued' together, then the flock is equivalent to cells with only a little collagen (extracellular matrix) in a cell-rich tissue (Figure 6.6b). This would be the sort of tissue we expect in an early wound repair tissue.

But notice how the *cells* dominate over the matrix in this situation (Figure 6.6b).

As cells are much weaker than the collagen material, such cell-rich tissues are not mechanically strong; or at least that is true until these cells push out *much* more collagen, as extracellular matrix (ECM), between themselves. As this process of collagen deposition continues, the proportion of matrix to cells rises until the collagen/ECM begins to dominate the mechanical properties. This corresponds to stages when a new connective tissue begins to mature into a strong, stable structure (Figure 6.6c).

Here our analogy becomes really bizarre, as we are unlikely to encounter sheep fleeces that are more than 20 cm thick. Yet adult tissues can have inter-cellular matrix several cell-diameters thick, which would translate into several sheep-diameters at our scale, implying that they can grow fleeces of some metres thick. However, this is a fantasy illustration – and cells are remarkable! In the case of collagen-producing cells, layer after layer of collagen is produced, pushing cells further and further apart and making the tissue stronger. It is as if there was no annual shearing of the fleece, or the collagen layer for cells, so that one flock eventually is comprised of lots of wool and only a few sheep. This sort of cell-to-matrix ratio is exactly what we see in mature connective tissues (without the wool!).

Surprisingly, this analogy is useful for one last lap. We can illustrate this in terms of events which follow an evening visit to the flock by a wolf. Where the flock/tissue is injured (red line in Figure 6.6c) and loses some of its bulk (viewed from the wolf's perspective, some sheep are taken out to dinner!). The resulting space will be filled by new sheep which have thin fleeces. The repaired defect will, for a while, again be sheep/cell rich and fleece/matrix poor, and therefore weak (Figure 6.6d). The process then repeats with new collagen (or wool) once more pushing apart the cells (sheep), until we return approximately to a joined-up, fleece-rich, sheep-poor flock (Figure 6.6c). And that is essentially how a wound repairs, if we leave out the sheep[10].

To conclude this section, have we, then, found a promising starting point for the composition of simple building units for layer engineering – cells in a collagen fibril network?

Collagen fibril networks (predominantly type I collagen) are at the heart of connective tissues, both in terms of tissue bulk and mechanical function.

Around 25 per cent of our protein tissue dry weight is collagen. This is a truly huge synthetic investment on the part of resident tissue fibroblasts, and something which, incidentally, they are not keen to do in culture. Clearly, then, collagen is a great candidate basic building material. The more of it we can engineer into the fabric of our simple tissue templates, the less the cells will have to labour to produce.

So we have reached a possible watershed. Using 'extreme tissue engineering' concepts and working from first principles, we have a candidate tissue fabrication approach:

1. Fabricate simple tissue layers out of (a) appropriate cells in (b) a thin sheet of collagen.
2. Build up both bulk and spatial complexity with many of these layers in 3D.

Two niggling practical problems hover over this stage:

- How do *we* make native fibrillar collagen materials with predictable properties, when at present only cells seem to be able to do this?
- How do we then get the layers to stick together?

But these questions are for later.

6.5 Making a virtue of hierarchical complexity: because we have to

So far we have tried to rationalize tissue 3D complexity, hierarchies and the asymmetrical cell-matrix structures which make them up. The aim has been to get beyond some of the traditional simplifications which are so ingrained in the world of isolated cell biology and '2D' cell culture.

This is perhaps the first nettle which it has been common to avoid grasping in much of tissue engineering. The bulk of what we know about, and how we think of, cells relates to monolayer cultures attached to flat plastic. Before the appearance of tissue engineering, few of us even bothered to grow cells in 3D, as it was considered impractically complex and largely irrelevant to the questions being asked. Clearly, the habit of some cell

---

[10]Note, though, that we have built up an analogy where the 'tissue' is represented not by the sheep but by the imaginary integrated flock. Note also that we have a flock/tissue-building unit which is very simple in composition (sheep + fleece or cells + collagen). However, sheep being sheep, with feet on the ground and backs to the sky, it is only one layer thick. So, our analogy leads us towards a promising-looking 'tissue building unit', i.e. a *thin layer* of cells in a fibrillar collagen mesh, **ready to repeat and layer-up into complex architectures**.

types to overgrow one another in culture, forming spontaneous multi-layers, can be regarded as a very simple form of 3D culture, but hopefully the reader will now understand how minimal this is in terms of the target complexity we really need to aim for. Once we claim that our vision is fabricating functional 3D tissues, this nettle can no longer be avoided.

In one respect, though, our understanding of 'simple sheets of cells' may turn out to be a bit of a bonus for tackling the second of our high(er) targets. In a way, this is the opposite sort of problem, and it relates to the way in which tissue structure just seems *so* complex when we look at it in our familiar histological slices.

This second blind-siding factor is perhaps even more deeply rooted in biological training. Although many modern imaging techniques can reconstruct structures in 3D, a large part of our understanding of tissue structure is still based on histological techniques where the information is essentially 2D, derived from very flat, very thin slices of tissue. The reason for the thinness is similar to that for

imaging cultured cells in monolayer – they produce the most beautiful images because the optical depth is optimal. How far this is from our target of complex 3D tissues is clear from a critical glance at Figure 6.7, an idealized diagram of skin. At its simplest, skin is regarded as at least three 'layers' (meaning in a plane parallel to its surface). These layers are the epidermis (keratinocyte layers), dermal (fibroblasts in a collagen-rich stroma) and subcutaneous or hypodermis of fat, plus or minus underlying muscle.

We take such images and μ-structural/anatomical analyses completely for granted now. But as a spatial analysis, they contain an assumption, they are selective in a very particular way. The plane in which those *slices are cut* is selected apparently for optimal teaching properties. They are ideal for showing off the main 'layers and interfaces', as in Figure 6.7. This means, though, that many tissues are commonly only really cut and illustrated in one or two planes, *not* three. Look in your text books at the structures of gut wall, blood vessels, oral tissues,

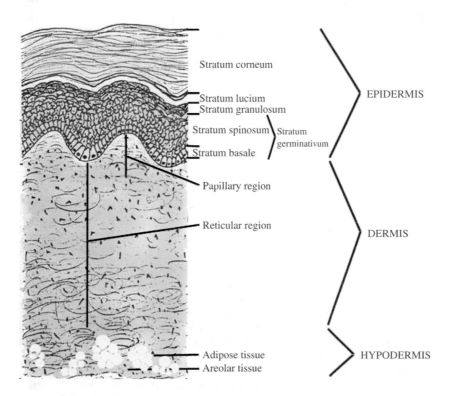

Figure 6.7 Three layers of skin, in 2D.

trachea, bladder, etc. They all resemble the skin shown here, in that the histological slices go across the maximum number of *layers*: surface to deep. So why do we so rarely see or consider the μ-structures of these tissues through or between the layers – that is, in the 3D plane parallel to the surface in skin?[11]

Simple, really: *if* you could get such sections cleanly through the layer (difficult as they are thin and not flat), they would be really boring. They would show an unappealingly constant structure over monotonous lengths of tissue. The dermal layer, for example, changes only slowly from one part of the body to another. We can, in fact, do this now with modern imaging – but we still don't often bother. However, once it is pointed out that there is *another* 'view' of these layered tissues (and almost all are 'layered' at some scale), we can suddenly see that making them need not be as complex as we thought. After all, this **third plane constancy** (monotony) means that only two of the three are different and complex (see Figure 6.2)!

To the extreme tissue engineer, this limited asymmetry is the source of great joy and promise. It means that, once we can identify a series of planes of *constant* structure, we can design simplified tissues as a series of layers (Figure 6.8(left)).

The epidermal and dermal layers are illustrated in Figure 6.8(left) in 3D. As we can see, these can be separated, arbitrarily, into at least three 'single layer' planes. The epidermis could, in theory, be split from the dermis in plane (a), while a parallel, but deeper plane (b) would peel apart the dermal and subcutaneous layers. However, there are clearly many more identifiable layers in this diagram, if we do the cake-type analysis. Indeed, there are more

planes to analyze than those which are parallel to the surface*.

* Also, not all of the planes and interfaces have to be flat (planar).

Tubes running through the tissue have concentric layers, and such radial layering is about as common as you can get in biology. These are indicated in Figure 6.8(right) as multi-layered, concentric cylinders running either from surface to deep or parallel to the surface, respectively ((c) and (d) in Figure 6.8(right)). Some of you will have noticed that with this step, we tend to have dropped one scale hierarchy, from mm of the dermis to tens of μm for a capillary or nerve. The example plane (c) includes structures which run down into the skin, such as hair follicles and sebaceous glands. Coaxial structures in the (d) plane might be blood vessels and nerves, or even small muscle groups and connecting tendons.

Contrary to non-specialist understanding, skilled plastic surgeons do not satisfy themselves with just concealing large facial defects with a layer of skin following trauma or tumour surgery. Where they can, their aim would be to replace/reconnect the complex facial muscle blocks and their fine nerve and vessel routes which give a balanced smile or a social blink. This is the target of *regeneration*, as opposed to *repair* with a covering skin graft. It also illustrates that analysis of layers and asymmetrical tissue hierarchies is no esoteric whim of micro-anatomy, but can hold the key to functional *regeneration* (as opposed to repair or 'filling the gap').

So, perhaps we can use these 3D hierarchies of structure – layers – as the basis for building up (fabricating) the *apparently* complex tissue structures that we take for granted as everyday users. We should be modest, though, as the idea of tissues as assemblies of precise, geometric layers in fact paints a highly stylized, grossly simplified version of natural reality. The point, though, is that even a simplified layer tissue would be a great advance on what we presently have. Furthermore, as we shall

---

[11] This piece of logic can take a few minutes to sink in if you are less familiar with spatial biology. It helps if you close this book and imagine you are a bookworm, munching through the pages. Now you can see that you would experience the same meal if you move in two of the three planes (between two leaves), but not when you try to eat through from page 13–133. The book is two-thirds homogeneous!

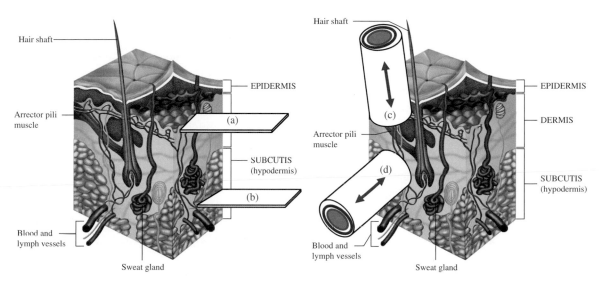

Figure 6.8 Contemplating the planar (left) and radial (right) layers of skin.

see, it also has great potential to be customized with greater and greater complex variation – just like real tissues.

In summarizing this section, we again glimpse the 'aiming low' tissue engineering paradox. The low target here comes from a naivety and limited cross-disciplinary understanding (e.g. the plastic surgeon's dream). It has led us too often to think we must engineer the *whole* tissue lump in one go, as it is too complex to sectionalize. Could it be, then, that tissue engineers who avoid the 'low target' of a single tissue lump might, in fact, fabricate much more controlled customized 3D structures by building in thin layers – deconstructed hierarchies, like cake-makers?

The process then becomes one of identifying and designing the necessary series of *layers* which could form our tissue template. To return to our Blue Mosque analogy, we should avoid the mistake of being overawed by complex beauty of the finished item and concentrate on finding the repetitive simplicity which is inevitably present. Completing the whole building is not yet the task of tissue engineers. We must first construct good, but simple, tissue templates. In the case of the Mosque, the ornate plaster and gold leaf is added later; and in

the case of facial reconstructions, that is the task for reconstructive surgeons.

6.6 Cell-layering and matrix-layering

What, then, do we need to do to fabricate such 3D structures through layer generation, even the simple ones? Is it really practical to use these natural planes of tissue layering and zoning as the basis for making complex, anisotropic bio-artificial tissues? In fact 'tissue layer engineering', or just **layer engineering**, is already in use. Perhaps even more surprising, there are two basic types of layer engineering and both are already available, though based on quite distinct technologies. These complementary approaches, cell-layer and matrix-layer engineering, neatly reflect the two basic forms of tissue we have identified already as our basic targets: cell-rich and matrix-rich tissues.

The first of these, cell-rich tissues, can be tackled by 'cell-layer engineering'. This uses many similar cell layers, grown as coherent sheets and then stacked to form multi-layers. The cells used for cell-layer engineering are commonly natural sheet-forming epithelia or cooperative cell types which work in the body at very high densities, such as muscle or liver cells.

The key enabling technology behind cell-layer engineering was the development of cell-culture substrates whose surface chemistry changes from hydrophilic to hydrophobic with temperature, commonly between 20–40° C. This transition in surface chemistry is key, as the hydrophobic or hydrophilic nature of a cell culture surface governs how, and if, cell attachment proteins stick to that surface (as we have seen before, cell adhesion proteins commonly include fibronectin and vitronectin from serum in the culture medium).

In this case, the poly N-isopropylacrylamide (p-NIPAAm) thermo-responsive coating which is applied to the tissue culture plastic supports protein-binds (and so cell adhesion) at temperatures around 37° C, at which cells are routinely cultured. However, if the culture temperature is reduced below say 30° C, its protein binding capacity flips. The adhesion proteins holding the cells down detach from the plastic, and take with them any cells which were using them.

The big plus about this system for cell detachment is how gentle it is. If cells are grown on (adherent to, at 37° C) such surfaces for some time – often in the order of (7–14 days) they form a confluent/continuous and stable sheet, incorporating

small amounts of extracellular matrix, commonly a basement membrane. Such cell-rich sheets are then detached from the plastic culture surface by reducing the temperature for a short period, and then they can be used individually or stacked into multilayers. The key point is that the cell-sheet remains intact as a floating, coherent layer of, for example, epithelium. This is unlike classical cell recovery by digestion of the cell monolayer with enzymes, which destroy the cell attachment and matrix proteins, so breaking up the sheet.

This technology has been used to produce a number of multi-layered tissues, including mucosa, cornea and muscle. Multi-layer engineering of muscle means that beating muscle sheets can be fabricated with reduced (and controllable) tendency to suffer serious hypoxic damage. Figure 6.9 summarizes the stages and mechanisms by which cell layers are fabricated on thermo-responsive surfaces.

The second type of layer engineering – this time producing *matrix-rich* tissues – aims to fabricate sheets containing cells within a mechanically strong extracellular matrix. The enabling technology supporting this matrix-layer engineering comes from technologies to construct native protein-based (often collagen) neo-tissues containing cells. A completely different approach, involving rapid shrinkage by fluid expulsion (collagen plastic

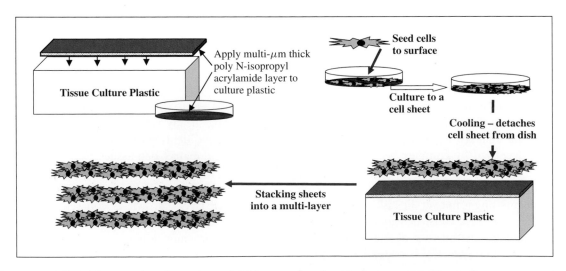

Figure 6.9 Cell-rich layer engineering: using poly N-isopropylacrylamide thermo-responsive surface, on conventional tissue culture plastic.

compression or CPC), has proved successful and avoids mechanical cell damage. The products of this CPC technology are thin, cellular sheets made predominantly of a dense, tough network of collagen nano-fibrils enclosing a closely enmeshed population of living cells. In other words, these are sheets of simple connective tissue.

It is worth explaining briefly just how CPC works to produce connective tissue building sheets. As we have seen, the two 'big problems' for engineered connective tissues are:

(i) production of collagen-fibril densities which even approach those of tissues; and
(ii) construction of even modest 3D tissue-like architectures, at the µm (i.e. cell-) scale, around the resident cells.

The trick with collagen plastic compression is that the basic tissue templates or replicas are formed initially out of hyper-hydrated, but native, collagen gels. The starting collagen for making these gels is already routinely extracted for clinical uses as acid-soluble tropocollagen, from animal tissues such as tendon and skin (see Chapter 4). Under physiological conditions (temperature and pH), the soluble collagen monomers aggregate to form nm-diameter fibrils. These nm-scale fibrils form around any cells which are suspended in the mix, enmeshing them, as they do in natural tissues. This 'fibrillogenesis' forms very soft cell-collagen gels, over approximately 10–15 minutes, comprising around 0.5 per cent collagen and about 99.5 per cent water (Figure 6.10).

Initially the excess of water, with resulting terrible mechanical properties, looks like big problems.

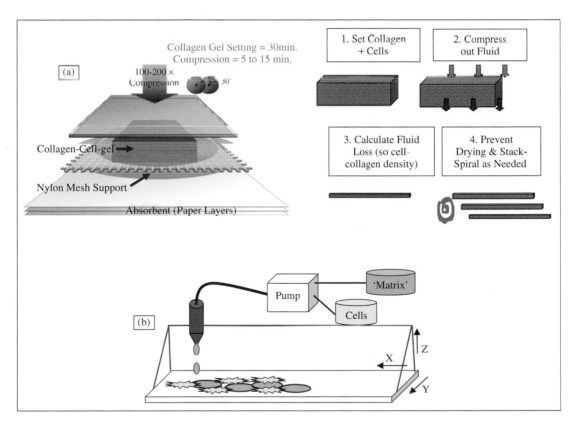

Figure 6.10 Enabling matrix layer engineering. [a] Collagen Plastic Compression: set-up and process shown on left (with thanks to Michael Anata for the diagram). Right hand side (1) to (4) show the main basic stages, including multi-layering. [b] Schematic of basic bio-printing. A delivery unit (e.g. a syringe), pump-fed with cells and a 'matrix' or cell support/gluing material, moves in X, Y and Z planes to deliver the desired patterns of cells and gluing matrix in 3D. Supporting 'matrix' must hold the cells in place, without damage, and mimic an extracellular matrix.

However, all this bulk means that we are dealing with an 'inflated' system, where the fine, μ-scale peri-cellular architecture can be fabricated at the mm-scale. This makes complex construction much simpler. Once the basic tissue template is gelled to a 'solid', most of this excess fluid can rapidly expelled (a) in a controlled manner and (b) in a single direction; with the following results:

(i) The original template structure remains largely intact, but *miniaturized* in one plane.

(ii) Miniaturization/compression is in the order of $\approx$100 fold, but only in the axis of compression, giving a thin, dense sheet of collagen matrix seeded with living cells.

(iii) Miraculously, any cells present, and now trapped in the fibril network, are undamaged by this rapid fluid outflow (apparently due to support of the nano-fibrils themselves).

(iv) Because fluid outflow is along a single axis, the collagen fibrils pack most densely at the fluid-leaving-surface (FLS) and in a series of parallel μm-scale thick lamellae above that. Cells find themselves enmeshed in a dense, laminated mesh without having 'done' anything.

(v) Typically, a 5 mm thick initial collagen gel is compressed to 50–100 μm. Since initial gel height, collagen concentration, cell density and total fluid removal are controlled, the final neo-tissue can be completely predicted. Typically, $\approx$25 million cells/ml lie in a tissue-like, anisotropic lamellar, native matrix of 10–15 per cent collagen.

In other words, we have solved our dual big problems by directly engineering with the materials we need in the end (native collagen plus cells). Furthermore, by building big and shrinking to the dimensions we want, we have made it easy to position and vary whatever cell and matrix additions we then insert. The apparent new problem we have, of having made a 50–100 μm thick tissue (a bit like moist cling-film), is not really a problem at all. In fact, this is perfect for layer engineering and heterogeneity building – rather like constructing micro-Swiss rolls.

Other approaches are under examination with the aim of producing similar **matrix-rich tissue layers**. There are, from other fields of rapid fabrication, a number of techniques which may be adapted to lay down together predetermined patterns of matrix protein and living cells. These include electrospinning and inkjet spray technologies, which becomes 'bio-printing' when used with cells.

Bio-printing is an area of increasing interest for rapid tissue fabrication, and it is a close relative of layer engineering. The idea has evolved from the mature field of print technology and rapid prototype engineering, especially in the automotive industries. The familiarity of inkjet printer technologies has made this an attractive idea in biological specialities. The aim is to print predetermined (μ-scale) patterns of support materials/scaffolds (with or without cells) in the same way that we would print a page of ink, but then to print another and another layer on top of this until a significant third dimension is built up. Again, progressive layers do not have to be identical, thus allowing for patterning in the 3D plane also. Clearly, this is a form of layer-by-layer assembly, at the multi-micro scale, but without the production of discrete tissue layers, as we had before in CPC and cell layering.

The automotive industry teaches us that this can work for selected materials, and indeed can make wonderfully complex and fine 3D structures, including apparently 'impossible' structures such as a sphere within a sphere. These first caught the imagination of bone tissue engineers for making hard-tissue implants with biomimetic micro-structure, customized to the particular patient defect or injury site. However, the challenges of this approach became more apparent as aspirations moved beyond making acellular replicas of hard tissue, in inorganic, calcium-phosphate-based materials.

First, the nature of the scaffold material is key. For rapid prototyping of engine parts, it was possible to use synthetic polymers and polymerize these as the process went on, then to cross-link or bond the various parts and μ-layers together with harsh, non-biological agents and condition. But these materials are unsuitable for tissue engineering, often not even undergoing simple dissolution (i.e.

'*biodegradability*' of conventional tissue engineering synthetic polymers).

There is also the logical driver, discussed previously (see Chapter 5), to place the required, living cells into the 3D micro-structure of the matrix *during* the bio-printing process[12]. However, the polymerization and cross-linking processes which are integral to bio-printing are normally incompatible with cell survival, at least for conventional synthetic polymers. Furthermore, the rapid transit of cells through tubes (or sprayed across air-gaps in the case of electro-spinning) poses considerable problems for cell survival. Reduction of fluid shear favours processes which are slower, but '*slower*' in soft, tissue-like structures makes it increasingly likely that each newly deposited layer will sink into, mix with or migrate through previous layers. In turn, this pushes the process designers more and more towards cross-linking the matrix or support-scaffold elements earlier and more tightly.

For all of these reasons, many research lines in this area are starting to use natural protein support materials, with natural cross-linking (e.g. fibrin or collagen: see Chapter 4), highlighting the family links with discrete layer engineering processes as described above (Figure 6.10b). These are high technical hurdles on their own, and are trickier still when combined with keeping the cells alive.

The problem is that the huge potential for fine control of 3D μ-structure offered by bio-printing and related fabrication methods does not help at all with the pressing problem of how we rapidly fabricate dense, strong natural tissue matrix around the cells. Indeed, the very advantages they bring for generating high resolution shape presently makes it technically harder to achieve this without killing the cells.

It is becoming clear that these approaches are no more 'technical-magical-bullets' for quick tissue engineering, than any other solutions we have looked at so far. Users of such print-related layering systems must tackle much the same big problems as everyone else: and these are not 'big problems' without reason. Focusing on the big problems which hold us all back is helpful in emphasizing the family links which exist between the various approaches to 'layer engineering'.

Building up of living tissues in layers, then, shares a great deal in concept terms across the field. The differences are more related to 'what is in', and 'how we make' the building blocks themselves. Some approaches fabricate discrete multi-micron-thick living tissue layers as a first stage, and then assemble-layer them together in a second stage. Bio-printing has a much higher hurdle, in that the whole process must be carried out in only one stage (though the rules stay much the same). In these circumstances, the history of biotech research teaches us that progress in both will be parallel. However, practical rapid tissue fabrication with simpler two-stage approaches is likely to come sooner than for single-stage techniques. In effect, the problems are the same, but we can allow ourselves more process time to tackle them.

6.7 No such thing as too many layers: theory and practice of tissue layer engineering

Here is the paradox of the Chinese restaurant. How is it that the most modest Chinese establishment boasts 6–8 pages of menu, containing perhaps hundreds of delicious-sounding dishes, yet these are inexpensive, quick to arrive, hot and (apparently) made by only a small staff? Are they super cooks? The answer, of course, is simple if you ever get a glimpse into the kitchen: rack after rack of tubs with different ingredients, thin and ready chopped, for rapid processing. Relatively modest numbers of these culinary building blocks (three racks full of vegetable tubs, two of chopped meats and one of seafood) can be combined quickly and with simple

---

[12]As we saw before (Chapter 5), dense, tissue-like 3D micro-structures are inherently rather too tightly packed together to allow cells to penetrate easily. As a result, where we make these *without* cells, our beautiful μ-structure will be compromised and potentially wrecked during the inescapable slow and protracted cell invasion-remodelling stages. Therefore, assembly of cells and matrix all in one step is highly desirable.

cooking sequences to produce literally hundreds of very different dishes. The result is a multiplicity of rapid, inexpensive dishes, needing only a fraction of the individual attention required by the skilled kitchen staff in the French or English restaurant next door, who train for years to individually assemble items on a menu of just eight savoury dishes and ten desserts, at twice the cost.

Applying this reductionist analysis, based on our previous identification of the main tissue planes and layers, skin can be broken down into a series of stylized layers. Using the three principle axes of the tissue, either planar or circumferential/coaxial, it becomes possible to generate complex 3D structures which resemble those of the natural tissue. This is illustrated in Figure 6.11 in the form of a model tissue. There are three separate levels of complex 3D structure built into this model:

(i) multiple, heterogeneous layers (they do not have to be flat, planar layers shown here);
(ii) 3D zones, running through a number of these layers,
(iii) concentric multi-layered tubes and channels.

It is important, though, for the reader to understand how high this aspiration actually is, compared with where we currently really are in cell and tissue engineering.

Current therapeutic strategies for spatial complexity include:

(i) the simplest possible (injection of a cell suspension into the vicinity of a lesion: e.g. stem/progenitor cells around a myocardial infarct);

(ii) random cell seeding of synthetic polymer sponges; or
(iii) pre-seeding of whole-tissue sheets;
(iv) bi-layers in which differentiating keratinocyte sheets are grown over a single collagen gel layer, seeded homogeneously with fibroblasts (e.g. Apligraf™, a clinically used skin substitute and cell therapy).

Some of these are illustrated in Figure 6.12, as a staged increase in complexity, from the familiar (2D) epithelial monolayer, through cell overgrowth and cell multi-layer formation, to true incorporation of different cell types into a fully 3D material. The greatest level of structural complexity (normally at the scale of 1–2 mm thick) is currently tackled using duplication of these cell-based structures, as in the case of Apligraf and other skin substitutes. It is important to note:

(i) the relative simplicity and gross scale of current non-homogeneous engineered tissues; and
(ii) their heavy dependence on cell activity (e.g. division and protein synthesis) to produce structure.

What new and ambitious technology can we adopt, then, for such a leap in target ambition? The answer comes from noting how easily we could generate complexity with the Swiss roll – in other words, build compositionally simple cell/collagen sheets (lots of slightly different components, as in the Chinese menu) and *then* introduce (pseudo-) complexity of 3D structure through the assembly process. This eliminates previous dependence on

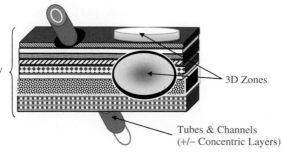

Multiple Heterogeneous Layers (NOT necessarily Flat or Parallel)

3D Zones

Tubes & Channels (+/– Concentric Layers)

Figure 6.11 Aspiration level: complex, non-homogeneous, 3D model 'tissue' – but built from many simple layers.

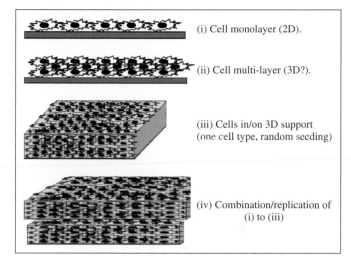

(i) Cell monolayer (2D).

(ii) Cell multi-layer (3D?).

(iii) Cells in/on 3D support (one cell type, random seeding)

(iv) Combination/replication of (i) to (iii)

Figure 6.12 Examples of current targets and strategies for engineering and assembly of 3D structure.

protein synthesis/assembly by cells. An attractive part of the strategy is the rapid assembly into multi-layer, thick tissues (Figure 6.11). The technical trick which makes layering such an attractive route to ever greater complexity turns out to be the ability to rapidly fabricate collagen networks with any number of different added particles, including cells.

6.7.1 Examples of layer engineering

We shall now look under the surface, into the detail of how we can 'make' layers and zones through use of examples. We can illustrate the principals through one main example, with a summary of the alternative possible technologies which follow the same track.

Let us develop this idea by looking in more detail at the building block layers which went into the structure in Figure 6.11. Figure 6.13 shows the nature of one such a matrix-rich building-layer for connective tissue engineering, with a scanning electron micrograph image of the dense collagen fibril matrix of a single $100\,\mu m$ PC layer.

If Figure 6.13a shows a cell-free collagen layer, the next layer(s) in the serial stack might be seeded with fibroblasts (Figure 6.13b). This fibroblast-seeded layer is mid-range cell density (perhaps more comparable to mature rather than repairing or growing tissues). In effect, cell density can be defined, layer

by layer, as millions of cells/unit volume of tissue (e.g. per mm^3 or per ml). This might, for example, be used to give a gradually reducing cell density deeper in the construct (i.e. further down the layer stack) by using a lower density in each successive layer. Such a structure would mimic dermis.

Clearly, any number of such separate sheets can be layered onto each other in the stack. Furthermore, they might be seeded with different cell types. The difference can be either as:

(i) one cell type in one layer and another in the next; or
(ii) two cell types in each layer, but in a ratio which changes gradually down the stack, giving a gradual transition.

Real tissue examples where these structures would be useful are easy to find. Larger blood vessels typically have an outer layer of fibroblasts which gradually grade into smooth muscle cell-rich layers which suddenly (innermost surface facing the blood) becomes a single-cell-thick layer of endothelial cells. Similarly (but this time a planar stack), the abdominal wall inside the skin changes suddenly from skeletal muscle sheets in different orientations, to a fibroblast in woven collagen layer (transversalis fascia), lined on the inner abdomen wall surface by a peritoneal cell layer.

Text Box 6.4 Well? *Can* we have too many layers?

Actually, there is a progressive advantage to tissue precision, and the ability to customize, in having as many layers as possible per millimetre of total tissue construct. This we can call '**resolution**'. It is a bit like the pixels in a digital photograph – the more there are, the better the image looks. This carries the predictable cost of more layers, more cost, and perhaps being slower to build up, but this another matter, as it is in digital photography.

There is a more pressing limitation which is easier to see if we look at a real example, say, skin. If we have only three layers – epidermis, dermis and fat, as mentioned above – each would be in the millimetre scale and the resulting tissue would be *very* simple. We could make the dermis ever more complex, with 10–20 layers of, say, 100 μm or (just) of 100 layers of 20 μm thick but, at much smaller than 20 μm, there is no space for the cells. This means that the theoretical limit for a cellular layer is a little over one cell diameter. There is, however, the possibility of adding in non-cellular layers, such as tough matrix, anti-adhesion or drug/protein carrying sheets, interspersed between cell layers or even cell-rich layers, such as epithelium.

So, there can be limits to the number and density of layers, but these can be predicted quite logically.

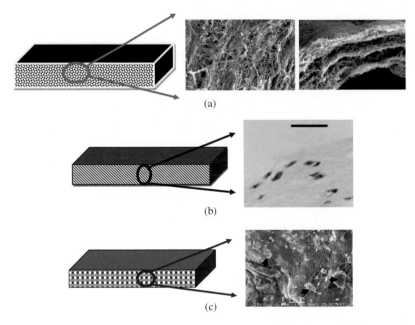

(a)

(b)

(c)

Figure 6.13 (a) Basic building-sheet for matrix layer engineering showing the dense fibrillar appearance of a 100 μm thick PC collagen layer (Figure 6.10). Left hand panel: bulk collagen. Right hand panel: cross-section of layer. (b) Fibroblast-rich collagen layer. H&E stained micrograph (inset) shows the distribution of (purple) cells within the orientated fibrous matrix of the single layer. (c) A hard tissue layer can be fabricated by the mixing of mineral particles with the collagen prior to gelling and compression. This leaves particles (in this case phosphate glass (arrows)) trapped interstitially within the fibril network, as in the inset scanning electron micrograph. With thanks to Dr. Tijna Alekseeva. (d) Assembly of layer series: Layers of differing, or similar composition can be stacked to give tissue-like structure at the multi-micron scale. The scanning electron micrograph (right) shows a stack of identical layers, formed by spiralling a single, large compressed collagen sheet (layers shown separate here for clarity, but would normally be in contact). Reproduced by kind permisson of Nelomi Anadagouda. (e) Local zones or depots can be inserted into position at either the layer-stacking or the gelling stages. (f) Micro-channels are formed to order through dense collagen tissues by co-compressing soluble (phosphate) 40 μm diameter glass fibres into the initial collagen gel. Far right inset shows fibres between rolled up collagen layers (arrows): the left hand SEM image shows a channel produced right through the construct when the glass dissolved. (e) and (f) reproduced with permission from Real-time monitoring of the setting reaction of brushite-forming cement using isothermal differential scanning calorimetry. M.P. Hofmann, S.N. Nazhat, U. Gbureck, J.E. Barralet. Journal of Biomedical Materials Research (Applied Biomaterials) 79B, (2006) 360–364. © Wiley. *Parts (d), (e) and (f) on the next page.*

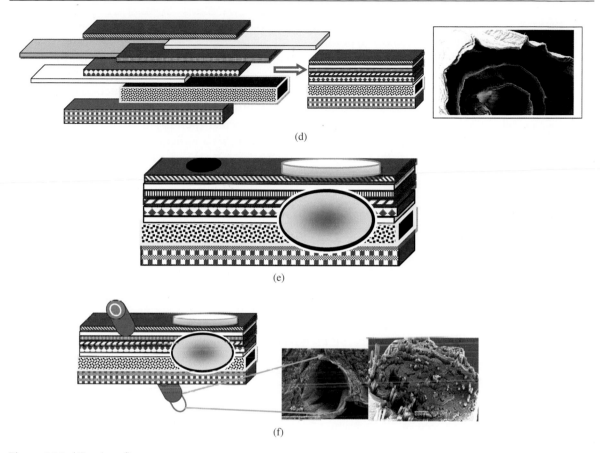

Figure 6.13 (*Continued*)

In fact, we can go yet further with the types of layering available; for example, the 'particles' trapped within the collagen network need not only be cells. Hard tissue layers can be formed by mixing mineral particles (again at any selected density and particle diameter) in with the initial collagen gel, with or without cells. Mineral particles can be, for example, hydroxyapatite, tricalcium phosphate or phosphate soluble glass across a size range of $\approx$100 nm to $\approx$20 μm (Figure 6.13c). This size range allows them to be retained by the fibrillar net at the lower end, and yet is a comfortable fit into the 50–100 μm thick collagen layer at the larger end.

Cells seeded along with the mineral particles largely survive co-compaction at modest mineral densities. This introduces the potential for fabricating hard tissues or hard-soft tissue interfaces. Such hard or hard-soft interfaces have a wide range of clinical and model tissue testing applications in the bone, calcified cartilage and dental fields. In particular, there is considerable current interest in engineering of the osteochondral[13] interface or junction. This is a major objective for surgical repair of joint damage. Tissue models of the osteochondral junction could help in understanding the origins of osteoarthritis, as it is suspected that changes in this layer could lead to some forms of osteoarthritis.

---

[13] e.g. the layer where the joint-end of a bone surface is bonded into the overlying articular cartilage.

By the same logic, this type of particle entrapment technology can be used to locate defined types and densities of cargo-carrying nano-micro vesicles or even carbon nanotubes. These can be pre-loaded with drugs, growth factors, gene-sequences, etc. and localized precisely within the final tissue to give new levels of controlled release and local biological control. A promising example of this has been described in the formation of ultra-high-density cell depots calculated to produce known levels of deep tissue hypoxia. This hypoxia leads to time-dependent generation of angiogenic growth factors, representing an entirely controllable pseudo-physiological therapy to promote local blood vessel in-growth. Such an ability to induce local re-vascularization at will is a long-standing clinical dream.

This train of research aims to develop a selection of both cell-rich and matrix-rich prefabricated 'layers' as building blocks. It only remains, then, to assemble these into stacks or layers *in the required sequence* for the simplified tissue of choice (Figure 6.13d). Integration of the layers of the stack (i.e. physical linkage) to form a single unit can be achieved either by co-compression of layers or subsequent cell action. Prevention of inter-layer adhesion, perhaps with insertion of a further *non-adherent* layer can be used to form gliding layers, similar to the synovial sheath in native tendon. Figure 6.13d illustrates the planar stacking of selected layers into intimate contact, as a complex tissue construct. The inset scanning electron micrograph shows an actual series of collagen layers.

As mentioned earlier, layering does not necessarily have to be parallel or flat to be biomimetic – far from it. Additional shapes and structures can be inserted into and across the layers to form localized 3D zones or depots in addition to the basic layer structure (Figure 6.13e). Depots of particles, fibres or cells can be inserted between the layers during layer assembly, or micro-injected after layering.

Alternatively, channel structures, either cutting across or running between existing planes, can be introduced as a single layer or as a co-axial tube and sheath. Figure 6.13f illustrates what is meant by layer-crossing micro-channels. These can be fabricated either as large diameter channels (>1 mm), formed relatively simply by rolling the collagen layers around a mandrel or by puncturing/drilling the constructs. However, the all-important micron scale channels, to mimic micro-vascular perfusion, or to guide capillary in-growth, are another matter. Once again, though, these can be achieved relatively simply by engineering of the soft collagen layer structure. Tissue examples of axial and co-axial layering can be found in blood or lymph vessels and nerves, as they run through other tissues, hair follicles in the skin, tear ducts and parts of hearing, uro-genital and airway systems.

One example of micro-scale channelling uses the plastic nature of the collagen PC process (Figure 6.13f). An example is shown (right hand inset) as two scanning electron micrographs. The right-hand (lower power) image shows a spiralled dense collagen construct, into which many parallel 40 μm diameter glass fibres have been compressed prior to spiralling. Such fibres can be made of any required size or composition. In this case, they were made of a soluble (phosphate) glass, where the familiar silica component of the glass is replaced by phosphate, making it fully soluble in a few hours. The glass dissolves and is flushed away soon after fabrication, to leave full-length patent channels right through the constructs.

This, then, is a 'lost-fibre' μ-channelling technique. Since dissolution products of the glass are generally not toxic, cells seeded into the collagen survive well. Indeed, recent work has shown that cells (in this case vascular endothelial cells) pre-seeded onto the glass fibres prior to plastic compaction come off the fibres and adhere to the walls of the 50 μm diameter channel, potentially forming an endothelial lined μ-channel. In nature, we would call this a simple capillary.

In a comparable approach, another group has seeded endothelial cells into 0.5 mm strands of agarose and cast this into a soft collagen gel to again produce an endothelial cell-lined channel through a gel matrix, in this case designed to

Text Box 6.5 More than one way to make a 'layer'

Perhaps it is time to take a small digression into the practical side of what interests us. We are starting to see a dazzling new collection of routes to 'assemble tissues and scaffolds' with such speed and detail that we might be forgiven for pinching ourselves. Readers and tissue engineering brethren who are less prone to swallowing whole fabulous tales of what could be coming, might wonder where the warts and the quicksand pits are hidden. The first thing, though, is to classify the different approaches and identify their family trees. After all, it is important to understand the origins of our ugly sisters before making any life-changing choices. Three seemingly loosely related examples of 'layer assembly' are:

(i) layer-by-layer nano-fabrication of surfaces on scaffolds and biomaterials
(ii) ink-jet, or 'bio-printing' of scaffolds and tissues,
(iii) tissue layer engineering (cell-layer and matrix-layer types).

All three aim to build up 3D structure by progressively adding layer onto layer in predefined sequence – clearly an attractive approach.

(i) Layer-by-layer fabrication of scaffolds involves coating surfaces with molecular (nm-scale) layers, bound by opposite charges (Figure 6.14a). This provides nm scale control of composition-structure in the z plane (red arrow), but mm or cm in the x and y planes.

Figure 6.14a

(ii) Ink-jet, or bio-printing, deposits μm-scale drops, layer after layer, as liquid plus cells plus scaffold, to build up 3D structure in the z plane (Figure 6.14b). In this case, spatial control is in the μm-scale, *but this time in all three planes* (x, y and z).

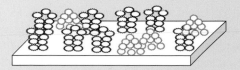

Figure 6.14b

(iii) Cell-layer and matrix-layer engineering (Figure 6.14c and described above) controls the z plane assembly at the tens of μm scale (arrow ≈50 μm). However, layers can be mm or cm in the x and y planes with little or no variation in structure.

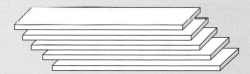

Figure 6.14c

As a result, layer-by-layer fabrication (i) gives the highest resolution in *one* plane, bio-printing (ii) provides the greatest resolution in *all three* planes, and cell-matrix layer engineering (iii) offers the lowest resolution and provides structure mainly in the z plane. However, because its scale reflects that of resident cells and its operation is simple and cell-friendly, (iii) is on track to produce living, functional 3D 'tissues' first.

mimic a larger vessel. In fact, there are now a number of versions of the lost fibre technique for rapid fabrication of microvascular-sized channels through solid implants. This has included the use of caramelized sucrose μ-strands (otherwise known as candy floss or cotton candy) to form channels.

Colourful as this is (in fact, pink!), the very rapid dissolution of method in this case is limited to making channels in synthetic polymers where it is not a problem that such sugar fibres are ultra-soluble in water, dissolving in a second or two in the gel.

6.8 Other forms of tissue fabrication in layers and zones

We have worked through one example here of *direct* layer engineering of tissues, in this case based on controlled fluid flow and shrinkage of native cell collagen gels. This use of fabrica ted living connective tissue sheets as building blocks for layer assembly was selected as it illustrates most of the desirable targets and approaches in a single system. Clearly, though, there are other examples of approaches following partially or wholly the same general track. Not least, as we saw earlier, there are mirror image systems designed to make cell layers for assembling layered constructs. These are designed for use where little extracellular matrix is needed, aside from small amounts around the cell (pericellular matrix) such as the basement membranes of epithelia or in muscle tissues. These cell layers are, of course, entirely 'grown' in culture over time (see Chapter 8), and are then assembled to give cell-rich multi-layer structures.

Other techniques, mostly for direct matrix assembly, have been adopted from:

- biomaterials production (electrospinning);
- rapid prototyping, such as in the automobile industries (layer-by-layer deposition and photo-cross linking);
- inkjet printing adapted to 3D deposition of layers containing proteins, synthetic polymers and, in some cases, cells.

Being adaptations from *non-bio* production technologies which we might regard as rather '**cell-brutal**', many of these struggle to operate with, or even around, living cells. Electrospinning as well as inkjet-type systems tend to be inherently cell-lethal due to both physical (shear and desiccation) and chemical solvent stages. Not surprisingly, then, most past and present uses aim to produce 'scaffolds' with complex and pre-determined 3D structures, but which are seeded with cells in a separate step. Electrospinning typically has been developed to produce nano-fibrous materials (protein or synthetic polymer), often with defined alignments which can then be layered (and cell-seeded). Inkjet technologies have been used to form complex structures rich in mineral to mimic hard tissues.

Controversially, some examples are now being developed where cell deposition is possible into the polymer 'scaffold' – the emerging cell-bioprinting technologies. These are beginning to tackle problems of μ-nozzle blocking and high shear but are moving towards production times of hours for significant tissue size and 3D (layered) complexity. However, it should be clear to the reader that successful examples of these methods will inevitably follow a similar concept and strategy route map to our main example. This is because they all aim to build up thick, complex tissue structures from many layers and zones made initially from simple building-blocks. In effect, they will all build up relatively simple compositions into disproportionally complex μ-structures on the model of Swiss rolls or Black Forest gateaux.

6.8.1 Section summary

The idea of this section has been to build up to a view of how we can (and do) develop a new approach to engineering of tissues. This extreme tissue engineering message, then, involves recalibrating our image of the target tissues. This sees the native target tissue as made up of a sequence of many similar layers and zones, in differing planes. It then becomes relatively straightforward to fabricate simple but mimetic collagen-based μ-layers which can be assembled, in sequence, to build up a cellular tissue mimic or template. If we are to make full use of this extreme new view, it is important to practise the deconstruction-reconstruction process using layers in all three planes. This can include radial/concentric, planar stacks and non-parallel layering. Such an approach to fabrication process design leads naturally on to the use of tedious repetition (and very thin layers) as a means of building up hierarchical structures across the major size scales.

Figure 6.15 Bridge construction analogy. Building up complex, long-range structure in 3D space is often best achieved using repetitive segments assembled in sequence. In effect this is the layer engineering principle, seen here during assembly of the Golden Gate Bridge, where decking segments (some identical, some different) are slung into position to be joined in a pre-determined sequence.

Finally, we should now be able to see, in best extreme TE traditions, that by aiming at the apparently impossibly high target of fabricating the tissue from hundreds of basic building block layers (instead of all in one go), we can achieve previously unthinkable levels of biomimesis in a fraction of the time. To appreciate how far this has taken us, recall that one of the most successful and complex engineered clinical skin equivalents at present is Apligraf™. This comprises only two layers, takes many days to grow (so is expensive) and delivers cells in a hyper-hydrated gel, acting more as a cell delivery than as a strong graft tissue. Now compare the complexity and bio-mimicry being contemplated in Figure 6.13. Paradoxically, then, by raising this particular bar, we are opening extreme new possibilities, often resembling those common in good process engineering and mass production.

An interesting aspect of this form of layer engineering and assembly is that of growing progressively in complexity from originally simple approaches to fabricating the layers. This is

beginning to spawn speedier or more effective device-based variants, but this evolution has a distinct direction, towards mechanization. It is not driven by a pre-existing device or technology which then has to be adapted to the biological demands of the cells and tissue, as is the case, for example, in bio-printing approaches.

6.9 Familiar asymmetrical construction components: everyday 'layer engineering'

'Segmental building' is a short-cut term for non-homogeneous, directional (or sequential) assembly of prefabricated segments into complex, anisotropic 3D structures. This, of course, is exactly what we are looking for in a rapid tissue assembly process. The bridge-building analogy (specifically *suspension* bridges) is useful here, as it emphasizes the potential for the rapid engineering of large, complex structures using layers and zones (in this case, decking segments). The way in which civil engineers go about erecting such structures can act as useful illustrations of the extreme tissue engineering solution which is beginning to emerge under the name of layer engineering.

It is important, of course, to keep the analogy in perspective, as there is much about bridge assembly which is anything but mimetic of cell-related assembly. However, the idea of pre-fabricating the large numbers of similar basic structural components and assembling them into position *in a strictly planned sequence* definitely has echoes of layer tissue engineering.

The key lesson here is just how valuable it is to plan the assembly sequence. Figure 6.15 shows San Francisco's Golden Gate Bridge under construction. The system of assembling prefabricated sections of deck in a directional, self-supporting sequence is clear. The direction of progression is forced by the necessity to work away from points of maximum support (i.e. the towers). Such basic mechanical logics can also dominate how we assemble layers and how resident cells work on natural tissues or template tissues.

Hanging and fixing together a long series of deck sections is a mechanically tricky process, and the bi-directional progression away from the towers is not only economical of effort and beautiful to see, but ensures a degree of balance and speed – in this case, between earthquakes!

6.10 Summary

What have we learned, then, about how to make the most of 3D hierarchical complexity? Or, to put it another way, have we answered our initial question, 'Is it worth the hassle'? Well, firstly, it is axiomatic that we have little choice but to use it and optimize it if we have any aspirations to be mimic natural tissues. It is clear, though, that once we understand how the natural hierarchical levels are built up in native tissues then, by repetitive positioning of similar parts in sequence, we can adapt well-tried strategies from other fields of construction engineering.

The take-home message seems to be to learn how to fabricate increasingly complex building 'segments' (layers when necessary). In this way, our designs for tissue assembly can be as **flexible** and **finely tuned**, as we undoubtedly need. This track is unavoidable if we aspire to mimic that most basic characteristic of natural tissues: *adaptability*.

After all, what sort of a visionary target have we produced if it is to fabricate a generic 'skin', for example? Imagine the reconstructive surgeon who needs skin for four successive patients to reconstruct (i) eyelids, (ii) a facial scar, (iii) post-tumour breast tissue and (iv) a post-burn thigh injury. Neither the surgeon nor the respective patients will be over-impressed by the ambition of our vision if they are offered a generic, lowest-common-denominator sheet of 'average skin' which looks like and behaves like the skin found at *none* of these sites!

Perhaps worse, patients needing eyelid reconstruction will be particularly unhappy with a skin equivalent which is functional and well adapted to work on the thigh. Less obvious, but no less important, is that in practice, the surgeon will also need different skin characteristics if the patient is 5 or 65 years old, or on chemotherapy, immunosuppression or anti-thrombotic drugs.

Surely, then, the 'generic tissue' targets we have adopted in the past are already too low to be called ambitious. In addressing this reality of biology, it is now becoming clear that we must either rely on the innate ability of our cells to *grow* this 3D complexity (as they do in embryos), or we must find ways to *fabricate* it for ourselves, as we have done for centuries, in other materials. This 'grow-fabricate' tension will occupy us for the remaining chapters.

But the honest answer to our 'is it worth the hassle?' question must be, YES. This is not only because, if we are frank, we don't really have a choice but also, surprisingly, that like so many other 'tall orders', *tackling the toughies head-on* can be the easiest long-term route anyway.

Further reading

1. Elloumi-Hannachi, I., Yamato, M. & Okano, T. (2010). Cell sheet engineering: a unique nanotechnology for scaffold-free tissue reconstruction with clinical applications in regenerative medicine. *Journal of Internal Medicine* **267**, 54–70.
2. Yang, J., Yamato, M., Nishida, K., Hayashida, Y., Shimizu, T., Kikuchi, A., Tano, Y. & Okano, T. (2006). Corneal epithelial stem cell delivery using cell sheet engineering: not lost in transplantation. *Journal of Drug Targeting* **14**, 471–482.
   [References 1 & 2: Example paper describing the making and uses of engineered cell sheets: cell-layer engineering.]
3. Phillips, J. B. & Brown, R. A. (2010). Micro structured materials and mechanical cues in 3D collagen gels. In: Haycock, J. (ed.) *3D Cell Culture: method and protocols.* Humana Press, Totawa, NJ.
4. Brown, R. A., Wiseman, M., Chuo, C.-B., Cheema, U. & Nazhat, S. N. (2005). Ultra-rapid engineering of biomimetic tissues: A plastic compression fabrication process for nano-micro structures. *Advanced Functional Materials* 15, 1762–1770.
   [References 3 & 4: Review and original description of matrix layer engineering using compressed cellular-collagen sheets.]
5. Guillotin, B. & Guillemot, F. (2011). Cell patterning technologies for organotypic tissue fabrication. *Trends in Biotechnology* **29**, 183–190.

[Review of cell, bio-printing approaches using a bottom-up, fabrication approach.]

6. Jakab, K., Norotte, C., Marga, F., Murphy, K., Vunjak-Novakovic, G. & Forgacs, G. (2010). Tissue engineering by self-assembly and bio-printing of living cells. *Biofabrication* 2(2), 022001.
   [Bio-printing from the opposite, 'morphogenesis' angle: position the cells in 3D and persuade them to finish it off.]

7. Nazhat S., Neel, E. A., Kidane, A., Ahmed, I., Hope, C., Kershaw, M., Lee, P. D., Stride, E., Saffari, N., Knowles, J. C. & Brown, R. A. (2007). Controlled micro-channeling in dense collagen gels by degrading glass fibres. *Biomacromolecules* 8, 543–551.
   [Channelling – vascularising into deep constructs is increasingly popular: (i) using sacrificial μ-fibres in cell-collagen constructs.]

8. Takei, T., Sakai, S., Yokonuma, T., Ijima, H. & Kawakami, K. (2007). Fabrication of artificial endothelialised tubes with predetermined 3D configuration from flexible cell-enclosing alginate fibres. *Biotechnology Progress* 23, 182–186.
   [Channelling with endothelial cells in agarose.]

9. Bellan, L. M., Singh, S. P., Henderson, P. W., Porri, T. J., Craighead, H. G. & Spectre, J. A. (2009). Fabrication of an artificial, 3D vascular network using sacrificial sugar structures. *Soft Matter* 5, 1354–1357.
   [Channelling with sacrificial candy-floss (cotton-candy in US) fibres set into cell-free polycaprolactone.]

In biological systems of tissue repair and growth, *tension* is a recurrent theme. *Tension balances* exist between the antagonistic growth factors, some of which promote tissue matrix production, while some promote matrix removal. Protein-digesting enzymes which degrade matrix components like collagen work *in tension* against protease inhibitors (e.g. TIMPs). Some cells specialize in matrix removal and are *in tension* with cells depositing matrix (e.g. bone osteoclasts and osteoblasts), but there is another, potent form of controlling tension in many of our tissues: the controlled balance of mechanical tension acting on the matrix-fabric of those tissues. How much better could our tissue engineering be if we understood how that tension control operated?

7

Other Ways to Grow Tissues?

7.1 General philosophies for repair, replacement and regeneration

As with so much of conventional tissue engineering, we tend to run helter-skelter past the basic concepts and potential clues which could be at the very centre of the squelchy problem. This is nowhere more likely than for the central question of '*How* do we intend to make tissues?' So, in the best traditions of extreme tissue engineering (i.e. doing it differently), this chapter will take a sedate walk around the question of whether there are any examples out there where we *already* 'make' tissues, particularly *in situ* (i.e. *in vivo*; see Text Box 7.1). Clearly, if we could identify any such examples *and understand how they*

work, we would gain considerably from the clues and parallels they would provide (and perhaps also develop a healthy modesty).

As discussed in Chapter 1, the idea of replacement of tissues is not exactly special to TERM. Indeed, we are entering a crowded field where it is sometimes easy to miss the occupants because they are so familiar. There is considerable advantage, then, in carefully identifying and analyzing the available options and existing cues before committing to any general strategy for making tissues. In this case it is helpful to revisit (from Chapter 1) one of the most basic divisions in implant logic, i.e. to identify where we definitely *do not* want to go. This is the division between replacing

Extreme Tissue Engineering: Concepts and Strategies for Tissue Fabrication, First Edition. Robert A. Brown.
© 2013 John Wiley & Sons, Ltd. Published 2013 by John Wiley & Sons, Ltd.

Text Box 7.1 Has someone out there already solved our problem?

When I was a young researcher, a pathologist-mentor passed on to me one of the great deflators of scientific pomposity: 'My boy, if you think you have a truly original idea, you probably just don't read German!' In fact, it can get worse. It might be that the answer is already out there, and *in English*, but you don't see it because it is in a dialect called 'Surgeon'. How many non-surgeons read surgical textbooks?

tissues with inanimate devices rather than living implants.

As discussed in Chapter 4, concepts based on the former, prosthetic type of replacement are, paradoxically, a more important influence on current tissue engineering than might be expected, given their position on the wrong side of this division. This is a consequence of their recent therapeutic success and the faulty aspiration for tissue engineering solutions to provide the same *immediate* benefits as a good prosthesis.

A characteristic feature of successful prosthetic implants is their almost immediate therapeutic benefit to the patient. In fact, as stated in Chapter 4, such implants never work better than on the day they are put in. The paradox here with tissue engineering is clear, since the whole aim of engineered tissues is that they take on a lifetime body function (i.e. they last as long as the surrounding tissues). The compromise we *must* accept is that engineered

tissues are unlikely to be as functionally effective as prosthetic devices *in the first instance*.

Consequently, it should, perhaps, be a core principle of extreme tissue engineering that we avoid strategies based on the logic of making prosthetic implants. A more honest strategy-building logic would accept the downside compromise of delayed optimal function and focus on making the most of the strengths of living implants: their bio-activities (Text Box 7.2).

Although blurring across the prosthetics/living tissue boundary is common and, in some cases, useful, it can also have very troublesome consequences. However, as a boundary which is easy to identify, it is well worth either avoiding it or clearly marking its crossing. A more difficult distinction to make is that between *engineering tissues* (from the ground up), and the *biological expansion of existing tissues*. Engineering, or modification, of existing tissues is routinely performed by surgeons

Text Box 7.2 Contrasting 'wide' with 'narrow' approaches to problem solving

A useful analogy to illustrate the decision-making implications of opposing problem-solving strategies is that of two solutions for keeping warm in winter. People can either wear thicker clothes and run around more (i.e. self-only, inward facing strategies) or turn up their central heating and decorate their homes in rich warm colours (non-self, unfocused, outward facing strategies). The two classes of solution to the problem operate in completely different directions. Neither one is necessarily correct in all cases and sometimes combinations of the two may produce the best result. The only problem comes when we mix up the two logics and expect things to work, for example:

- I wear a thick sweater and expect this to make my party guests feel better.
- I paint the walls a warm colour and hope that the dog* is more comfortable.

  * **Trivia note:** Dogs seem to be red/green colourblind, although they probably see brighter images than humans, with better night vision and detection of movement. So your dog probably would not see your lounge walls as a 'warm' colour even if he cared. '*The results of the colour vision tests are all consistent with the conclusion that dogs have dichromatic colour vision.*' Neitz, J., Geist, T. & Jacobs, G.H. (1989). Color vision in the dog. *Visual Neuroscience* **3**, 119–125.

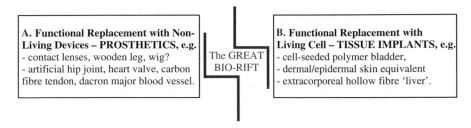

Figure 7.1 The basic non-living/living division ([A] and [B]) between tissue replacement strategies.

as they repair, reconstruct or remodel their patient's tissues. Wherever (or indeed, *if* ever) this interface is finally identified, it is clearly a good idea to learn what we can from surgical tissue modifying skills which have been developed over many decades. In particular, clues from this sector can teach us a great deal about the tissue-like processes and controls which we are keen to build on, as the benefits of the living/non-living compromise (Figure 7.1).

The purpose of this chapter, then, is to look more closely at one particular bio-process which is used, often incidentally, by reconstructive surgeons and repair biologists. The hope here is that we can copy or harness it. This is the process of **connective tissue 'growth'**, particularly where it occurs in adults. After all, many strategies implicitly accept that our engineered constructs will (magically) grow. This assumption is made on the simplistic basis that they are living and so will 'grow' up to the physical dimensions they reach naturally. For example, this is implicit in the concept of the *in vivo* bioreactor and extended to our *culture* bioreactors (see Chapter 8). Yet, without understanding what makes adult (post-embryonic) tissues grow and how that growth switches on and off, we really are going on very little else besides instinct. If we don't understand:

(i) what limits the size of natural tissues;
(ii) why they grow symmetrically (same left and right arm and index finger lengths); or
(iii) what makes wound tissues stop growing,

then what makes us so sure that our constructs will:

(i) grow at all;
(ii) stop growing, *ever*; or
(iii) not grow *smaller*?

7.1.1 What does reconstructive surgery have to teach us?

In its simplest form, surgical reconstruction uses the surgeon's skills and anatomical knowledge to recover part or all of the function of a tissue with the help of *natural tissue repair processes*. This involves repositioning some tissue-parts and joining others together, then encouraging the tissue repair process to 'paper over' the gaps that this leaves. For those from the non-bio tissue engineering tribes, the real extent of 'the gaps' can be illustrated by a plumbing analogy. Imagine that a plumber fitted up your new kitchen sink by pushing the pipes together and holding them together with a fine silk threads, delicately sewing a loose joint together. You might reasonably consider the inter-stitch gaps where water sprays out to be a terminal problem (particularly when the plumber's bill arrives), but this is exactly how a vascular surgeon would fit in a cardiac graft to your failing heart. And it works: the gaps fill!

We have naturally tended to focus so far on the rebuilding and repair of tissue or the movement of small living tissue spare parts from one body site to another. However, such reconstructions also rely on off-focus events which attract far less attention. These are the remodelling, reshaping and restructuring (indeed, growth or shrinkage) of tissues which *surround and support* the primary reconstruction. In fact, it is clear that the remodelling processes that we are so focused on during tissue repair are accompanied by a (much slower) remodelling of surrounding, but connected, body structures which were otherwise untouched.

The tensions across sutured skin injuries or around skin donor sites gradually decrease over

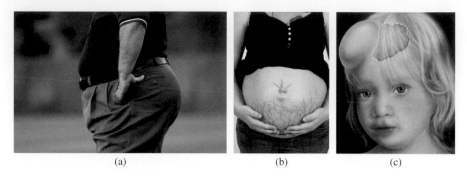

(a)   (b)   (c)

Figure 7.2 Two examples of skin growth in adults (a and b) and one in a child (c), where mechanical forces drive the tissue expansion – but how? In obesity, the retaining skin clearly grows (a), rather than stretches, under tensions from the accumulating underlying tissue. The same seems to be true in pregnancy (b), often at such a rate that scarring is common. Tissue expanders (c) are used by surgeons to 'create' new skin needed elsewhere on the body for graft-repairs. This is 'mechano-engineering' of skin *in situ*. If only we knew how it works! Credits: (a) © iStockphoto.com/Rob Friedman. (b) © iStockphoto.com/Steven Frame.

time, as surrounding skin 'remodels'. Comparable tensions are set up where cut nerves are repaired, for example following hand injuries, but these also seem to reduce in time where functional recovery is successful. Less helpful, but certainly instructive, the process can work against us. Facelifts inevitably sag again as months and years go by. More of a problem is the connective tissue contractures (shortening) which can occur in adhesion tissues around many surgical interventions. These are common, for example, between body wall layers (fascia) and internal organs, around abdominal hernia repairs or around injured tendons in the hand.

Such contracture is a process where (sometimes following repair, sometimes for reasons unknown) the affected or surrounding tissues become geometrically smaller – they 'contract'. This is a strong process, and any moveable structures which are attached get pulled together. If it is a finger, the digit gets stuck in a bent position; if it is around a section of gut wall or urethra, it can lead to dire and painful blockages.

Could we learn a thing or two, then, about how these shrinkage and sagging processes occur. Better yet, could we harness them, by looking more closely at what goes on during reconstructive surgery, to help us make and expand tissues in the lab? Alternatively, is it possible to persuade the body to make extra little bits of this-or-that tissue *in situ*, for use as grafts elsewhere in the body? This latter idea is

basically the *in vivo* bioreactor strategy proposed by some tissue engineers. In reality, these techniques may already be in use by orthopaedic and plastic surgeons in the form of bone distraction (lengthening) and skin tissue expander procedures.

7.1.2 Clues from the natural growth of tissues

While there is a purist argument that tissue engineering is distinct from these surgical procedures, it is a pragmatic no-brainer that we must use whatever lessons they can teach us. Tissue engineering can be considered as aiming to grow *substantial amounts* of *new* functional tissue (Figure 7.2). The keywords in this phrase are underlined. First, there will be general agreement that we aim to engineer *substantial* pieces of tissue, in other words in the millimetre rather than in the micron scale (or below). This is important, as it distinguishes tissue engineering from basic repair biology, say at a wound margin, where tiny amounts of new tissue are formed by local cell action.

Second, there would be general agreement that in engineering of tissues, the aim is to work through processes which implicitly form *new* tissue. This point distinguishes tissue engineering from the successful areas of transplant biology/surgery, where any amount of modifications are developed to enhancement and improve the function of a previously used (i.e. not new) tissues and organs.

Text Box 7.3 Balanced tension in the skin of the hand

Take a look at the palm of your hand. The skin is folded over the simple hinge articulations at the front of your finger joints (Figure 7.3) but it is smooth and crease-free over the finger bones. Now turn over your hand, with your fingers pushed back and extended as far as they will go, and take a look at the back of your hand. The skin is now most likely wrinkled and in folds – clearly neither tight nor under tension. But you should now be able to feel the reason why.

This is definitely *not* a neutral, at-rest position for your hand. In this position, you have stretched or hyper-extended (hence the 'feeling' you have) the skin of your palm, so much so that the blood circulation changes (Wright *et al.*, 2006). In fact, if you find the

resting or neutral position for your hand by letting the whole arm go limp, you will find that your fingers are slightly flexed into a partial 'claw'. In this position, the skin of both your palm and back of your hand should be in *balanced tension* – smooth except for over the joints. This is a visible manifestation of the effect, noted by many human anatomists, that once structures such as a major artery are dissected out of a limb, they seem too short to fit. In other words, *in situ*, they are constantly extended under a tension.

Reference:

Wright T. C., Green E., Phillips, J. B., Kostyuk, O. & Brown, R. A. (2006). Characterization of a 'blanch-blush' mechano-response in palmar skin. *Journal of Investigative Dermatology* **126**, 220–226.

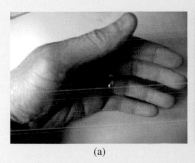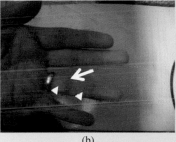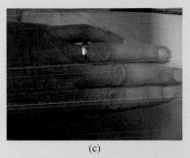

(a)　　　(b)　　　(c)

Figure 7.3 Hands: (a) in neutral, relaxed position; (b) showing extended-stretched palm skin: between (arrow heads) and over the joints (arrow); and (c) extended, folded skin on the back of the hand.

Finally, this leads us to the 'grow' word. Of the three keywords, this one is perhaps where we may learn most and build better tissue engineering concepts. Understanding this would lead to the greatest advances, making the other two seem useful but technical caveats.

To conclude this section, then, we should definitely take on board the idea that THE key objective in tissue engineering is to grow tissue structure, with the emphasis on **grow**. Since we are all-too-familiar with the mechanisms by which man-made structures are made to 'grow' by engineering and fabrication processes, the next section will consider natural examples from which we might learn how natural tissues grow biologically.

7.2 What part of *grow* do we not understand?

'Grow' is an interesting word in its ambiguity and multiple meanings. In biology and medicine it tends to have a rather specific but also, to be honest, somewhat poorly understood series of meanings. In reality, it is more widely used to describe any form of geometric increase in size or proportion. For example, we like to think of children *growing* in height and salamanders *growing* new limbs after injury as rather special (i.e. biological). Yet the same word is used for both processes, even though they are clearly very different. Similarly, we are perfectly happy with the idea that the Eiffel Tower grew over a

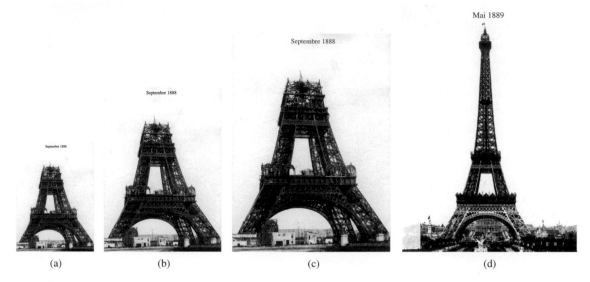

Figure 7.4 The Eiffel Tower grows in dimensions in images (a) to (c) – but this clearly is different to that growth which occurs between (b) and (d). Interestingly, (a) to (c) is analogous to soft tissue (interstitial) growth, but (b) to (d) better represents bone/hard tissue (appositional) growth. Equally interesting, (a) to (c) is a photo-trick which we cannot achieve by engineering.

period of months (Figure 7.4) or that the Himalayas grew by seismic activity (as do tsunami waves on a different timescale!). Intriguingly, we even consider that a tunnel (i.e. tubular void) *grows*, although in this case as a result of excavators *removing* material.

In other words, growth can be the appearance and extension of three-dimensional structure by almost any animate or inanimate means. As we shall see in the next chapter, the diversity of how we achieve growth may be at the very core of our thinking about how to engineer tissues. In particular, we shall wrestle with the tension between growth as a biological cell-driven process and growth of structure that we can achieve in the human world by routine engineering and fabrication.

First, though, let us make the distinction we have used before between cell-rich and matrix-rich tissues. In effect, most of embryology and early mammalian development can be seen as a series of cell rich tissue organisations, template formation and growth processes, under relatively tight gene control. Clearly, the major cell-rich organs (liver, kidney, lung, heart) continue to grow in size by the addition of more cell-rich tissue throughout childhood, halting mostly (though not completely) in adulthood.

The major matrix-rich connective tissues have, in the main, been laid down in their basic template shape and form by the time of birth. These templates grow in size during childhood and adolescence by addition of new matrix-rich tissue. In fact, the proportion of cells to matrix falls precipitously as they take on increasingly the mechanical support functions necessary for a growing, active body.

In a way, this shift, through childhood, is completely inevitable. On the one hand, cell-rich tissues inevitably have only limited mechanical strength. On the other hand, the rapidly increasing mass of a growing child demands increasingly strong, stiff structures for support of its shape and movements. Consequently, to achieve adult load-bearing function, such support-tissues must undergo a dramatic *reduction* in their content of cellular (weaker) material, if only to make way for the ever-increasing proportion of strong, load-bearing extracellular matrix (ECM) protein material.

The extent and pattern to which this happens differs between body tissues and their age, as each one adapts to the load-carrying demands placed upon it. If we are interested in hunting clues, then, our question must be '**How does that happen naturally**?'

7.2.1 Childhood growth of soft connective tissues: a good focus?

Using the analysis above, we can see that human embryology and development represent the formation and patterning of **cell-rich** tissues. In this case we have a neat distinction, because the years between birth and adolescence, where childhood growth is greatest, is the period where the vast majority of connective (**matrix-rich**) tissues are laid down. Interestingly, while entire journals are dedicated to the study of development, far less is known about childhood tissue growth. Although this growth is 'only' simple geometric extension of the embryonic patterns, understanding the processes that control and drive it would be key to engineering *new* adult tissues. The section heading here specifically focuses on soft tissue growth, as we already have a good

knowledge of how bone grows – and paradoxically, that is where we must start.

Long bones (i.e. most of the skeleton) grow during childhood through cell and matrix addition in small tissue areas known as physeal growth plates (Figure 7.5). Physeal growth plates are thin strips of specialized cartilage towards each end of each long bone. Such growth plates are directional bone-producing machines, comprising three basic parts in a sequence of three layers. The first (closest to the end of the bone) is a depot of uncommitted progenitor (stem) cells. The second is a layer of rapidly dividing chondrocytes, and under this is the third layer, where cells undergo rapid swelling. These swelling cells eventually die to be replaced by bone and vascular tissue.

In other words, this bone-making machine operates by briefly producing a swelling cartilage tissue, which is rapidly removed and replaced by hard bone.

The reason that the growth plate is important for us to consider here is that it is a mechanism to

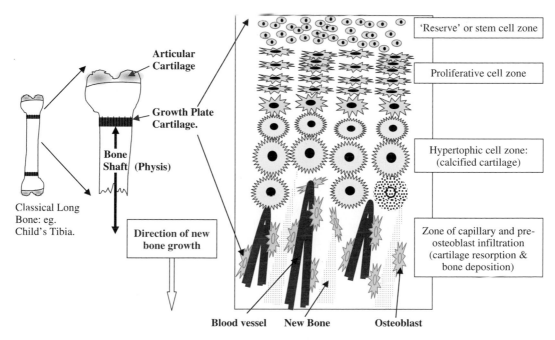

Figure 7.5 Diagram illustrating the main components of a long-bone growth plate – the 'motor' for soft tissue growth. Right hand detail shows stem cells (top) proliferating, then undergoing hypertrophy, death and remodelling by incoming vessels and osteoblasts. Expansion comes from the swelling hypertrophy. Notice that there is strong direction to the growth, downwards here, as columns of cells and, later, cartilage and bone matrix are deposited.

develop (grow) new tissue very rapidly *in a principal direction*. The rapid division of cells in the growth plate (second, proliferative zone) is under tight hormonal control by the body – in fact, by the pituitary gland. This means that throughout childhood, most of the growth plates undergo concerted spurts of action and inaction. These are seen as childhood growth bursts, with eventual cessation of growth and loss of the growth plates after puberty. These bursts of cell division, followed by cell swelling or hypertrophy (the third, hypertrophic zone, see Figure 7.5), inevitably generate compressive loads down onto the adjacent shaft of the bone. Since the bone shaft cannot compress significantly, the ends of the bone are progressively pushed apart and the bone-shaft elongates.

To summarize, this means that we can see long bone extension as a mechanically driven cyclical process. In each cycle, a tiny new layer of cartilage is swollen with water and is replaced by a new layer of bone matrix. The cycle repeats regularly, producing a form of ratchet-extension of the long bones. In rapidly growing vertebrates, this can generate 1–2 mm or more of new bone per day. Derivative versions of this process increase the dimensions (widths and diameters) of the articulating joints at the end of the bones, although of course at *much* slower rates.

But why the focus here on bone/hard tissue extension, when we were supposed to be considering *soft tissue* growth mechanisms? The answer is simple: 'mechanisms'. Almost all of our soft tissues (skin, tendons, vessels, nerves, etc.) are firmly attached to, or stretched over, our bony frame and around its articulations. Thus, the expansion/growth of that stiff frame inevitably becomes the *directional motor* and the *rate-limiting accelerator* which drives expansion of the soft bits that are attached.

Suddenly, it becomes a whole lot simpler to understand the apparent miracle of this biological control process, which leaves our body shapes so symmetrical, wrinkle-free and predictable. In effect, all of these diverse (soft) parts react to adjacent growth-plate driving motors, which are themselves under hormone, and so central, synchronized control.

Exercise Box 7.1

Exception: As always, exceptions are *so* informative. As a short test, list the accessible parts of your body that are *not* stretched over some part of your skeleton. This exercise does not need a formal solution – there are few such body parts. Interestingly, they tend to have much more size-shape variation between people, as well as asymmetry and size variability within the same individual, than those which are attached to bones. Equally interestingly, these also tend to be the parts which attract most angst and cruel attention during adolescence and in our later sex lives – but this is more to do with psychology than structure!

7.2.2 Mechanically induced 'growth' of tissues in children

But, what about *soft* connective tissues? After all, this is the subject of the present section. For soft tissues, there can be no single point of growth as there is in the growth plates of bones. Soft tissues grow at all points throughout their mass. This is called *interstitial* growth, as opposed to *appositional* growth, at the outer surfaces, in bones (illustrated in Figure 7.4). So what is it that controls and drives soft tissue growth? As we hinted at in the last section, the mechanism is apparently both simple and plausible. The directional motor for soft connective tissue growth can be thought of as growth plate expansion during long bone growth, centrally synchronized by pituitary and sex hormones.

The basic hypothesis which explains this linkage demands only that the cells within soft connective tissues can be stimulated to extend and grow their surrounding matrix when they are subjected to an appropriate tensile load. If we accept this working hypothesis for a moment, it is possible to understand how the concerted complexity of limb and trunk growth in childhood can occur. Extension of the spine and long bones through their growth plate machinery must inevitably place *directional* tensile loads onto all of the soft surrounding connective tissue that are attached. Our tensile load hypothesis then explains how all of these surrounding soft tissues (skin, blood vessels, nerves, fascia, ligaments) grow in a progressive manner linked in space and time with growth of the skeleton.

This is a catch-up growth system in which the soft, elastic connective tissues are under a constant, background tension, sufficient to stimulate growth. This growth would not only follow, directly on bone-extension in time and rate, but also spatially, in shape. In fact, cells would then also respond to the principal directions, or *vectors*, of tensile force set up by the patterns of bone extension.

Not least, this would explain why children and young people, over progressive growth cycles, never seem to have unpleasantly wrinkly and then painfully tight-stretched skin, as a result of some independent or partially synchronous control of growth between the hard and the soft tissues.

7.2.3 Mechanically induced 'growth' of adult tissue

Clearly, such a **tension-driven** growth process would be an excellent source of clues and controls to mimic in our tissue engineering, if we understood it. On the other hand, all of this would be rather incidental to tissue engineers if it only worked for cells and tissues *in children*. Happily, the extreme tissue engineer can be encouraged by indications that the basic process of tension-driven soft tissue growth is present, and indeed is ready to go, in *adults*.

We are actually quite familiar with some forms of adult tension-driven growth of soft connective tissues. For example, lots of us know someone who is seriously overweight but, on occasions, diets back down to more reasonable proportions. Such obese-slimming cycles (Figure 7.2a) are not uncommon and, especially amongst younger individuals, they are accompanied by something which is remarkable for its absence (so easy to miss).

The abdominal circumference, notably the skin, might conservatively increase in length round the waist by 25–50 per cent in large numbers of people, particularly in the USA (Text Box 7.4). Yet these are substantially greater increases than could be expected if the skin were just stretching. This is particularly striking, as we do not see any of the expected signs of skin thinning, such as increased visibility of blood vessels, changes in tone or texture or tendencies to split, burst,

---

Text Box 7.4 Obesity and waist measurements

According to Ford *et al.* (2003) and data from 1999–2000, average US male waist measurements at that time were 97 cm (all ages and ethnic backgrounds). The 95 and 99 percentile figures were 126 and 143 cm, respectively. This suggests that even if an individual *started* at an average waist circumference and increased their girth until it reached a level of 1 in 20 of the population (the 95 percentile level), their skin and abdominal wall circumference would have increased by 25 per cent. Indeed, if they reached the heady heights of one out of 100 Americans, then their skin and tissue would have increased in length, in that plane, by 50 per cent.

These are two useful points for our ball-park extrapolation. First, 50 per cent linear increase is, at 1 in 100, not an uncommon level of expansion (albeit this is the USA). It implies that 1.5 million males would have this waist measurement, and that at least some of them were originally average girth. Second, try, if you can, to stretch your own waist skin by anything approaching 50 per cent in that plane. With a struggle, you may manage around 10 per cent extension. In fact, abdominal tissues are twice as stiff in this (the transverse), as opposed to the axial plane (Kureshi *et al.*, 2008). The take-away message, then is that millions of adults apparently manage to extend their skin and body walls by 3–4 times as much as might be reasonably expected by tissue stretching alone.

References

Ford, E.S., Mokdad, A. H. & Giles, W. H. (2003). Trends in Waist Circumference among U.S. Adults. *Obesity Research* **11**, 1223–1231.

Kureshi, A., Vaiude, P., Nazhat, S. N., Petrie, A. & Brown, R. A. (2008). Matrix mechanical properties of transversalis fascia in inguinal herniation as a model for tissue expansion. *Journal of Biomechanics* **41**, 3462–3468.

bruise or rupture[14]. Slimming is more variable, and perhaps age-dependent, but it is equally striking that these lucky individuals are happy to show off their new shape. This slimmer shape does not normally include hanging layers of loose skin, free to sag without its adipose support (except in some extreme examples). It certainly looks very much as if, in most cases, the excess bulk of skin has been removed (remodelled away), leaving the skin as smooth and tight as it was when it held an extra 10 kg of adipose. Although the common assumption might be that our skin *stretches and recoils* as we get fatter and slimmer, in fact we can now see that this stretch-and-recoil idea does not really stack up, either for weight gain or for weight loss.

A more commonplace, though equally dramatic, example occurs in pregnancy (Figure 7.2b). For its extent, rate and focal nature of skin extension, this must exceed that of almost all obese-slimming cycles. Of course, unlike obesity, it always occurs in a complete 'growth-shrink' cycle and, by definition, is restricted to adults. Again, the skin grows to accommodate the expanding abdomen and uterus and then resorbs back after the birth (Figure 7.2b). This can often be so close its original dimensions (again without excess tissue) that even self-conscious mothers are perfectly happy to play with their babies on the beach in last year's bikini. In this case, the tissue expansion/growth can be so rapid that there are residual signs of the process. These are the dreaded 'stretch marks', again indicating that the skin was not as stretchy as we might have thought and could not, in this case, grow fast enough. Stretch marks are scar-like tissue structures which seem to form as a result of minor local connective tissue injuries. In this case, they may form where local strains generated by the expanding foetus exceeds the material properties of the overlying dermis and its growth potential.

These, then, represent two of the natural examples of adult, tension-driven soft tissue growth.

7.2.4 Growth has a mirror image – 'ungrowth' or shrinkage-remodelling

Hopefully, most of you will already have spotted the extra enigma of both the pregnancy and the obesity-slimming cycle examples (do not worry if you missed this though, as it is so obvious and familiar as to be largely invisible). The extra enigma here is that these forms of tension-driven adult tissue growth are *reversible*. They can act in cycles. In the case of obesity, additional tension is generated by the localized deposition of adipose tissue. Obviously, in pregnancy the skin over the abdomen must accommodate rapid intra-uterine growth of the foetus. Both inevitably act as sources of progressive tensile loading on the surrounding soft tissues, which can be seen as comparable with that of extending bones in children*.

* Happily, *shrinkage* of children after they have grown is uncommon.

So, there seems to be a mirror image of tension-driven adult tissue growth, namely the loss of bulk tissue when the tensile load across it falls. This can be interpreted as meaning that the loss of basal tension in soft tissues *normally*[15] stimulates tissue-matrix contraction and resorption such that the tissue becomes smaller. Clearly such a geometric shrinkage (or tissue contraction) would also tend to restore the pre-existing tissue tension, so signalling the end of this shrinkage episode of the process. In other words, cellular activity seems to change tissue dimensions by deposition (or in this case *removal*) of bulk connective tissue to minimize changes in the basal tissue tension. This, in effect, is the description of a cell-based, tensional homeostasis system – implying that tissues work within (and maintain) an overall, background resting tension

---

[14]In fact, many patients with inguinal hernia ('ruptures') are not obese, and the body wall layer does *not* rupture (as in burst)!

[15]Tissue tension is so 'normal' that we are generally not aware of it until, for some reason, the tension is lost – see Figure 7.6.

Figure 7.6 Tissue tension is so 'taken for granted' that it only shows when it is not there, as in the skin of this oversized Shar Pei dog. © iStockphoto.com/Vitaly Titov.

(Figure 7.6). The presence of such a tension would explain why skin is taut, rather than wrinkly, following loss of body fat – and in pregnant mothers, following child-birth.

In everyday life, we take completely for granted this 'taking-out and adding-back' of material in the connective tissues and we barely notice it. In connective tissue and repair biology, we call this amazing and constant effect 'remodelling' and leave it at that. However, by any analysis, this is a specific, mechanically driven form of material restructuring, apparently controlled through a basal tissue tension. Doubtless we would give it a great deal more attention, and discussion, if some bright anatomist a century ago had named it 'ungrowth'.

7.3 If growth and ungrowth maintain a tensional homeostasis, what are its controls?

At this point, we need to ask ourselves if is it likely that we are looking at two control processes held in approximate balance. One would involve a stimulus going to cells, perhaps circulating hormones to push connective tissue cells to make more matrix (so growth) every Tuesday, Wednesday and Saturday until the hormone is switched off. A completely separate stimulus, perhaps loss of background tensile load or a second hormone, might bring about tissue contraction or shrinkage at other times by matrix removal.

Such balanced systems, based on opposing mechanisms working in Yin-Yang fashion, are beloved of biologists. For example we think of the coagulation cascade balanced against the clot lysis (fibrinolytic) system of the blood. However, such antagonist systems have to have intimate and interlinked controls, and even then there are inevitably common pathologies where the balance fails, such as haemophilia on one hand and thrombosis/stroke on the other. Though it is always hard to rule out the existence of complex control systems, in this case the idea of a complicated two-process system, with interlocked controls, seems less attractive. For one thing, there are natural examples (Text Box 7.5) where one would expect a two-process system to produce consequences which we do not, in fact, see, but this cannot be conclusive either way.

The cell-mechanics theory, though, is more persuasive in this case, since it represents a single, unifying control mechanism for all tissue remodelling. After all, such a 'material tension' control system *is* available to cells anyway. So, while it is possible that an *additional* and complex biological control process has evolved *as well*, this simple mechanical system is available whether cells use it or not. It is based in fundamental mechanics: if tension in a material rises, insertion of more material (generating greater 'length') *must* reduce that tension. If tension falls, then removal of material bulk (shortened length) *must* increase load on its attachment points, increasing tension.

It is tempting, then, to expect that the 'KIS' principle (keep it simple) operates here. In effect, a simple control system is available anyway, based on the relationship between material geometry and tensions within its material components. So why would a more complex Yin-Yang system need to evolve? So the theory remains.

An important consequence of the material-mechanics control mechanism theory for tissue growth and remodelling is that a background tension needs to exist in soft tissues. Furthermore, the resident connective tissue cells would need to actively maintain that tension. That such a background tension exists in soft connective tissues, even at rest, is evident in our most visible of soft

---

Text Box 7.5 Author's reflection on the conundrum – can childhood growth reverse?

In old age, osteoporosis can cause loss of skeletal height/length, and wasting or loss of muscle-mass leaves many overlying tissues loose and sagging. The question is, does this represent an example of control failure or an uncoupling of the controls of tissue growth and shrinkage? Unfortunately, in this case the answer is tricky, as *all* cell processes slow down with age anyway, so they may just not keep up with external changes any more. Also, 'oldies' are inherently wrinkly anyway! So this test of ungrowth is ambiguous.

At the other end of the age scale, children with inherited bone growth disorders are left as very short adults, but their soft tissues are appropriately sized for their short stature. Their skin and vessels are suitably smooth and tight, like the rest of us. Similarly, where a bone growth plate is lost in childhood (due to a bone tumour) the bone cannot extend but the overlying skin, nerves and vessels stop growing also. These soft tissues do not grow into folds, even though the contra-lateral leg (bone and skin) continues to grow normally.

The Shar Pei dog skin (Figure 7.6) presents an interesting, if weird, exception. Our hypothesis (yet to be tested) must be that this strain of dog has an unfortunate (some consider cute) mutation of a gene which is key to the mechano-responsiveness of skin. Tissue engineers and repair biologists alike may have much to learn from this dog.

---

tissues – the skin. This is remarkably smooth and tensioned over the main bones, where it is under constant load. It only sags into folds over joints where regular variation of loading is unavoidable (see Text Box 7.3). As suggested in Text Box 7.5, the gradual failure of this remodelling control system with increasing age may only highlight its dependence on mechano-responsive cells. We can assume that the number and activity of these decline with increasing age, reducing tissue remodelling rates (adaptive growth and ungrowth) and leading to loss of tissue tension.

7.3.1 Tension-driven growth and tensional homeostasis – the cell's perspective?

Support for the idea that connective tissues grow in response to tensional forces acting on resident cells comes from the description of a fibroblast 'tensional homeostasis' system. As discussed previously, many cells – most commonly fibroblasts – are known to generate their own tensile forces on the collagen fibril network in which they are embedded. This is most easily seen where cells are cultured in native 3D collagen gels which are allowed to float freely in medium. Over the following hours and days, the resident cells attach to the collagen fibrils which make up the gel and apply their own tiny motor forces on the gel. As a result, the gel visibly shrinks in diameter and thickness.

In effect, in this system, we can 'see' the forces generated by resident cells as a change in gross dimension of the floating gel. The experimentalist measures the reduction in diameter ever 2–6 hours, as a percentage of starting diameter. Based on the 'simplest-theory-possible' principle, the observer then assumes that the change in shape reflects the sum of all the cell contraction forces generated independently and locally. Obviously, there are other, less simple possibilities. For example, not all of the cells in the gel need to have the same contraction abilities (the different-populations interpretation). There may be 'tough guys', or even 'thug-type' cell populations mixed in with the 'softies'. In modern mechano-biology, it seems all too likely that 'independent and local' is an assumption too far. Nevertheless, clusters of cells could act together, as a cooperative, coordinated by biochemical or mechanical signals.

Rather less obviously, we might question the assumption that cell contraction is isometric, with forces of the same magnitude applied in all direction, 360 degrees around the spherical cell. The isometric tension assumption arises because the actin-myosin (muscle-like) movements come from the cell cytoskeleton, and these can act in any and all axes. While it might be that cell-forces are isometric in the first instance, this would be a very special, extreme position which would be unlikely to persist

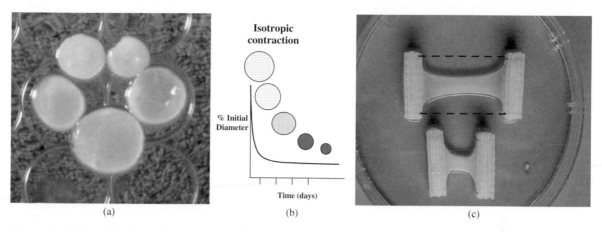

Figure 7.7 (a) Free-floating collagen gels plus fibroblasts contracting (left to right). (b) Graph indicating the contraction rate over time. (c) Same gels uniaxially tethered between porous bars. Top shows gel narrowing, perpendicular to the tethered axis and the original shape, as the dotted line. Lower gel was released just before photo was taken.

for long in any natural system. In particular, for the cytoskeleton to generate loads onto the extra cellular material, it must first *attach* to that material.

In practice, and especially for materials made of long thin, highly asymmetric fibrils, such as collagen, this step will inevitably end up producing very *an*isotropic forces as the fibril component reorganizes and realigns. In fact, the assumptions surrounding the fibroblasts-in-collagen contraction model were developed more than four decades ago and, as we shall see (Figure 7.7), are ripe for revisiting.

Quite by chance, this model is also remarkable in being an extremely low-force, low-resistance system. Forces generated by the cell cytoskeleton are very small, even when magnified by many cells – normally far too small to produce measurable gross-scale movements, least of all those we can measure with a cm ruler. However, the first requirement is that such 3D cell gels are free floating in their culture medium, so friction to movement is minimal. Secondly, the gels are nearly all water, with 0.2 per cent collagen fibres: hence 99.8 per cent water!

Consequently, when cell-cytoskeletal fibres attach to and pull on collagen fibrils through their integrin membrane receptors, there is almost no resistance to their movement. What is more, the gels are so hyper-hydrated that cells are able to contract them

to 50–80 per cent of their initial size before the collagen stiffness increases to a level which stops further shrinkage (Figure 7.7).

The story of the collagen-gel-contraction model, with its indirect, visual read-out of contraction could have continued on much the same lines but for the development of direct, real-time measurement of force generation. This changed many of the basic assumptions of the system (mostly beyond the scope of this analysis). One example of this direct cell force measurement involved tethering rectangular gels at either end of porous bars. Sensitive strain-gauges could then monitor the contraction forces, as they were generated, in the single tethered axis (Figure 7.7c). At the same time simple uniaxial loads could be applied to the cell gel tissue through a computer-controlled motor. The precision and rate measurement of these systems, with the ability to monitor cell reactions to applied loads, has gradually changed how we understand these cell-matrix forces.

It turns out that forces generate by cells on the collagen network increase gradually, up to a relatively constant level, over 2–10 hours, depending on the cell type (Figure 7.8a,b). This relatively constant matrix tension is normally maintained by the resident cells over many hours or days (as long as

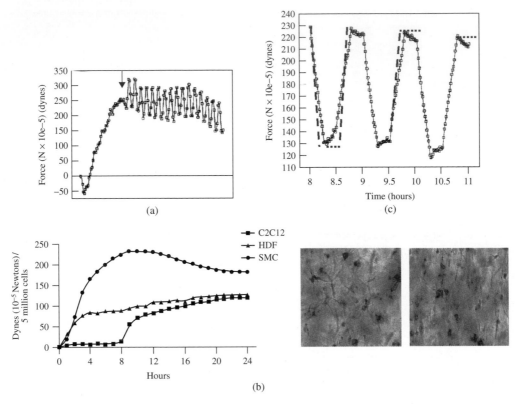

Figure 7.8 Tensional homeostasis graph. (a) Force generated by fibroblasts in collagen (24 hrs) rises to a maximum over ≈4 hours and remains relatively constant, maintaining at a characteristic 'equilibrium'. In this case, at the stage where cells began to maintain 'tensional homeostasis', a external cyclic load was applied via a motor (red arrow). Source: Brown, R.A., Pajapati, R. McGrouther, D.A., Yannas, I.V. & Eastwood, M. (1998). Tensional homeostasis in dermal fibroblasts: Mechanical Responses to mechanical loading in 3-dimensional substrates. *Journal of Cellular Physiology* **175**, 323–332. (b) Force-time responses (24 hrs) for three cell types: smooth muscle cells (greatest), dermal fibroblasts (same rate, lower max), myoblasts (delayed onset). Inset: myoblasts in multi-axial (left) and uniaxial loaded zones (right, aligned parallel to the load). (c) Detailed expansion of the four-hour loading cycle with the cell response to external load. Actual applied loads are shown by the dotted extension lines. Cells seem to work to maintain a constant endogenous tension, increasing and decreasing their loading in reaction to each part of the applied load cycle, so distorting the cell-free pattern.

is measured so far, or until another process intervenes).

Interestingly, one such 'other process' can be the application of an external load to the mini-tissue in the same plane. When this happens, there is almost immediately a mechanical reaction from the resident cells to reduce the external change. In other words, where a small additional load is added by the motor, there is a reduction in cell contraction until the baseline returns. Where the baseline tension is reduced by the system motor, there is a corresponding adjustment in endogenous contraction from the cells (Figure 7.8c).

This behaviour represents a tensional homeostasis at the cellular level. Cells are maintaining a constant tensile load against changes in external loads on their matrix. They seem to do this from inside the gel, by altering the cytoskeletal loads they generate in reaction to the changing external loads in an equal and opposite manner.

At first glance, this behaviour appears to be a futile reaction to the changes in the outside world/environment. As mentioned previously, the cell forces involved are absolutely *minute* (and cell numbers small), and this is particularly true in relation to the relatively huge external loads acting on native skin and tendon tissues. However, tensional homeostasis behaviour is exactly what would be predicted from the concepts of mechanically reactive remodelling that we have been discussing. In other words, perhaps these fibroblasts are *not* involved in some heroic but futile tussle of the few and the weak against the irresistible. This small band of matrix material maintenance cells is not trying to correct short-term shape distortions of their parent tissue. Rather, this may represent the detectable signs of their constant vigil to maintain the material's properties and its 3D dimensions by ECM remodelling.

In fact, since these first observations, further evidence of mechanically reactive cell collagen remodelling has been reported. These cell responses seem to be part of the mysterious remodelling process we are so interested in – the same remodelling process by which tissues acquire and adapt their mechanical function (and 3D architecture) against the ravages of a changing external environment (including calorie-intake!). It is almost as if the fabric of our tissues were maintained by *a thousand blind but constant tailors*.

7.3.2 Mechanically reactive collagen remodelling – the 'constant tailor' theory

We may, then, have reached the point where it is possible to put some meaning and mechanism to a whole series of bio-observation and tissue behaviour anecdotes. It seems plausible, at the very least, to say that connective tissue cells constantly adapt their matrix stiffness by increasing or decreasing collagen deposition in response to changes in the basal matrix tension. If this is correct, it would resemble how a blind craftsman repairs or reshapes a piece of fabric, testing the strength by pulling across the materials, then adding stitches or patches until it can hold the tensions that are routinely applied. We can think of this as the '***constant tailor***' theory, helping to define the relationship between cells (e.g. fibroblasts) and the extracellular matrix which makes up the fabric material of our tissues. In this case, the tailors are our matrix-maintaining cells.

The idea here is that their maintenance programme is to constantly apply a small but steady tension on the fabric in which they live. Where the x, y or z planes of the material are stiff, the cell itself will be distorted. Where they are less stiff, the matrix will be pulled together and *it* will deform. Either way, the cell-tailor will receive an unmistakable message and presumably deposit or remove collagen matrix (a few stitches) in that location and plane.

The constant tailor theory presents us with an interesting consequence, as it makes it clear why the tailor-cell needs to maintain a tensional homeostasis. This is an essential component, in that it provides the constant baseline against which to measure material stiffness. In fact, it is far easier and much more accurate to monitor changes in dimensions or material properties where the material is under a constant, *known* load, rather than hanging loose.

This can be pictured in one of three forms: in one plane (1D) for a rope or chain, in two planes (2D) for a fabric and all three planes (3D) for a bulk structure such as an office-block or a body-tissue (Figure 7.9). Without a background tension in the tethering rope between a kite and the flyer, information from the loose cords is ambiguous. Has the rope broken/stretched, is the kite being blown towards the flyer or is the kite falling? Indeed, could the flyer himself (i.e. the cell) be moving? If the kite is falling, for instance, a slack string provides absolutely zero information about direction or rate of fall. When the rope or fabric is tight we can quickly detect decreases or increases in background tension, but much of this information-stream is lost when the material or structure is loose.

There is a second (obvious) consequence of the constant tailor theory hinted at repeatedly throughout this section. This is that the *direction* or **plane matters**. As shown in Figure 7.9, this system can provide unambiguous information to the cell-tailor about which direction its material is more or less stiff. We must presume, at present, that this allows the cell to deposit or remove matrix material (e.g. collagen fibres) in a corresponding plane, to keep the tissue strongest in the direction it is loaded most.

Though there is still much to learn about the detail of this process, the importance of the main

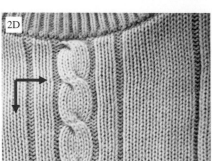

Figure 7.9 Illustrations of useful background tensions: (1D) in a kite or a boat-tethering rope; (2D) 2D planes in clothing fabric; (3D) a block of flats, where all three matter (arrows).

direction of loading, or primary force vector, cannot be overstated. The fact that many tissues are adapted to function under regular and repetitive load directions may be particularly important, as the constant tailor control process is particularly sensitive to changes in *force vector*. This leads on to the idea that the cells responsible for adaptive mechanical remodelling may respond most to local changes in force vector. This is the idea of 'OOPS', or 'Out Of Plane Stimulation', discussed in the appendix to this chapter.

7.4 Can we already generate tension-driven growth in *in vivo* tissue engineering?

It may be that we must endure some modest wobble of our tissue engineering self-esteem. To use a heroic image, we may have struggled to reach high into the foothills of the 'Sierra Tissue Engineering' only to find someone else's flag above us!

7.4.1 Mechanical loading of existing tissues

As we have seen earlier, plastic and reconstructive surgeons commonly use tissue expanders, inflated under patient's dermis, to 'grow' new skin, for grafting. This involves inserting a silicone bag under the skin and forcing in saline under pressure, with top-ups every few days over a period of weeks. The result is 'new' skin, which can be grafted elsewhere once the bag is deflated. In this case, it is clear that the tissue has grown under the applied load in dimensions and mass. Consider the potential for engineering of new tissues if the triggering process for such localized adult tissue growth could be understood *and applied* when and where we need it.

Orthopaedic surgeons have, for many years, been able to extend the long bones of patients with growth deficiencies by osteogenic distraction. This involves breaking the femur, tibia or humerus in question and allowing a soft fracture callus to form naturally. This is the normal bone tissue repair process, in which the callus eventually mineralizes, joining the opposing bone fracture ends. However, in osteogenic distraction, a strong frame is implanted around the fractured bone and this frame is expanded to pull against and stretch the fracture callus tissue. This stretching (i.e. the distraction) is carried out at a standardized rate, usually a few mm per day, in the axis of the bone. In effect, the soft fracture callus grows in length under (and in the plane of) this applied load.

The clever part here is that the rate of extension of the frame is designed to *just* match the rate at which the stretching fracture callus is mineralized and turned to bone. As a result, distracted callus cannot fully mineralize as it would in a normal facture repair. However, neither can the soft callus get any longer. Rather, the new material becomes calcified as part of the patient's bone, which progressively elongates until the process is stopped and the frame removed.

A third surgical example, the stretched scar, is rather less a treatment than a form of damage limitation. In many skin incision wounds (i.e. surgical) it is necessary to suture the wound margins under a tension. Such tensions are normally an inevitable consequence of the patient's anatomy around or beneath the injury site. For example, tensions form across the abdominal wall, near to major arm or leg joints, or over the face under the effects of jaw movements. In such cases, the sutures carry most of these external loadings in order to *unload* the wound margins while new tissue and scar forms. Unfortunately, it is often necessary to remove these sutures at relatively early stages, which throws the load onto the new repair tissue in the same way that the distraction frame stretches the new fracture callus. The result is similar, producing growth of the scar tissue across and between the original wound margins, which produces a characteristic 'V' shaped stretched scar by lateral growth of new skin (scar).

7.5 Conclusions: what can we learn from engineered growth?

It is quite clear that these clinical examples can be regarded as forms of *in vivo* tissue engineering or, in some cases, *in vivo* bioreactors. It is even possible that some are already in use for questionable 'engineering' of skeletal geometry, for example to enhance sports performance such as finger action or jump height. Paradoxically, it is difficult to envisage how such processes, legitimate or nefarious, can progress far beyond their present empirical base without the sort of detailed understanding which is coming from new, *ex vivo* tissue engineering. However, in the best 'Looking-Glass' traditions, extreme tissue engineering can undoubtedly teach us much and throw up critical clues through careful analysis of its tissue-production success stories.

The big question now facing extreme tissue engineers in this field is how can we change the external loading demands on tissues (engineered or natural) to trick the resident cells into growing their extracellular matrix, when and where *we* need that growth?

Appendix to Chapter 7

Note: this appendix is designed to stimulate (where appropriate) an insecurity-driven desire to learn more about cell-matrix mechanics, rather than to answer all the questions you may have.

Forces on cells are a bit like bagpipes on people: position and direction are *critical*.

Those of you who have ever heard Scottish or Irish bagpipes being played live may appreciate this analogy very well. Heard in the open air, especially massed, or across distances over an open hillside, the bagpipes must produce one of the most haunting, enchanting and stirring sounds imaginable. However, experienced at close range in small rooms, especially with soft furnishings, they can represent a banned form of torture. So it must be when mechanical forces act on cells embedded within their 3D support material. How they affect cells

---

Text Box 7.6 Monolayer cell culture on plastic is not a control for mechanical 'stimulation'

Cells culture in monolayer on plastic is, in fact, a very special and extreme form of cell-mechanical loading because, when cells attach to plastic, they rapidly assemble a contractile (actin-myosin) cytoskeleton which generates very high intra- and inter-cellular loads on their trans-membrane attachment points. Since the plastic is so unnaturally stiff and completely symmetrical in its stiffness, in all three planes, it represents a cell environment which is both high-load and poorly, or potentially, non-physiological.

Text Box 7.7 Reflecting back on the value and vulnerability of the tribes of tissue engineering

As we saw in Chapter 1, the constituent tribes of tissue engineering are its unique and strongest feature, but they also bring special needs and vulnerabilities. One of these is the question of the shared, pseudo-shared and downright misinterpreted language we use. It is easier to see this effect where peoples of different cultures speak different languages, as in Figure 7.10. In this case, it is necessary to understand the French language to be struck by a cultural and philosophical difference concerning where and when it is appropriate to urinate in public. To other (more repressed?) cultures, it might seem surprisingly inappropriate *ever* to have to indicate where *public* urination is unacceptable (the assumption being 'nowhere, *ever*'). Having an industry which makes and sells the official signs is a further surprise!

The rich variety of approaches brought to tissue engineering by its tribes can be obscured by our *apparently* common scientific English. An example of this was experienced by the author. After talking for 15 minutes to an eminent, if patient, biological colleague about cell guidance and force directions, he finally tired and enquired how *gene* vectors could possibly come into the story . . .

Figure 7.10

depends on the materials they pass through and the direction they come from to reach those cells. To our colleagues from the physical science tribes of tissue engineering, this is completely obvious – knowledge of a force vector is at least as important as its magnitude or frequency. Within the biological and medical-related tribes, however, this is not so obvious and can be a real surprise (Text Boxes 7.6 and 7.7).

It is important to understand from the start that the nature of these '*load carrying demands*' is not at all simple and merits a detailed analysis. An idea common in the biological literature is that cells or tissues are either mechanically *stimulated* or not. But this does not come close to adequate. It leads to the disarming idea that if cells can be 'stimulated' by mechanical forces, we absolutely must have a 'control'. In familiar cell-molecular systems, a baseline or 'control' output comes from using a zero concentration of the compound in question. Unfortunately, this is not appropriate in the case of mechanical controls, mainly as it embarrassingly misunderstands the initial concept. *Mechanical load* is not a single, simple, dose-dependent form of 'stimulation' which can be switched on, or more particularly, *off*. Specifically, cells in monolayer culture on plastic are most definitely *not* a zero-force or baseline 'control' system, as sometimes thought.

In the first place, on this planet (g = 1.0), we can reliably assume that pretty well all tissues are constantly under some form of mechanical loading as a result of their mass. Thus, unless we are employing large amounts of expensive equipment or an experiment on the International Space Station, the very idea of 'mechanically stimulated' (or unstimulated) is a mistake.

Interestingly, where cells respond to micro- or zero gravity experimentally, this may not be due to the direct effects of removing the cell *itself* from gravity. Given the tiny mass of a single cell, it may not be able to detect the effects of gravitational forces directly. However, pretty well all mammalian cells are part of, and firmly attached to, a great mass of other cells and extracellular matrix. Since such tissues have substantial mass, the resident cells will inevitably be deformed by the resultant compressive, shear and tensile forces. However, in isolated cell experimental terms, this is a very different *indirect* response.

Secondly, it is a dangerous simplification to look at the main form of loading on a tissue and conclude that such a tissue is simply compression-, tension- or shear-responsive. Although an artery wall is under strong shear loading at the blood/vessel interface, this will also generate directional (axial) tensions below the vessel surface, compression of the mid-layers, turning to circumferential tensions in the outer, elastic retaining wall layers. Similarly, compression on the articular cartilage of our joints (e.g. as we stand on our leg bones) generates tensile loads at both the cartilage margins and menisci, as well as shear at the articulating surfaces as the joints roll and slide. The reality is that combinations or *patterns* of loads act on different micro-sites and cell zones

or layers, and that these patterns change during the period over which loads are applied (Text Box 7.8).

The detail of these patterns may be characteristic (potentially like a fingerprint), with one or two 'principal elements' dominating during any regularly performed movement. Different points in a tissue, then, are subjected to combinations of mechanical stress, in tension, compression or shear, with one of these predominating at any given site or time in the loading-relaxation cycle.

As we have seen, some of these loads can be described as 'static' or anatomical. These are tensions *built into* the physical dimensions of our stiff skeletons and less stiff but attached soft tissues. Other loads, commonly those generated by skeletal muscles, will have patterns which are complex and dynamic. These are generally very *repetitious* (e.g. cyclical) and *constant*, such as when we run. The cyclic repetition of this loading is again a consequence of the 'constant' anatomy around which the muscles act. This is because our anatomies (muscle position/bulk, joint locations and bone-length levers) tend to remain fairly constant (aside from injury and aging).

As a result, we can predict that cells within any of the main tissues will be subject to complex

---

[16] Actually, this still leaves us with a serious paradox – that the cells, in this case, also manufacture the ECM which they so constantly and diligently adapt.

Text Box 7.8 Paradox in the ranks of the cell-mechanic-biologists

As in previous chapters, we find so much to learn by tackling the big paradoxes head on. In cell mechanics (biology of cytomechanics) there have now been many, many works demonstrating that application of changing mechanical loads to cells in culture 'stimulates' them. The nature of the stimulation depends on what is being asked. The 'elephant in the room', though, is the question of 'how come all our bodies are not growing all the time, expanding with

warts, tumours or just new mass in all sorts of directions?' If this were the case, then sports stars should be completely bizarre in physical form and hugely different between, say, swimmers and golfers.

Yet, when our cells are in our bodies (rather than naked on a culture dish), they become almost completely deaf to the vast majority of loads our tissues experience. The big difference is that cytomechanics is *not* the same as tissue mechanics. Cells in a physiological tissue are surrounded by a material (the extracellular matrix – ECM[16]), and this completely dominates how cells respond.

Exercise Box 7.2

Discuss the idea that 'zero mechanical loading is not really an option for physiological cell systems', including the use of zero gravity!

and highly dynamic **load patterns**. These are, in effect, localized 'load signatures' characterized by the main direction of loading (the principal force vectors) as well as their magnitude and frequency. Most importantly, however, these signatures will be remarkably constant for any given group of cells over considerable periods. This makes it possible for such cells to adapt the shape, composition and load-carrying characteristics of their extracellular matrix to resist each particular local load signature.

All we must propose, then to make such a system work, is that stromal cells are programmed to replace existing matrix with new materials of a type, density and orientation which minimizes their own (cytoskeletal) deformation during this routine (signature) pattern of loading. In proposing this, we are effectively describing a mechanism for what is generally assumed to be **mechano-dependent tissue remodelling**. In bone, this has been known for well over a century as Wolff's law.[17]

This suggests that connective tissue cells lay down an extracellular matrix which optimally stress-shields those same cells from the prevailing pattern of external loading. Any changes to that pattern of loading will tend to elicit a cellular response, which is to remodel that matrix material to again minimize the strain on local cells. In effect, it is a completely cyclic and interdependent relationship between connective tissue cells and the mechanical properties of the matrix material they lay down. The cells at the centre of this relationship (fibroblasts, chondrocytes, smooth muscle cells, etc.) are the 'constant tailors' of our earlier theory.

To conclude, if this analysis is correct, one potent way to influence the constant tailor's work-pattern will be to apply our new 'activating' or controlling forces along *vectors* which are outside the signature pattern. In other words, according to the 'OOPS' principle mentioned earlier (**O**ut **O**f **P**lane **S**timulation), loading tissues out of plane of the existing collagen fibre alignment is most likely to 'stimulate' cell adaptive mechano-remodelling.

Further reading

1. Perrotta, S. & Lentini, S. (2010). In patients with severe active aortic valve endocarditis, is a stentless valve as good as the homograft ? *Interactive Cardiovascular and Thoracic Surgery* **11**, 309–313.
[An uncommon direct comparison example review of grafts versus prosthetic implants, in this case heart valve . . .]
2. Meswania, J. M., Taylor, S. J. & Blunn, G. W. (2008). Design and characterization of a novel permanent magnet synchronous motor used in a growing prosthesis for young patients with bone cancer. *Proceedings of The Institution Of Mechanical Engineers. Part H, Journal Of Engineering In Medicine* **222**, 393–402.
[. . . but making prostheses which grow and remodel in time/space, in the way that a graft can, is very rare and tricky.]
3. Proff, P. & Römer, P. (2009). The molecular mechanism behind bone remodelling: a review. *Clinical Oral Investigations* **13**, 355–362.
[Hard tissues renew, remodel and adapt].
4. Mackey, A. L., Heinemeier, K. M., Koskinen, S. O. & Kjaer, M. (2008). Dynamic adaptation of tendon and muscle connective tissue to mechanical loading (review). *Connective Tissue Research* **49**, 165–168.
[Soft tissues renew, remodel and adapt]
5. Sabharwal, S. (2011). Enhancement of bone formation during distraction osteogenesis: pediatric applications (review). *Journal of the American Academy of Orthopaedic Surgeons* **19**, 101–111.
[Osteogenic distraction . . . growth of bone by stretching a fracture-repair site (callus): engineering tissue *in situ* with mechanics].
6. Page-McCaw, A., Ewald, A. J. & Werb, Z. (2007). Matrix metalloproteinases and regulation of tissue remodelling. *Nature Reviews Molecular Cell Biology* **8**, 221–233.

---

[17]Wolff's Law: the theory that healthy bone will adapt its composition and architecture to the loads it is placed under (Julius Wolff, surgeon, 1836–1902).

[Definitive review of connective tissue remodelling – view from (only) the perspective of the enzymes that break down the tissue].

7. Brown, R. A. (2006). Cytomechanics in connective tissue repair and engineering. In: Chaponnier, C., Desmouliere, A. & Gabbiani, G. (eds.) *Tissue Repair Contraction and the Myofibroblast*, pp. 7–24. Landes Bioscience, Georgetown, TX.
[Analysis of how mechanical forces, at the cell level, can and do affect tissue remodelling – including the 'oops' principle.]

8. Docampo, M. J., Zanna, G., Fondevila, D., Cabrera, J., López-Iglesias, C., Carvalho, A., Cerrato, S., Ferrer, L. & Bassols, A. (2011). Increased HAS2-driven hyaluronic acid synthesis in shar-pei dogs with hereditary cutaneous hyaluronosis (mucinosis). *Veterinary Dermatology* **22**, 535–545.
[Wrinkly (Shar Pei) dogs have 5 × normal levels of hyaluronan, with HAS2 gene over-expression: but why should a surplus of charged polysaccharide prevent Shar Pei fibroblasts maintaining a skin tension?]

Bioreactor origins

Although 'the bioreactor' has been a major part of tissue engineering thinking and aspiration since its beginnings, it cannot be claimed as special to the subject. Fermentation bioreactors predate us by quite a margin and, if we are to borrow their name, the least we can do is to understand what it *already* means. The simple wine-beer fermentation jar shown here is a classic – if simple – bioreactor in which biological organisms (yeast, in this case) are fed with sugar and other nutrients under controlled, often dynamic, conditions to promote production of useful materials, such as alcoholic drinks. Classically, this involves biochemical reactions and changes, (i) mostly *in solution*, (ii) mostly carried out by *whole organisms*. Interestingly, neither of these features ranks high in the 3D tissue-bioreactors now envisaged for the growth of engineered tissues.

8 Bioreactors and All That Bio-Engineering Jazz

Extreme Tissue Engineering: Concepts and Strategies for Tissue Fabrication, First Edition. Robert A. Brown.
© 2013 John Wiley & Sons, Ltd. Published 2013 by John Wiley & Sons, Ltd.

8.1 What *are* 'tissue bioreactors' and why do we need them?

The central place of 'the bioreactor' in traditional tissue engineering thinking is suspiciously elastic and ambiguous. The idea of the bioreactor is that the basic building-blocks of the tissue, once assembled, can be brought together into a 3D tissue facsimile by a culture stage. However, these variables seem to have worryingly large ranges, requiring anything from tight cell control and brief culture periods down to those supplying only minimal cell controls over long culture periods. Consequently, the complexity, duration and even the timing of this *bioreactor culture stage* has been difficult to define and understand.

In this case, to be suitably analytical, we should ask: 'Can we be sure that *tissue bioreactors* really are likely to do all that is being asked, or is this a tissue engineering 'fig-leaf' used to conceal an embarrassing area of uncertainty?'

8.1.1 Rumblings of unease in the smaller communities

To start to answer this question (as in other chapters), let us look more closely at some of the really fundamental assumptions of the 'tissue bioreactor story'. Indeed, there is a huge assumption buried deep within the original blueprints for tissue engineering. For shorthand, we can call it the 'architecture-control assumption', which proposes that our bioreactor conditions will control the tissue micro-architecture of the constructs we grow. It now seems that most of us have, at one time or another, signed up to this assumption without really checking the small print. And the small print, not surprisingly, includes a sub-section which states '. . . but do remember, this is an assumption!'

To recap, the assumption is that: 'given a suitable array of control cues and raw materials, the cells we grow will make a new and functional tissue where there previously was none' (Text Box 8.1; also discussed in Chapter 7).

The optimistic view, that this is a safe assumption, comes from:

- Long-held developmental biology theory, and its understanding of how embryonic tissues come together from small cell-balls and
- Some epithelial, sheet-cell types, which can self, assemble themselves into simple but reasonably faithful tissue-layer replicas.

In the same vein, we can see that there has been a significant expansion in structural complexity which can be generated as we moved from '2D' to '3D' culture systems.

In this climate, it is easy to understand the buy-in of much of the community, especially 3D biomaterials scientists, cell and developmental biologists and engineers. However, one of the smaller tribes of tissue engineering has consistently shuffled their feet and kept returning to the 'assumption' word. These are the tissue repair biologists.

Repair biologists (including some surgeons) are also impressed by this vision and they are just as keen for it to become possible. However, they have also been suturing, pouring and pushing every conceivable potion and composition into tissue injury sites that 'most people do not want to know about' for the best part of 2.5 millennia.

Text Box 8.1 Two types of cell-control cue

Viewed from a process technology standpoint, cell control cues here fall into two broad categories. The first type of cue regulates all things related to the composition of the fabricated tissue, including the order, rate and concentration in which components are incorporated. The 'components' will include cell types (with proliferation and differentiation) and soluble and insoluble (e.g. extracellular matrix) export products. This is complex but familiar, especially to the cell biology and medical communities.

However, the second family is far less familiar. These are cues which are needed to control the 'where' – the spatial/temporal organisation cues critical for *functional* 3D architecture.

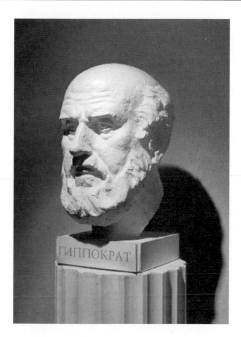

Figure 8.1 Hippocrates of Kos (*ca.* 460 BC – *ca.* 370 BC; Greek: Ἱπποκράτης; Hippokrátēs).
© iStockphoto.com/Philip Sigin-Lavdanski

Indeed, Hippocrates (Figure 8.1) did not seem to be averse to filling tissue voids with a range of exotic materials. But despite every possible motivation (and surface wounds are *extremely* emotive) and plenty of imagination, success has been strictly modest in getting usable new tissue to form where and when required. This rather does suggest that the 'architecture-control' assumption might not be quite as sound as we would like. Indeed, even where inspirational approaches (including bee-fluids or tropical tree-bark) have been replaced by modern gene and cellular mechanistic therapies, this unease persists.

As discussed in Chapter 1, significantly sized (non-fatal) defects in vertebrates normally fill with tissues which are rarely as functional as we would like. Scars appear in almost every body location. So the repair-biologists logic goes, if molecular and cellular cues *in the correct body-site* cannot, after many millennia of evolution, persuade natural systems to rebuild tissue as it once was (i.e. regeneration, see Chapter 1), what makes us think that farming them

in a laboratory bioreactor will do any better? The suspicion here is that the enabling environment of at least some clean wound-beds (post 50 million years of evolution) should be better than a reaction chamber (evolution time ≈15 years). Looked at from a philosophical perspective, the ability to make omelettes with *random structure* out of highly *structured and symmetrical* eggs implies nothing about the feasibility of the reverse process (Dumpty *et al.*, 1835)[18].

8.1.2 *Hunting for special cells or special cues*

The assumption that our technology is up to the task of persuading cells to do just what we want is a difficult one to break out from. Figure 8.2 shows a logic gradient along which biotech scientists can move freely, working diligently towards whichever end their vision takes them. The logic suggests that we can produce our tissues either by:

1. prodding our rather everyday cells into the right action with subtle environmental instructions/cues; or
2. finding those special precursor cells (which currently seem to exist in embryos) which can produce tissues on their own with very little help from us (a recent sub-plot to this suggests that we can go one step further back, de-programming adult-committed cells, then reprogramming them to something else – in other words, inducible programmable stem cells).

On one hand (a), this implies a breakthrough in the understanding of how to control normal cellar processes, especially 3D spatial controls. On the other (b), it hopes that, through much trawling and good fortune, we will locate cell types needing little or no external control. Clearly, either of these is dangerously open-ended, but the availability of both options apparently makes it possible to offset (or to 'fudge') the risks. In effect, as one approach

---

[18]Note: this is a joke reference. Please do not be tempted to look it up. It refers to an old British Nursery rhyme about the futility of trying to repair broken eggs . . .

Figure 8.2 Logic spectrum for how to achieve the fabrication of functional tissues through culturing of cells.

hits its inevitable problems, we can move our effort to the other.

However, justification for labouring away at the problem (either end of the logic gradient) is only reasonable where there are plausible approaches or new knowledge with which to assault that problem. The idea of just 'trying' every available version or combination is at best risky, and at worst futile. Just like Hippocrates, tissue repair biologists have been here before – fishing for solutions too far ahead of their understanding. They are now rightly cautious about committing to the same approach for the next 2.5 millennia. After all, although the solution may be only around the corner, scarless healing is *still* a dream.

8.1.3 Farming – culture or engineered fabrication

Although the need for tissue bioreactors in this scheme (Figure 8.2) rises (left) and falls (right), the underlying assumption at both ends of the logic is that cells will be the primary producers of both tissue substance *and* structure. The only remaining question is who (or what) supplies the controls that regulate production rate/sequence, component type and spatial accumulation of material. This reflects the traditional biological belief that only cells can make complex bio-systems. It assumes that the role of engineering systems in the process will be restricted to monitoring instruments, nifty labour-saving culture chambers and cryo-storage. In effect, then, current tissue engineering has an implied assumption that tissue will be made through ever-better cell 'farming' (like salmon farming), rather than engineered fabrication (like mobile phone production).

The role of new science in generating useful products from the culture of living systems has a long and pragmatic history. We can trace this progress from the explosion in agricultural production, ranging from GM products, marine farming, hydroponics and enhanced animal welfare, to safe product storage, cryopreservation and meat traceability. Our analogy here, then, is agricultural.

Despite a long history of scientific benefits in agriculture, many in tissue engineering may not appreciate the drawing of this parallel, yet it is all too valid, as we can see from the evolution of conventional bioreactors. For example, the expansion of cells (especially stem cells) for regenerative medicine involves their growth (proliferation) in nutrient media. This is directly analogous to hydroponic culture of plants or aquatic farming of fish fry in hatcheries (also known as aqua-culture). The aim in each case is to develop defined nutrients, blended with control/stimulant and anti-microbial agents within tightly controlled protocols of temperature, pH, sunlight, etc. These are designed to produce reproducible biological products, from plants or plant products to young fish and fish fillets (Figure 8.3).

Success is, in part, limited by the complexity of the biological system (i.e. the whole organism and the material we intend to produce). In the case of farmed salmon fry (or, later, their maturation into adults), the organism itself holds all the necessary information for production and control to generate new fry, subject to basic conditions such as water content, flow/mixing, temperature and light quality. So, for whole-organism culture (such as salmon), the complexity of the organism and the fabrication controls are not an issue for us as the organism itself comes comprehensively equipped to carry out the *full* process where suitable *enabling* conditions are provided.

In the same way, stem cell expansion, though complex in its detail, is simplified by the fact that the cells themselves come ready-programmed to divide without too much outside control. Clearly, there are questions (particularly for stem cell expansion) of

Figure 8.3 Farmed fish fry. © Gwynnbrook Farm

maintaining a particular differentiation state of the cells. But this is much the same as maintaining good inbred salmon stock over a number of salmon generations despite mutation, genetic drift or infection. In other words, we tend to be generally happy with what the bio-system will produce for us. Much of agricultural science and cell culture, then, is about adapting or enhancing the enabling conditions in order to shift or speed up what we get as a crop towards what we need.

This may seem like a small, useless fragment of pedantic logic shuffling around the similarities of culture and farming, but it is really far more. Once we accept that our concepts have shifted from construct *engineering* (i.e. fabrication) to cell-dependant culture or farming, the more astute reader will see that we are also expecting to break one of the most basic rules of farming. Although the retail value of fish lips may, in some countries, be attractively high, and pumpkin fruit is the only saleable part of that plant, no one is yet suggesting that we farm *just* lips or *only* the fruit (Figure 8.4). The bone-filled fish and the tough leaves are not essential as a product, but they are as much a part of the production process (the culture) as the bioreactor is in tissue engineering. They are not part of the high-value product, but equally there is no product without them.

The complex 3D tissues that we require here cannot be made in isolation, without the control and production machinery. This makes our dream of the tissue bioreactor a deceptively high hurdle, as we have set ourselves, for the first time, the task of producing meat without bone or brain, or

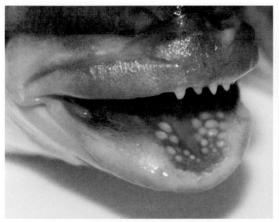

Figure 8.4 'Isolated farm produce'. Pan-ready fish lips (left) and leaf-free pumpkin (right). Clearly, these are agricultural fantasies. We farm *whole organisms*.

fruit without root or leaf. This is the equivalent of a cow-free steak farm – an idea we shall need to revisit.

At this point, it is possible to hear the distant sounds of all-out rebellion from the host of tissue engineers who are keen to assert that culture bioreactors ARE the answer and, in time, we shall find the right conditions. We can in time devise advanced tissue bioreactor technologies which can replace all the missing cues and controls. After all, it is early days yet, as Hippocrates of Kos might have claimed. However, as a result of the **Hippocrates caveat**, it might be wise to run this aspiration through a reality check. To analyze how well this stacks up, we need to look under the lid of 'the history and origins of tissue bioreactors'.

8.2 Bioreactors: origins of tissue bioreactor logic, and its problems

8.2.1 What have tissue engineers ever done for bioreactor technology?

Mammalian cell biologists are sometimes surprised to find that 'bioreactor' is a word which engineers and bio-process chemists commonly consider their own. Rather inconsiderately (for us), we must admit that *tissue* engineers did not invent '**bioreactors**'. This is a term used for many decades (indeed, centuries in the case of fermentation) to describe vessels, frequently stirred, in which suspended cells – from yeast and bacteria to genetically modified mammalian cells – grow and produce materials that we want. Clearly (and happily), the earliest examples of these were the wine and beer products of alcohol brewing chambers. Their modern equivalents now churn out (literally, in the case of milk-derived products) everything from washing powder enzymes to medical antibodies, all with exceptional efficiency and precision.

So, the only special contribution that tissue engineers seem to have made is to adapt this to the culture of 3D (i.e. multi-cellular, spatially organized) structures. In other words, the most important difference is that the product of tissue-bioreactors is intended to be a complex 3D structure, whereas conventional bioreactors produce soluble molecular mixtures. In fact, we can now see that conventional biochemical bioreactors have been developed into highly effective production units in which complex, sophisticated product mixtures are made with 'engineering precision'. While fermentation bioreactors can make highly complex chemical mixes such as fine wines, and perfectly reproducible antibodies genetically modified cells, these products are essentially *soluble*. They certainly have minimal to zero 3D spatial organisation across length-scales greater than a few nanometres – in other words, emphatically *not* what we need for tissues (see Chapter 5).

So, what are the similarities between tissue bioreactors and their conventional relatives? In both types of bioreactors, the function is to 'control biological output' from large numbers of cells, so they both ultimately depend on getting cells to generate product. This culture-dependent production can sit anywhere along a spectrum which runs from biochemical engineering (closely controlled process stages) to more traditional, farming-style culture systems. In the latter case, controls are light and aimed chiefly at enhancing the native biological processes that are inbuilt in the farmed organism. The more controls are applied to such systems as environmental factors, the closer they become to (bio)chemical engineering processes. However, the further down the engineering line the process goes, the louder is the demand for detailed knowledge of the innate cellular controls.

As we move away from systems that depend on good Pilsner beer yeasts and towards those where the producer-cells wall themselves into collagen-elastin tombs (i.e. into a matrix-rich tissue), this detailed knowledge rapidly becomes the limiting factor. Here we reach the big difference. While many decades of research have made it possible to control intracellular processing to produce refined *soluble* products, we are a long way yet from the same understanding where the cell product is a spatially organized 3D tissue (Figure 8.5).

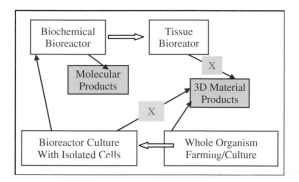

Figure 8.5 Illustration of the bioreactor logic-loop which needs either engineering-level control of material synthesis (not achieved yet) or cultivation-type fabrication without whole organisms (cell-only; again not really achieved yet).

8.2.2 The 3D caveat

Oddly enough, although the transition from biochemical to tissue engineering bioreactor may sound modest, it unfortunately contains a critical caveat which changes everything. This is buried in the quantitative detail and can be glimpsed from two directions. In the first view, we can see that we have quietly moved a long way from the target of making soluble (definitely non-3D, non-material) products. However, our assumption has been that this can (easily) be extended to the *solid* 3D structures required from tissue engineering bioreactors.

Worse still, these 3D structures are not simple to form or easy to maintain using living cells. Target tissues are frequently large in overall dimensions and made of dense material. This leads to inherently poor perfusion (mass transport) properties, especially for the deeper layers. Put another way (but the same caveat, really) in tissue bioreactors the solid material product is deposited like prison walls around the cells which produce it, rather than being pushed out into solution and swept away by the stirrer. This change means that, for success, the quantitatively dominant product will be the bulk *solid* support matrix which makes up the tissue. In matrix-rich constructs (skin, tendon, vessels), this commonly takes the form of new extracellular matrix material (see Chapter 3). In cell-rich constructs, the product is more cells, but in the same structural mass. This is the *prison wall* caveat

However gently we try to pass over the prison wall caveat, its effect on the bioreactor logic is thunderous. By taking the option that *cells will make what we need*, this logic forces us to a heavy dependence on inbuilt biological responses and controls. These are the innate controls of the producer elements (i.e. the cells). But without a detailed understanding of how they operate, our tissue bioreactor aspirations begin to look a very, very long way from our comfort zone of conventional biochemical bioreactors.

Text Box 8.2 Between a rock and a hard place

The whole idea of tissue bioreactors seems to lie precisely at the impact point of a rock-like assumption and a caveat hard place. Could it be that we have here an attractive concept which is shockingly far ahead of its time? The prison wall caveat is that the cell-product *must* end up being an organized, substantial 3D material (i.e. a tissue). This makes the science and engineering needed to control production extremely difficult to understand. The daunting scientific problem, in turn, forces ever greater reliance on cell-dependent (farming-like) processing.

However, as if to lock the trap, we find that such approaches are traditionally only effective where they 'culture' (farm) whole organisms which include a *complete* control system (the fish-lips conundrum). Our only major success so far in fabricating useful products from *part*-organisms or cell-masses lies in the use of conventional engineering bioreactors, which generate soluble, non-material products.

Figures 8.5 and 8.6 illustrate this rather irritating logic-loop which bogs down our dream. From where we stand now, either route out of this paradox would seem to require decades of basic research, not **system optimization**, as often suggested. While this is not a universally welcome analysis, it should deflect us from futile cycles into more balanced, coherent strategies.

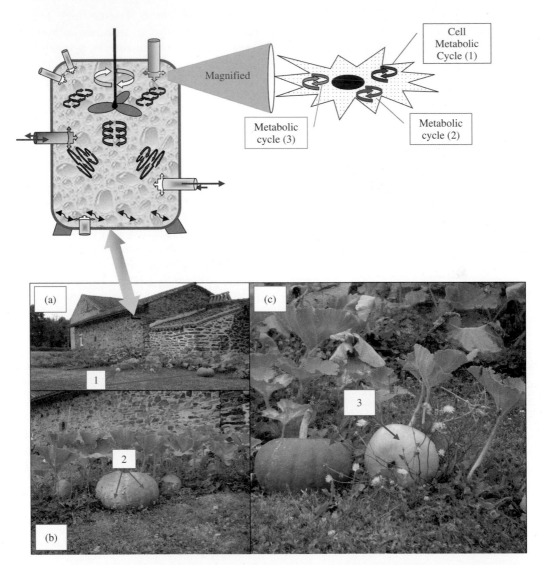

Figure 8.6 Diagram illustrating (by exaggeration) the extremes of culture control, which form the 'bioreactor spectrum'. The top panel shows a diagrammatic form of engineering bioreactor for tight biochemical control of single-cell-suspension production of useful soluble products (e.g. protein). The cell-medium suspension is stirred (local mixing/stagnation/turbulence can be modelled mathematically) for controlled nutrient/gas exchange with cells. Input of metabolites, nutrients, etc., as well as out-take of product and wastes, can be batched or continuous, under tight control. Conditions are monitored (e.g. pH, ionic strength, temperature) for feedback correction. Meanwhile, the main biochemical cycles and enzyme efficiencies (right hand panel; cell cycles 1–3) are known in detail and can be controlled. The lower panel (a to c), in contrast, shows a simple agricultural culture system for production of pumpkin vegetable *material* in large, useful 3D lumps. Though simple, the culture system still supplies essential nutrients and microenvironment (as far as the requirements are known). It monitors basic changes in conditions and reacts to correct these as they develop. However, the precise demands of the system are not (cannot be?) well understood, so the system is based primarily on reliance on the innate ability of pumpkin plants (i.e. the whole organism) to fabricate pumpkins. The controls are very light but, equally, heterogeneities and variance in the system and the product are easily tolerated (arrows 1 to 3).

8.2.3 Fundamental difference between biochemical and tissue bioreactors: 3D solid material fabrication

In effect, all of the tight control and understanding which goes into conventionally engineered bioreactors can be regarded as manipulation of the organic and physical chemistry of two *soluble* compartments, the first is the intracellular cytosol; the second is the extracellular culture medium. Nutrients, gases, waste metabolites, catalytic enzymes and product/by-products/contaminants generated by the cells are in solution, either inside or outside the cell, in the culture medium. This is commonly acknowledged by the idea that 'pools' of this or that metabolite are located within intra- or extra-cellular compartments. There is 'traffic' between pools with characteristic dynamics and rate constants.

This reflects the sophisticated level of understanding of intracellular biochemistry, with its many well-mapped metabolic pathways, predictable kinetics and enzyme control points. This is so much so that the chemical processing can be mathematically modelled down to the level where stirred, sluggish and turbulent flow of fluids can be used to regulate mass transport of products between different parts of the system (extracellular/intracellular pools, compartments or zones of the bioreactor.

Figure 8.7a illustrates this predictability, based on dynamic (bio)chemical processing of solutes in closed chambers, These have characteristically well understood, separated pools of reactants, within a *two-compartment system*. Unfortunately, it is completely unreasonable at present to expect this level of

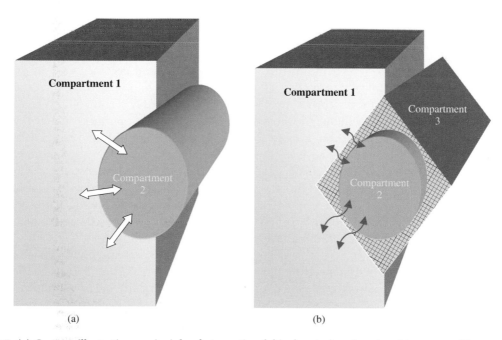

(a)   (b)

Figure 8.7 (a) Cartoon illustrating a principle of conventional, biochemical engineering, bioreactors. These can be seen as two compartment systems – intracellular and extracellular fluids – separated by the cell membrane. The equations for mass transport of reactants and various cell products can be calculated on assumptions based on access between the two compartments, with mixing or stagnation influencing diffusion times in each. (b) With the addition of a third compartment (extracellular matrix, ECM material) enclosing the cell-compartment, all of those equations are void. First, the third compartment/ECM changes *not only* the mass transport, but its predictability as ECM structure is dynamic, anisotropic and heterogeneous. Second, the main bulk product of the cell chemistry goes into the ECM (so out of solution), altering the chemical equilibria in unknown, relatively unpredictable ways.

control in 3D tissue bioreactors, because the main products, by definition, cannot be soluble. In order to become a solid tissue, the cell products from tissue bioreactors must be packed, in great bulk and at high densities, between the producer cells.

This means that the knowledge base and conventional bioreactor rules, no matter how successful, cannot be applied, because we have slipped gently into what is now a *three-compartment system* (Figure 8.7b). There are, as before, the intracellular and extracellular (medium) compartments with their distinctive, intermeshing chemistries, but physically *between* them is an extracellular (tissue matrix) material – a solid. This changes everything. Not only do the main products *not* mix, traffic or diffuse according to any of the previous rules, but this build-up of solid product increasingly dominates mass transport in the *other* two compartments. Worse still, it does not seem likely that we will ever side-step this third compartment solid matrix 'problem', as it represents the very thing we want to make. This is one issue we *have* to fix.

Far from being a pedantic piece of classification, this understanding is critical. It means that the 'bioreactor' concept is unlikely to get us off the equally difficult farming process problem of expecting to grow isolated tissues without the whole organism (the fish-lip conundrum). Is it possible, then, that tissue engineering has inadvertently set itself an enormously high target by mixing its exemplars? This seems to be a fatal concoction of:

- top-down culture (farming) of isolated tissue-parts, *but* without the luxury, of the entire organism; and
- bottom-up bioreactor synthesis, *but* with the previously unattainable target of controlled 3D architecture.

The difficulty is that the 'but' caveat in both examples makes that route look worryingly implausible.

This, perhaps, finally nails down the niggling feeling from Section 8.1 that we are being held back by a questionable assumption (the Hippocrates caveat). It comes down to our old friend '3D architecture and how we make or grow it'. Next, we should look at the special consequences and needs that develop from this implausibility, which perhaps might allow us to creep up on some new (extreme TE) solutions.

8.2.4 Why should a little thing like 'matrix' change so much?

Predictable biochemical reactions

Those readers who have experience of organic and biochemical processing will understand the dominating importance of:

(a) reaction equilibria on rates of production/consumption; and
(b) mass transport of reactants and products (for large or dense systems).

At almost all stages, the important metabolic enzymes will drive reactions at rates which relate to the concentration of the reactants and products *in solution*. But there's the rub: *in solution*. When a dominant proportion of the soluble cell product *leaves* solution and becomes solid, as extracellular matrix, it ceases to play the same predictable games, and new rules apply (Figure 8.7). Where a reaction product is taken out of the system by leaving solution, we might (simplistically) expect the reaction to be accelerated. There are many simple examples of this in conventional chemistry. For example, where the product of a slow reaction is poorly soluble and precipitates, this loss of product from the soluble phase can pull the reaction faster, due to the removal of product inhibition.

However, the protein-polysaccharide composite materials which make up the ECM considered here are *not* low-solubility salts with known solubility coefficients. They aggregate through complex, time-dependent bond formation, following cell processing, intermolecular recognition and cross-linking. Just for good measure, many undergo maturation through water exclusion.

These processes are currently not well understood, and are certainly not often predictable at a mathematical level. This is especially true for ECM, which is an inherently heterogeneous asymmetrical material in 3D space. This is not to say that these reactions and production processes will not happen – just that they are transformed from simple and predictable into 'too complex to predict with current technologies'. To coin a phrase to sum up this position: *no predictability, no controlled processing* (or the '**NoPre-NoCon**' principle if you need a mnemonic).

Predictable mass transport of reactants in the 3D space between compartments

In terms of the effects on mass transport, it is important to recall that:

(a) the ECM is the *bulk* material to be produced;
(b) the aim is to promote ECM deposition *between* the producing cells;
(c) good mimics of native ECM will be fibre-anisotropic, with locally distinct zones and layers.

This represents almost the opposite of the separating structure in the two-compartment system – that is, the cell membrane. Cell membrane is ultra-thin and biochemically constant in 3D space and in time – a paradigm of predictability. Any local in-homogeneities of extracellular culture fluid – for example between stirred and stagnant zones in Figure 8.7a – can be monitored, controlled and predicted.

The number one purpose of *tissue* bioreactors is to increase the mass and heterogeneity of ECM material, with properties which are completely different to cell membranes. So the drift (or should that be 'headlong dive'?) away from 'good bioreactor' conditions (predictability of nutrient/waste and product transport) is both inevitable and progressive. The 'progressive' point grows from our requirement that the tissue structure should get more and more complex (i.e. tissue-like).

The important take-home message here is that, on this analysis, the longer our bioreactors are cultured for, the less predictably they will work! Hence, suggestions that our bioreactors will give us the tissues we need if we could just run them longer and longer sound like ever-bolder strides out

into the valley of death (complete with whistling a jolly tune). Since we often do not understand the timing of cell-mediated ECM deposition, we cannot yet hope to formulate the new equations that might lead us forward safely. Once again, this is not to say that these processes cannot occur to make ECM and tissues in bioreactors – clearly, they do. It is just that two-compartment thinking cannot be used to *predict* the process (and **NoPre-NoCon**).

8.2.5 *The place of tissue bioreactors in tissue engineering logic: what happened to all the good analogies?*

We might expect that a good analogy of the fabrication (engineering type) systems considered here could be the manufacture and assembly of cars. Basically, the process sequence resembles our tissue fabrication process in that engine parts, seats, wheels and bodywork need to be made and gathered together, along with the workers and conveyor assembly line. Once the components are collected, the growth stage can begin, where cars are 'grown', refined and finally come off from the end of the production line (**). Cars would then be painted, polished and finished for shipping *directly* to the users. Alternatively, rough car shells *plus* a 'finishing kit' (polish, go-faster stripes, extra spot lamps, etc.**) could be shipped (i.e. *indirectly*) to a dealer network. In this case, the intermediate stage customizes and completes the car to the needs of the final user.

By analogy, the tissue engineering system might aim to make a fully finished, customized tissue graft in the bioreactor. Alternatively, the strategy might be to use the bioreactor to make a rough template in which the cells and matrix would mature and remodel after it being implanted to the patient.

Readers may now be feeling a little uncomfortable with this analogy. It is, in fact, not such a good match for the process we envisage, in particular at the two points marked (**). Specifically, cars do not themselves 'grow', of course. The inanimate object increases in size and complexity as parts are assembled by the workers. In addition, the 'finishing kit' supplied to the dealers might be a cute idea, but it

Text Box 8.3 'Bioreactor' caveats – the differences between bioreactor morphs

1. First of all, within the area of tissue engineering, many workers refer to bioreactors in the context of cell expansion systems. In practical terms, these are designed and operated quite differently from 3D tissue-generating bioreactors. Although cells in a natural repair site *appear* to do all things at once, (i) this is, in fact, an illusion, and (ii) 'control' is the real issue anyway. In fact, most cell types are hard enough to control when they are doing just one thing at a time (e.g. dividing or fabricating new tissue. When we aim to prepare as many cells as possible (for cell seeding or injectable cell therapies) the number one target is to optimize *proliferation* in a cell expansion bioreactor. Clearly, then, some of the points in this chapter do not apply to cell expansion bioreactors, where the number one aim is to persuade cells to fabricate as much 3D material as possible. Simplification is essential, so it is best that we concentrate on producing *either* cell expansion *or* matrix production systems. Indeed, it is a brave tissue engineer who aims to do both in a single system with our present knowledge base.

2. Much of the discussion here relates to *matrix-rich* tissue engineering, as opposed to *cell-rich* (such as engineering organs like liver or kidney, made up largely of dense cell aggregates). Again, these have been discussed elsewhere and raise somewhat different problems. Whatever the 3D volume available in your particular construct, it will be filled with both (a) living (cells) and (b) non-living elements (extra-cell material, hard or soft, and water) in some tissue-specific ratio. Where (a) above becomes a greater and greater percentage of total volume, energy/nutrient consumption rises dramatically. Where (b) increases, consumption falls in proportion (subject to metabolic activity). This simple balance dominates the difference between cell-and matrix-rich tissue engineering bioreactors.

3. Among the more traditional biochemical engineering types of bioreactor there are a bewildering range of approaches, applications and technologies designed to provide efficient, predictable operation over long periods. For example, there are:
   (a) mammalian cell bioreactors (sometimes transformed cells for reproducibility and enhanced activity);
   (b) plant cell systems;
   (c) bacterial cell systems;
   (d) fungal cell systems.

   There are also simpler, non-cellular bioreactors where whole cells have been replaced by specific biochemical elements required for the reactions carried out. At the other extreme, examples of bioreactor processes are in use with whole (admittedly small) organisms, nematode worms and plants.

   Inevitably, there are also many, many technologies developed for achieving the basics such as mixing, separation and recharging of the bioreactor contents. Some examples are given in Chaudhuri & Al-Rubeai (2005).

Reference:

Chaudhuri, J. B. & Al-Rubeai, M. (eds, 2005) *Bioreactors for Tissue Engineering. Principles, Design and Operation.* Springer, Dordrecht, The Netherlands.

---

hardly parallels the massive tissue-tissue integration and maturation which happens when an implant is sutured into the body. No, this analogy is flawed, as it relies on a human engineering assembly model where the products are non-living systems.

Perhaps a better analogy might be salmon farming. The first stage is the fabrication and collection of the component parts, fish-fry, food, hormone pellets, net pens, circulation pumps and salinity meters. We might envisage an assembly stage where the fish pens are towed out into the bay, the necessary pumps are bolted into position (**) and the cages filled with fish, etc., ready to farm. Then the fish farmer has the choice of whether to take out lots of small salmon for sale to wholesalers, for fattening and packaging for supermarkets, or to grow them to full size for direct supply to the restaurant table and customer.

The flaw in this analogy is a little harder to see, but a closer look shows that the assembly stage process is, in fact, *not* designed to assemble fish, but the fish farm itself. In fact, the part of this process we control and engineer is the assembly of pens and monitoring equipment, etc. The salmon largely assemble themselves, providing they are housed well out in the bay. So again, this is *not* a great analogy for a controlled process to generate fish. In this case, the bioreactor may more accurately be more identified as the bay where the fish pens are anchored.

The fact that we are having trouble generating good parallels with familiar engineering or farming processes may be an indication that we are missing an important element of what we expect to happen as a tissue is engineered. At the engineering-fabrication end of this spectrum (car assembly), the analogies are clearly rather thin and flawed. A possible miss-match here is that in human fabrication processes, we almost always add new materials or parts *to the surfaces* of a growing structure. We pretty well *never* design assembly processes where the new parts are inserted *into* the existing structures which were made at earlier stages of the process. This is so obvious that it is easy to miss, because it would be too disruptive, in the human assembly world, even to attempt to disturb the 'inside-out' sequence which we might call layer-by-layer or appositional assembly. After all, we (the human producers of the structure) are on the outside and we would like to stay that way after assembly is completed (a rare exception might have occurred in the final assembly stages of the Egyptian pyramids: this is called entombment).

Some readers will be thinking that large buildings are an example of humans constructing from the inside, but this is to miss the point. The 'outer' fabricated surface would extend to the inside of hollow structures. For example, plasterers come in sequence to put a series of smooth layers over the initial bricklayer's work. In turn, these are followed by the painters and decorators, who add ever more cosmetic 'final' layers. The actual building really only gets bigger when more masonry is added to the outer edges of what is already present. No one would suggest inserting an extra row of bricks every metre up the height of an *existing* outer wall to make the building taller.

Not least, assembling structures in this way would mean that all of the previous parts of the structure, laid down at earlier times, would be spatially disturbed, compressed, stretched or distorted, so they would themselves need to be modified continuously during the growth (see Chapter 7). But this is precisely what happens when cells fabricate (assemble) soft tissues during bio-growth. That is, while new parts of the tissue bulk are being added in one area, other areas are being 'remodelled' to accommodate the resulting shape-space changes. A key difference from the human standpoint is that the fabricator-cells *live*, full time, within and *surrounded by* the structures they fabricate. In fact, a much more accurate human-world analogy for tissue-growth would be the rather nebulous process of how we 'grow' our towns and cities. This is not a process dominated by how we construct houses, office blocks, bus stations and football stadia. Rather, it is the process by which we knock down and reshape old, existing districts, roads, business zones, etc. to accommodate the building of new additional city parts (Figure 8.8).

Figure 8.8 Picture of a city from the air (Boston-Cambridge, USA). Imagine that the city fathers want to insert a new Olympic rowing facility on either side of the bridge. This would narrow the river and require the demolition of housing and parts of MIT (red and green circles, respectively). These lost facilities could be moved out by a few kilometres, displacing old factories (yellow circle) to the edge of town (yellow arrow). Meanwhile, the bridge (now too long) is demolished and remodelled to be higher but shorter. This is urban remodelling.

Using the 'urban expansion-redevelopment' analogy, we can suddenly get a clearer view of what makes cell-based tissue fabrication so different from human factories and building sites. Cells are multi-micron-scale factories in their own right, permanently sat *inside* the fabric of their own production. The cells are the 'factories', producing tissue structures inside-out. We can now see why the automotive assembly analogy was flawed. In that case, workers always lay down a steel plate, coat it with protective plastic, then layers of paints and decorative stripes – *strictly in that sequence*. They would never go back at some mid-stage (e.g. after the first coats of paint) to add ribs or fluted shapes into the steel base-structure. The difficulty of imagining how this might be done only illustrates how ingrained our human-scale thinking is (e.g. injecting in more liquid steel in the last example is clearly silly). Equally, revisiting our house-building process, any builders who try to inject extra layers of plaster *behind* the decorative wallpaper would clearly have a short career (not to mention needing possible medical attention).

But if we consider our tissue assembly process to be more like that of city development, things get better. Now, we start to see that the process is really a *combination*, where addition of the car assembly plant and group of workers houses is only a the first part. Expansion goes hand in hand with remodelling of the surrounding buildings (to make space), addition of new road and rail links to bring in parts and workers, telephone lines and offices for the new district government (and tax officials!). In other words, there are *two* sections to what we are aiming to do, and only one of them (construction of new simple structures) can be based on the principles of human engineering and fabrication. The second part of natural growth (progressing all the time, hand-in-hand) is the reshaping of the previously fabricated structures to accommodate the new – i.e. *remodelling*.

This also has implications for our analogies at the tissue-farming end of this spectrum, in particular the difficulties with drawing out good analogies. The fault line here is obvious in retrospect, but its recognition profoundly affects tissue bioreactor

logic. It is the problem touched on earlier, of growing fish lips or pumpkin fruit *in isolation*. We never normally even try to farm or culture isolated parts (e.g. single tissues or organs) of our domesticated animal or plant crops. Rumour has it that there were early attempts to genetically engineer chickens with four legs, replacing the much lower meat quality wings. This was never likely to catch on, for reasons of ethics or taste, but it would still not have broken our rule, in that, however many limbs the bird has, whole living chickens would have carried out the tissue fabrication.

Frankly, if there was a realistic possibility of developing processes to successfully farm isolated tissues, we might expect to hear of work on culture processes to grow 100 per cent fillet steaks in huge, sterile vats. Perhaps an early plan might be to grow the most valued Kobe beef steak at anything up to $600 per kg. It is possible to question the validity of this example on the grounds that muscle tissue is complex in structure, must be highly vascular and so is too high a target even for its market value – so perhaps we should instead expect to see a beluga caviar farming bioreactor. Fish eggs, at least are relatively simple in structure and the market value is even greater, currently $7,000 to $10,000 per kg. This is around a third the value of gold ($\approx$$33/gm) but, as a non-durable, consumable item, it could be a marketing dream.

However, the elephant in this room is that farming-type fabrication depends on the inbuilt controls of a whole viable organism. In tissue engineering bioreactor logic, it is rarely argued that we can use the whole organism. Yet, without any real precedent, the tissue engineer's vision of 'growing' functional 3D tissues can start to look like an early alchemist's claims to gold production.

So, to conclude this section, our engineering analogies for tissue fabrication fall down because human fabrication systems are fundamentally different in scale and mechanism to those by which tissues grow naturally. This is aggravated by the realization that hopes of developing successful tissue bioreactors out of biochemical bioreactor (fermentor) technology may not be realistic. Worse still, the widespread assumption that tissue bioreactors could still work by relying on farming-like process controls looks increasingly rose-tinted. It must at least await serious progress in understanding stem cell and developmental biology, until it reaches the level where single tissues can be grown free of the parent organism.

But before the reader gives up in gloomy despair, it is worth using what we have learned, including the negatives and no-go areas, to help us plot a new 'bioreactor concept'. Key to this is the city planning analogy and the glimpse it provides of a two-part process:

(i) Expansion.
(ii) Remodelling.

In particular, we can see that that our engineering/fabrication skills may be sufficient for assembling simple tissue replicas, but this will not take us further because of our limited understanding of the natural tissue remodelling that needs to happen in parallel with tissue expansion. Therefore, our new concept is a *two*-part process in which the technologies are better fitted to our capabilities:

- The first part involves engineering and assembly of relatively simple tissue replicas or templates.
- The second part would be based on bioreactor-based growth processes, where this template is remodelled, expanded (grown) and provided with bio-complexity by its resident cells.

The problem now is in coming to terms with the concept that we are now *not* dealing with either cultivation or engineering, but *both*. Engineer the simple bulk, then cultivate this such that resident cells complete the second, growth and remodelling process.

Shock, horror, gasp – it's a *sequence*, not a choice.

8.3 Current strategies for tissue bioreactor process control: views of Christmas past and present

This is the point where we examine the stages reached by tissue bioreactor engineering at the present time, irrespective of the logic-analysis in previous sections. At its base, we are focused towards developing systems, chambers and associated processes which will successfully maintain '3D tissue cultures' for extended periods. The purpose of this extended maintenance is to persuade the 3D template or constructs both to mature in composition and architecture, and to grow in functionality, including size, strength or biofunction. For matrix-rich tissue applications, this most commonly requires the accumulation of large amounts of dense, organized ECM (extracellular matrix). For cell-rich tissues (typically organ engineering), there is a much reduced drive for mechanical support and a dense ECM material, but a more pressing need for differentiated cell sheets, blocks or tubes. These would commonly need an appropriate 3D organisation to produce, for example, ducts, tubules, filtration surfaces or vascular integration.

Despite the apparent diversity when viewed in terms of the target tissues, many tissue bioreactors and the associated systems developed so far have much in common, since their resident construct cells must be:

(a) kept alive and highly productive. This involves supplying all the raw materials they need, plus oxygen, and removal of wastes at (i) appropriately rapid rates, and (ii) throughout the 3D volume of the construct, minimizing gradients, except where such gradients are functionally useful (later). '*Construct Perfusion*'
(b) kept active, i.e. doing or making what we need them to do or/produce! This involves '*Cell-Control*'.
(c) kept free of infection from any form of exogenous micro-organism (i.e. sterile, despite all the other comings and goings of nutrient media, measuring probes and additional components). '*Sterility*'.

Many other (perhaps less fundamental) demands have been added in some cases, including maintenance/direction/reprogramming of cell phenotype, maintenance of stem cell de-differentiation, and generation of cell stratification or (vascular) micro-tube formation. However, satisfying these three general backbone requirements encompasses most current tissue bioreactor targets.

8.3.1 Bioreactor enabling factors

Construct perfusion

Control or optimization of nutrient and waste mass transport to/from suspended cells has been the central success of conventional fermentor-type bioreactor engineering. It has been one of the central targets to date, with the aim of predicting and then regulating overall bioreactor performance. It is, perhaps, possible to get an unclouded glimpse of how early we are on the *tissue* bioreactor learning curve by examining two of the more basic assumptions for their level of wobbliness:

1. Control of deep cell perfusion is a dangerously low target if we are serious about producing functional 3D tissue architecture. Functional perfusion is only likely to be an enabling factor, either preventing cell death or switching on (or at best speeding up) tissue production. While this is an important basic, as we have seen before, the *speed* of matrix production alone contributes very little to control of 3D spatial organisation.

2. So far, most attention has been on monitoring and controlling nutrient, waste or oxygen levels in the *external medium* in which the construct is bathed, largely because it is technically a simple matter to measure the external fluid. However, recent systems in which micro-monitor probes have been used to measure real-time levels at fixed depths in the tissue constructs have highlighted how dangerously over-simple this approach can be. As already discussed, resident cell *consumption*, rather than diffusion, is frequently the determining factor, but changes in location and activity of cell clusters deep within 3D constructs are not yet predictable. Added to the poor predictability of diffusion properties in the different planes of anisotropic tissues, it become clear that we cannot realistically hope to exert meaningful control over deep-cell perfusion when we only monitor the external culture

medium. Indeed, it is worse, of course, as it leads us to imagine that all is well below the surface, so there is no need for further work.

As we have discussed previously, the aim of controlling 3D architecture *or* composition of deep tissue cells or matrix-zones, based on information taken from the construct surface or culture fluid, is either a seriously modest strategy or a massively optimistic target. In biological circles, it is generally assumed that diffusion path length (surface to core) and material density are the factors that govern where and when damaging gradients and deficiencies occur.

However, it is commonly the high density of active cells (in pockets or multi-layers), rather than simple diffusion, which is the determining factor. In other words, it is cell consumption of nutrient/oxygen and production of wastes (discussed in Chapters 3 and 4) which dominates the formation of tissue gradients (actually, extracellular matrix is surprisingly nutrient-permeable). For example, it will be common for the surface one or two hundred microns of cells to be active enough to deplete the lower cell layers much earlier than would be expected by diffusion alone.

Clearly, monitoring the culture medium alone can presently give little indication and very little measurement of such effects. Until our understanding of the dynamics of 3D tissue structure and localized cell consumption are improved, it will be necessary to directly monitor deeper 3D construct layers.

Aside from *measuring* these gradients, current efforts at improving deep perfusion (reducing such gradients) are focused on biomimetic approaches such as incorporation of μ-channelling or blood capillary mimics, with or without cyclic mechanical loading to drive fluid movements. However, as we have learned from past TE strategies, such approaches remain inefficient, random stabs at the problem if they are not closely coupled with quantitative monitoring of actual deep perfusion. This means that we should add to our list of **good TE practice** the need to ensure that analysis of mass transport within the 3D constructs is an integral

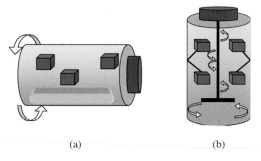

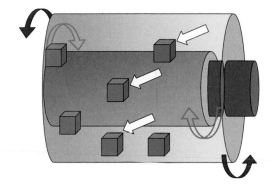

Figure 8.9 Diagrams showing the general principles of (a) roller bottle and (b) spinner flask culture systems. Constructs are suspended in media within the cylindrical bottles on a controlled speed roller mixer. Once they are rotating, an internal ridge helps generate fluid motion, which keeps the constructs moving and suspended. Spinner flasks (b) have an internal free-spinning magnetic bar, turned by a magnetic stirrer under the flask. This generates a gentle (sub-vortex) rotation of the media, around the constructs, held static within this flow on rods.

Figure 8.10 Rotating wall bioreactors. Highly simplified diagram illustrating the principle of action. 3D constructs (white arrows) under culture are maintained 'floating' and relatively statically in culture medium by the independent rotation of the inner (green arrows) and outer (red arrows) chamber walls. For a commercially available rotating wall bioreactor, see Figure 8.13.

part of biomaterials or cell biology initiatives to control perfusion. This is likely to become in future, one of the basic norms for 3D-bioreactor targets.

Modern works reviewing tissue bioreactor technology are available, describing recent approaches to the 'construct perfusion' question (see reading list). The earliest of these systems used existing cell culture technologies, which basically agitated or circulated the culture medium crudely around the growing constructs (Figure 8.9). The two basic forms for this were spinner flask and roller bottle cultures. The problem here is that mammalian cells are easily killed or damaged by fluid-shear. Unfortunately, these systems generate relatively high and, more importantly, uncontrolled fluid shear levels, especially around corners and angles of 3D constructs, resulting in local necrosis.

Other forms of fluid exchange can be used. These resemble examples found in conventional suspension bioreactors, where media is exchanged gradually and mixed at the same time by low flow pumps. However, mixing is most critical for the fluid layer directly adjacent to the 3D construct surface, where nutrient depletion and waste accumulation is most pronounced. This layer governs diffusion gradient formation into the construct but, in many

cell-seeded constructs, it also tends to impact on large numbers of cells. Consequently, attempts to stir this layer must be highly controlled if damaging fluid shear is to be avoided.

This problem was effectively solved by the development of rotating wall bioreactors (Figure 8.10). These comprise a culture fluid-filled chamber formed between two independently rotating walls. By setting the device walls to rotate differentially at suitable rates, adapted to the construct characteristics, it is possible to generate controlled gentle fluid movements, which keep the constructs suspended at the same height as the construct apparently falls in the opposite direction. In effect, the rate the construct falls is offset by the rotation, such that it maintains its relative position. This also has some appearances of 'culturing under microgravity' (though the accuracy of this idea is contested, it is less important in the present context). The key point here is that the damaging concentration gradients at the construct surfaces can be decreased without generating lethal shear. The result is that cell growth and activity rates in these chambers are widely reported to be excellent.

As we shall see later, other biomimetic approaches to control are now under development, such as the introduction of perfusion μ-channels and forced interstitial fluid movement. However, these involve tackling the problem at a new (less basic) level, where the construct architecture is designed and prefabricated as part of the process package rather than aiming to produce generic bioreactors that will grow a range of ill-defined tissue 'lumps'. A feature of these perfusion solutions is that they make the bioreactor more complex (within limits) but do not tackle the question of mechano-regulation of key cells within the constructs.

Attempts to control any given mechanical micro-environment on the 3D tissue cells have been reported, and this forms the subject of a later section. The key point here is that this normally involves some form of physical contact – clamping or restraint of the constructs–which inevitably increases the design complexity. Magnetically driven construct loading systems only partly dodge this 'direct contact' point, and they certainly introduce many new variables, such as distance from the magnetic source.

Interestingly, in the case of rotating wall bioreactors, direct physical contact with constructs would be pretty well incompatible with their operation, in effect defining the limits of the use of such bioreactors. But these, in any event, may be better described as minimal-mechanics (rather than microgravity) bioreactors – and this, paradoxically, is not presently a common design target.

Sterility (and scale-up)

Although sterility is included in our listing (albeit under 'enabling factors'), this is essentially a technical driver rather than a concept of tissue engineering (Text Box 8.4). Indeed, many of the basic requirements and solutions are pretty well known in advance. Consequently, despite the critical importance of this aspect and the time commonly expended on its design, it is already well understood from other, traditional disciplines and generally does not need special tissue engineering attention.

However, one key enabling factor has sterile operation at its core, and this is the topic of scale-up. The importance of scale-up is not so much that it is special to tissue engineering – more that tissue engineering is especially susceptible to its application (or lack of). The cross-disciplinary nature of tissue engineering makes its translation to practical applications (including the development of bioreactors) particularly vulnerable to late or poor scale-up design. Since culture systems are initially developed in biological or academic bioengineering labs, it has proved all too possible for them to be developed (too) far down the translation path, to industry or clinic, at the scale of a *cottage industry*.

Text Box 8.4 Sterile and aseptic

Sterile is a very specific term in cell biology. Its absolute significance is worth appreciating in full, particularly to those coming from outside biology. This is simple when we understand how different it is from the term *aseptic*. When we discuss sterility in microbiology, be certain that it means the *absolute* absence of living micro-organisms (i.e. bacteria, fungi, (or their microspores), mycoplasma, viruses or any other non-mammalian cell capable of division). If *any* contamination occurs in culture systems (including bioreactors), the contaminating organisms will eventually overgrow and take over the culture we want to survive, simply because they divide quicker and kill off mammalian cells. The key here is that there are *no* half measures: one bacterium or yeast effectively cancels 'sterility'.

Clearly, non-biologists may have more direct, personal experience of infections, and in such cases then maintaining aseptic conditions are enough. This means that infections are OK as long as they do not overwhelm the *organism*. This last word contains the key factor. The human bioreactor (i.e. your body) has an immune system – the tissue bioreactor does not. As immuno-competent organisms, we can carry substantial loads of 'exogenous bugs' with no bad effect. However, our cultures cannot fight even a single infective bug. Therefore, 'sterile' is the term needed here, *and it is an absolute*.

In other words, the original manual culture process developed in the research lab just grew larger.

'Growing larger', as a cottage industry production, is not the same as undergoing the translation into a scaled up *production* process. This has, in the past, become a fatal problem (notably for skin implant applications reaching the clinic) because regulatory authorities demand that 'the production process' must be rigidly constant once products go for testing. Where implants have gone to trial based on labour-intensive lab scale processing, it is that cottage production process (and *only* that) which the regulators have approved. Once this position is reached, the process cannot easily be redesigned or streamlined for scale-up without entirely re-starting the clinical testing, at major cost. This has resulted in making it near-impossible to generate economies of scale, as the lab scale bio-processing is fossilized into the system.

The ever-present 'sterility-driver' effectively represents a special, and particularly high-profile, aspect of this issue. Indeed, the sterility question can be particularly useful as a warning sign for the wider question of scale-up. Once the question of how to maintain sterility raises its head for any particular bioreactor design, it is probably time to start looking for collaborators in the field of production scale bio-processing. So, we can now identify a new tissue engineering rule for good practice. This proposes that as soon as a bioreactor process becomes interesting in the biologist's lab, and needs special sterilization treatments, it should also attract the involvement of engineering colleagues in order to introduce good scalable concepts.

8.3.2 Cell and architecture control

Control of cell function and tissue architecture during 3D bioreactor operation is the central *active* aim of any bioreactor process (perfusion and sterility being enabling functions). The most important modes of direct cell control tend to fall into two areas, based on the following:

(i)   Administration of cell-regulating biomolecules such as growth factors, hormones,

cytokines and gene sequences. Not surprisingly, these tend to be the focus of the cell and molecular biology tribes of tissue engineering.

(ii) Controls which use mechano-regulation. These tend to be more in the domain of the engineering and bio-engineering tribes.

Clearly, for integrated cell control, these groups will eventually need to work in closer conjunction.

Control using bio-molecular factors

Concepts and strategies for regulation of cell and culture activities by delivering growth factors, nucleotide sequences and other bioactive agents are essentially the same as those applied in general cell and molecular biotechnology. The selected growth factors, antibodies, gene transfection or antisense sequences (for example) are provided, at suitable times, to elicit cell responses previously described most commonly in 2D systems. Using combinations of these, the intended options available are almost endless.

The limiting factors here are practicality (bioreagent cost/availability and regulatory hurdles) plus knowledge of what they really do, and *how they do it*. Strategies for bio-molecular regulation, therefore swiftly leave the scope of tissue engineering bioreactors, except in one area – *delivery*. Delivery of such molecules to deeper (3D core) parts of our constructs may well be seriously affected by the construct properties, especially the structure of any ECM or support material. This comes about for one quite simple reason: the bio-molecular factors in question range from fairly large to absolutely huge.

Even smaller growth factors or hormones would be in the range of 8–30 KDa in molecular weight, or around 40–160 times larger than glucose. Larger factors and nucleotide sequences would be many times this, making their passage slow at best across even modest path-lengths of dense materials (see Chapter 5, Figure 5.3). Consequently, where such factors are delivered simply, via the culture medium (the most common approach) the material properties of the construct *at the time of delivery* (and not the bio-activities of the factor) will dominate where and how fast they work.

What is more, this confounding factor will change dramatically with increasing culture period in many constructs. The common micro-porous polymer scaffolds will allow relatively rapid and directionally homogeneous diffusion to deeper zones in the early stages, where cell and ECM densities are low. However, with increasing time in culture, this will change in a manner which is both spatio-temporally complex and unpredictable. Worryingly, it will be most pronounced in the most *successful** construct

*Note: An additional, highlighted entry in the tissue engineering bioreactor good practice rulebook is: beware any systems whose success have downstream failure built in.

and bioreactor systems. In other words, there are time-dependent consequences of the cell culture process itself on the rate and directions of mass transport in constructs, which (a) cannot be ignored past a few days of culture, and (b) will be *more severe* and have earlier onset for large proteins, including growth factors. This is why the phrase 'at the time of delivery' is highlighted above.

Immediately after first assembly, constructs are likely to be either micro-porous ($>50\,\mu m$ diameter) or nano-porous in their basic architectures (the latter often being the natural fibrous protein-based scaffolds, e.g. collagen and fibrin). Also, they may vary from highly anisotropic (typically fibrous) to largely random-pore in structure. These extremes will dictate the long-path diffusion rates and (critically) any favoured direction for protein movement when the molecular diameter is a limiting factor. Each 'scaffold' material will have specific properties in this area, and these are usually well documented in the literature. However, even where diffusion rates are rapid and multi-directional *in the early stages* (for example, through large pore isotropic materials), successful deposition of anisotropic ECM and dense cell layers must lead to slower and less predictable directions of macromolecular movement. Thus, long and successful bioreactor operation produces greater unpredictability.

In particular, this unpredictability will affect where and when bio-molecular agents act in the construct. Importantly, this represents a loss of *control*, not loss of action of these agents, as they will still affect the cells they do contact. Where such factors are delivered by addition to the external culture medium (the simplest and most common means), they will increasingly act on the surface cells of the construct and less on core or deeper layers. Where the construct surface is uneven, such as the presence of an incomplete cell layer or physical defects, clefts or channels, there will be zones that allow faster local access.

Experienced construct fabricators will instantly recognize that variations like these in the surface structure are the norm, rather than the exception. They vary between different regions on the same surface, between constructs and, most particularly, between different surfaces of the same construct, commonly as a result of support material fabrication. Such local zoning effects *could* be used to positive effect to generate tissue-like local structure; after all, native tissues are almost never symmetric and homogeneous. However, they are presently uncontrolled – and even unrecognized – variables, and so they further exaggerate the fall-off of process regulation. When the aim is to develop a bioprocess system for controlled fabrication, any strategy resulting in *loss of control* deserves close scrutiny (See tip bubble above, and then Chapter 5).

Control by mechanical conditioning

In recent years, the idea that tissue bioreactors (at least for the connective tissues) should incorporate mechanical cues has become increasingly familiar. Clearly, such a critical, yet imprecise, term as 'mechanical conditioning' deserves a detailed analysis in our understanding of route maps and strategies. Understanding the impact of external mechanical loading on the growth and function of mechanical (i.e. connective) tissues is both a basic need and a characteristic of tissue engineering bioreactors. For now, though, we shall confine the discussion to an analysis of the current forms of 'mechano-conditioning' bioreactors.

For the biological scientists it is important to emphasize that the basic, three-way divide of applied mechanical forces is between tensile, compressive

Text Box 8.5 Applying the force is one thing ...

Although it is not difficult to *apply* simple forces (tension, compression or shear, in one or more axes) to the target material or construct, what happens to them as they pass through the material is a completely different matter. The more deformable the material of the tissue construct, the more that our 'clean', definable *applied load* is converted into 'other forms' of local forces. We can call these the μ-load patterns, where tissue mechanics gives way to cytomechanics.

Also, in softer materials, more of these complex loads will be translated into deformation of the resident cells. In other words, a uniaxial tensile load applied to a soft,

extensible material will generate complex mixtures of compression, shear and tension at different points, even though the dominant overall load is tensile. What is more, the effect of *scale hierarchy* comes in here and these patterns of load are changed dramatically as we slip down a scale to the meso-scale of our cells. External loads applied through stiff materials are transmitted to cells in quite different ways, commonly with much less overall cell deformation, so there is little direct mechano-stimulation.

In other words, as discussed before, cells adherent to stiff substrates are more stress-shielded. The good news is that all of these effects have been understood since Newton's time. They are predictable and calculable – *if* we choose to predict and calculate them!

and shear forces. This allows us to divide tissue bioreactor types along the same basic lines, based on the *principal* class of loading which it is intended to apply (Text Box 8.5). For example, cartilage tissue engineers will generally design systems to apply principally compressive loads for bulk cartilage function with, in some more ambitious cases, shear-loads where the surface is considered. Vascular tissue engineers concentrate on fluid-shear forces for the inner lumen and (pulsed/cyclic) tensile loads on the outer wall. Tendon and ligament engineers tend to concentrate on uniaxial tensile loading. These are based on fairly simple concepts linked to the native properties of the *gross*** target tissue function, reaches *at maturity***.[19]

Examples of mechano-bioreactors are now fairly common and diverse. Figure 8.11 shows two basic types: the pulsing-flow format and one applying uniaxial tensile loads (either static or cyclical). Flow-type culture systems tend to focus on the production of hollow, tube-like tissues such as blood vessels, gut or uro-genital tract tissues. Compressive loading systems, for example in cartilage constructs, are commonly based on commercially available

compression test rigs (e.g. Instron-type), applying cyclic loads directly through a conventional incubator wall on to cell-biomaterial constructs in modified culture dishes (Figure 8.12). There is a huge range of possible variants for the means and pattern by which external loads can be applied through even these relatively simple systems. However, it is increasingly clear that next-generation approaches will benefit more from improved concepts of exactly what applied loads are doing at the *cell*** level and in *early*** developmental stage tissues.

The means by which motor power is applied to constructs and their resident cells has involved numerous mechanical loading regimes. These systems have generated much imagination, at least among the biological community. They are perhaps less interesting to the engineering community, where such choices are more everyday-quantitative than imaginative. The most common examples (though not in any order) would be:

- fluid flow (e.g. liquid driven from a peristaltic or syringe pump);
- gas pressure;
- stepper motor linked to screw-driven, lever or cam actuators;
- permanent-magnet impellers.

Each of these can be conveniently controlled (greater or lesser precision) via simple computer

[19]Note: Compare the ** paired terms above to derive our next bioreactor concept-rule, which is that, 'We need to look in the appropriate places when we look to identify mechanical cues directed at the control of *cell* function in *early* stage constructs'.

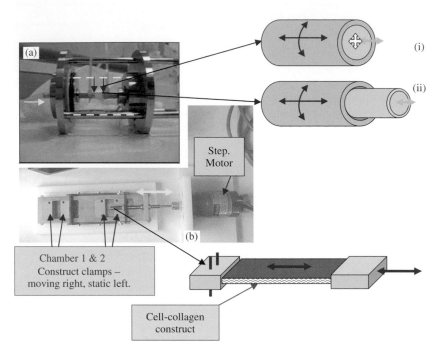

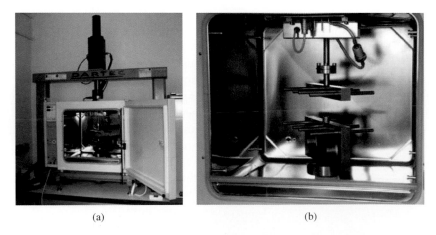

Figure 8.11 (a) Pulsing flow bioreactor comprising a flexible tube construct (red arrows) through which a heavily pulsed flow is driven (yellow arrow). The culture chamber surrounds the outer surface of the construct (glass chamber, white dotted lines, removed here). Set-up diagrams: (i) has the cylindrical construct *only* into the flow line, for lumen shear *and* wall tension (blue arrows); (ii) a flexible silicon lining inside the construct gives wall tension without shear. (b) Stepper motor (labelled 'Step. Motor') and screw-drive provide computer-controlled uniaxial cyclical tension (yellow arrow) onto constructs clamped within two tandem culture chambers. Set-up diagram shows one chamber with (pink) biomaterial-construct, clamped and anchored at one end and load to the opposite clamp.

Figure 8.12 Instron-based compression bioreactor for 'cartilage' culture. (a) General view showing the external 'Instron' computer-controlled loading frame operating through the incubator wall. (b) Internal detail of the incubator, showing the fittings to load the constructs within a multi-well culture plate.

systems to deliver most patterns of cyclic, incremental or static loading. As a general rule, those in which the motive power is applied most directly (e.g. motor through screw) allow the greatest precision. Non-contact magnetic drives, for example, are exquistily sensitive to the distance betweeen magnet and impeller (variable and difficult to control). Similarly, those where the applied loads are simplest (e.g. unaxial cylic tension) provide the most predictable cell-tissue loading patterns. However, these are made much more complex as they pass through the support materials. In contrast, radial pulse-cyclic loading, generated by peristaltic flow (Figure 8.11a) applied to soft materials, *starts* as dynamic and multi-axial even before it is made *really complex* by the radial, soft architecture of the tube that it acts on.

One of the most easily overlooked motor sources is that of the resident (adherent) cells themselves. Almost all living, adherent cells generate small contractile forces on their substrate which increases over hours after attachment. The fact that this is always present makes it an important factor. It is also important because cells seem to use externally induced alterations to this cell-substrate tension to monitor their mechanical surroundings (Chapter 7). However, in quantitative terms, the forces produced are small and relatively static. They usually have significant affects only on very soft, compliant substrate materials such as weak protein gels or synthetic hydrogels, and then only when cells can *attach* to the gel fibre network.

The concept that the growth of (connective) tissue constructs in culture 'can be controlled and enhanced by the external loads' is, therefore, now well established. This conflicts somewhat with earlier suggestions that micro-gravity culture favours tissue formation in some cases. Micro-gravity, in fact, represents the minimal possible theoretical level of mechanical loading, and it is as far from mimetic, at least of earthbound mechano-biology, as it seems possible to get.

This brings us to an interesting point of logical asymmetry. While it is clear that perfusion-led bioreactor design, such as rotating wall systems

Figure 8.13 NASA-inspired rotating wall bioreactor, said to mimic micro-gravity culture conditions. Constructs continuously 'fall' within slowly rotating culture chambers (one of four) mounted vertically on the blue base-plate.

(Figure 8.13), can make mechanical loading more difficult to engineer, the reverse is true for biomimetic mechano-bioreactors. Where the applied load is designed to generate physiological deformation (i.e. strain) in a repetitive (i.e. cyclical) pattern, these movements can be extremely effective in promoting directional movement of fluids within the substance of 3D constructs. Such *interstitial fluid* movement can be critical in breaking down diffusion and consumption gradients. Consequently, by coupling biomimetic loading regimes with biomimetic 'scaffold' anisotropy, it becomes possible to promote directional fluid movements, so improving deep zone perfusion while delivering cell-organizing mechanical loads. In this way, mechano-bioreactors can become self-circulating and self-perfusing.

To conclude this section, it is important to recap on a number of basic lessons from the development of tissue bioreactor technology (Text Box 8.6). It has clearly been common for biological specialists entering the bioreactor field to simplify the force patterns and perfusion controls needed for their

Text Box 8.6 Balloons in plastic mesh versus cells in collagen fibril network

In terms of tensile properties of constructs made from a number of different components, it is important to understand that there is a logical 'pecking order' for their relative contribution to overall mechanical properties. For example, let us consider a series of small balloons trapped inside a plastic fibre mesh. Most of the stiffness and the ultimate strength (to failure) of the *whole bulk* mixed material comes from the fibrous mesh, which is the stiffest, strongest element. If the fibres are more numerous in one plane than another (e.g. perpendicular to applied load in Figure 8.14a), then the construct will have anisotropic tensile properties (i.e. different in different planes, so *not* isotropic).

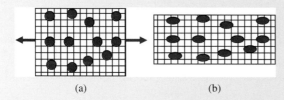

(a)   (b)

Figure 8.14 (a) Uniaxial tensile load. (b) After loading.

On the other hand, the more balloons we have, the less stiff and easier it is to deform (strain) the material.

Since we have only two (active) components, then the greater the proportion of the overall blend is occupied by balloons, the greater its compliance (i.e. strain or deformation produced by a load), as they *must* replace more of the stiffer (less deformable) plastic fibres. Again, if any plane of the construct has more fibres or more balloons than another, then the overall construct will have asymmetrical (anisotropic) mechanical properties. Shifting the balloon : fibre ratio, especially locally, will also tend to change that anisotropy.

In addition, even symmetrically applied external loads can deform the less stiff balloons (cells) asymmetrically, if the gross shape of the construct is asymmetrical (e.g. long and thin). The longer and thinner the construct (called by engineers the 'aspect ratio' – simply length : width), the more the balloons will deform in the long axis, as in Figure 8.14b. As a result, the deformation of soft particles (balloons or cells) will be anisotropic and far more complex than the loads applied. In native connective tissues, collagen fibres are the 'plastic mesh' and cells are the 'balloons'. The strongest, stiffest element is the fibrillar collagen, and groups of cells become the 'weakening' elements. Hence, the greater the cell density in an ECM, the less stiff it will be (in that plane).

Loading analysis of natural (tissue-like) materials, then, should be considered mainly in terms of its collagen fibre content *and its orientation* relative to the applied loads.

tissue applications, but inclusion of engineering specialities seem to help in the long term. There is clearly a set of basic enabling factors for bioreactor operation which must be addressed as housekeeping activities (essential but not enough to achieve our targets). These include sterility, capacity for scale-up and effective deep-zone perfusion. In addition, active factors are needed to control or locally restrict cell activity and architecture. Critically, these active factors *must* be capable of dynamic adaptation to time-dependent changing properties in the constructs.

Nowhere in extreme tissue engineering is the need for engineering and biological cooperation more important than in the hunt for successful tissue bioreactor systems. Separately they are naïve; linkage brings sophistication (Text Box 8.7).

To summarize some of the principles of tissue bioreactor good practice identified so far, we can construct a list:

(i) If it matters, measure it *directly*, at an appropriate scale and time.
(ii) Questions of bioreactor sterility are good markers of the need for scale-up.
(iii) Beware 3D bioreactors with a tendency to generate damaging conditions where they are successful.
(iv) Selection of control cues is scale- and stage-dependent (we work mainly at the 'cell' and 'early').
(v) 3D bioreactors designed by only one tribe are likely to be naïve in one major sector or another.

Text Box 8.7 Some clues and mnemonics for mechano-bioreactor function

(a) In Mechanical Loading and Connective Tissue Engineering: **Newton Rules!**
(b) In *Connective* tissues (almost) everything is connected, so forces *transmit*.
(c) Material randomness/anisotropy governs force transmission.

(d) *Force Vectors* interact with *Material Anisotropies* to produce **change & complexity**.
(e) **OOPS** (out-of-plane stimulation) and stress-shielding are key to cell mechanics.
(f) Bioreactor culture *duration* alters the transmission of both load and growth factors.

Look up Newton's Three Laws of Motion: how do they apply to the diagrams in Figure 8.14 (Text Box 8.6)?

8.4 Extreme tissue engineering solutions to the tissue bioreactor paradox: a view of Christmas future?

At the end of Section 8.2, we left things a on a bit of a cliffhanger. Where should we go next if basic tissue bioreactor strategies have so many flaws? In a nutshell, tissue bioreactor logic sits uncomfortably at the crossroads of two other strategies:

- Conventional engineering-type biochemical bioreactor technology.
- Advanced (whole organism) cultivation-type technologies.

This seems to have led to strategic targets for *tissue* bioreactors which seem unlikely to prove realistic.

So, to pick up the story again, our task from this position must be to plot out fresh forward strategies and identify the log-jam points that they inevitably contain. Options include:

1. Perhaps we can progress with the use of engineering-type fabrication processes by minimizing our dependance on controlled cell activity (i.e. sticking to what we are good at). This would involve applying strictly human-type fabrication processes to assemble simple, bulk material templates of the tissue. Any cell-dependent processing would be used as little and as late as possible. Put another way, this would involve translating how cells fabricate matrix and tissues into human factory-fabrication systems – more like Ford or Sony and less like salmon and pumpkin.

2. A second, almost 'looking-glass' approach exists for those who favour reliance on natural biological processing. This maintains that ways *really can* be found to cultivate tissues in 3D if we work at it. The first exteme version of this route involves actually growing constructs within whole organisms (i.e. in the patient). This is the so-called 'human bioreactor'.

3. The opposite extreme of bio-cultivation strategies proposes that we learn how to recapitulate how tissues developed *in utero*, then translate that into growth *in isolation*. In other words, the target here is to engineer developmental biology and (parts of) foetal growth.

Although the last of these sounds dramatically ambitious, it has its advocates. It also comes with an obvious, but essential, first requirement – a catalogue of very special cells. These cell types and cell stages would be comparable in their programming to the tissue producers in a developing embryo. Hence, the hunt is on to mimic embryonic stem cells, or to reprogram adult-derived cells and to determine the means to control them. The reader will recognize this as the logic-stream which underpins present stem cell, regenerative medicine resarch.

8.4.1 In vivo *versus* in vitro *tissue bioreactors: the new 'nature versus nurture' question?*

In traditional biology and bio-philosophy, there is an interminable argument about the proportion of the total adult organism function, which comes from inbuilt genetic information, and how much is dictated by the environment. Psychologists have refined this for human consumption into the nurture-nature debate for child development.

We now have a comparable discussion in tissue engineering. In this case, the 'nature' idea suggests that cells with the right programming, and a good 3D template, will grow into a functional tissue in the best and fastest manner, in an *in vivo* implantation site. In this logic, engineering 'control' is minimal and innate biological processing is maximized ('minimal bioreactor' input: Figure 8.15) The idea of directly engineering or fabricating tissues represents the opposite of this, or the 'nurture' track where the environmental factors are controlled and engineered (i.e. 'maximal bioreactor' input: Figure 8.15).

8.4.2 Do we need tissue bioreactors at all?

As we have seen, there is a widely held idea that it will be possible to use the natural growth process and to delegate control to the cells. This draws on the logic of tissue farming and embryo development. In effect, the concept of '*in vivo* bioreactors' suggests that even less basic knowledge or biological control is needed if we can combine innate tissue-building behaviour of our seeded cells with the support provided by the implant site itself. In effect, providing the human recipient (bioreactor) of the engineered implant with three meals a day and a warm room could be a more efficient growth machine than existing costly and imperfect bioreactors.

Indeed, conditions for growth and repair in normal adults are extremely favourable for bulk tissue formation, although, admittedly, this can produce scar tissues and is far less effective in the sick and or aging, who are mostly the patients. In this respect then, *in vivo* bioreactors, where they work, will have attractive growth rates, perfusion and matrix deposition rates. Sadly, where they do not produce suitable tissues (see Section 8.1), there is little to be done, because the tissue 'engineer' can exert little real process control.

Nursery implant sites

The concept of an implant nursery site is a version of the *in vivo* bioreactor, though, in this case the surgical nursery site itself is designed to allow a partially formed tissue template to mature and develop function. This has been suggested and tried, particularly where strong vascular in-growth is important (e.g. in bone). The nursery site can be quite remote and very different to where the implant will ultimately be used. This allows for the selection of sites

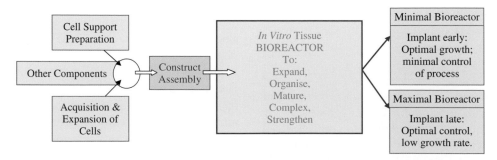

Figure 8.15 Diagram summarizing a standard process sequence assumed to be needed to fabricate tissue constructs. The left-hand cluster of boxes represents the acquisition and preparation of the basic components. These various component elements are then assembled into the initial construct (cells seeded, growth factors/drug depot inserted, etc. – small central box). The main box represents the 3D tissue bioreactor operation stage, which can be brief, allowing minimal cell integration/attachment prior to implantation ('minimal bioreactor' effort) or long, to allow the resident cells to fabricate tissue bulk ('maximal bioreactor' effort). Versions of this can range between maximum and minimum bioreactor effort. Maximal effort involves retaining control of the process conditions, but at high operating cost and slow construct growth rates. Alternatively, tight process control can be given up early in exchange for economy and speed.

which favour good tissue growth with less chance of damage to fragile early tissue templates.

For example, a template construct might first be implanted into a highly vascular tissue bed of choice (e.g. a large skeletal muscle), rather than the ultimate functional site, say in a bone. At a later stage, the robust, mature construct would be moved (with its vascular supply) in a second surgical operation. This option adds something to the speed of construct growth and integration, but these factors must be balanced against the cost, pain and risk of extra surgery on the patient.

The balance between in vivo and in vitro bioreactor (early versus late implantation)

Aside from the patients' view on extra surgery, the problem with *in vivo* bioreactors is that they largely give away operator control and so become skill-based, one-off surgical events rather than controlled bio-engineering processes (Figure 8.16). There can be little process control and systematic, predictable tailoring to the needs of the patient (age, disease status, gender, etc.). The task of the tissue engineer in such early implant processes is to fabricate initial templates which carry with them inbuilt controls and cues that direct local repair, with minimal scar tissue formation. These need to operate long after the construct is implanted (i.e. prolonged bio-control processes need to be inbuilt during 'construct assembly', to operate long after

the implantation, when it is unlikely that external controls can be effectively applied – Figures 8.15 and 8.16). This key tension is illustrated in Figure 8.15.

The early implantation of simple stage constructs has plenty of attractions, but it gives away control to the very vigorous *in vivo* repair processes. Retaining these controls with a prolonged *in vitro* bioreactor process offers the possibility of eventually getting the tissue quality and function that we need. However, the longer the *in vitro* stage, the more technically challenging are the control processes, and the greater the cost, complexity and time taken to achieve any function.

Currently, conventional 3D tissue bioreactors require weeks or months of culture time, and even then construct function is frequently poor or well below that of mature tissue. The compromise is to implant as soon as possible *after* we have cultured a basically functional 3D tissue. This is easy to say, but difficult to achieve. The question, then, is: 'What is the minimal culture period needed to get a useable graft template?' We can be sure that the answer will not be satisfactory until it is reduced to hours, rather than days or months.

The answer to our question here – 'Can we replace *in vitro* bioreactors with early implantation to an *in vivo* bioreactor?' – would seem to be 'no' in most cases so far tried. However, the question itself is almost certainly faulty, as we probably cannot

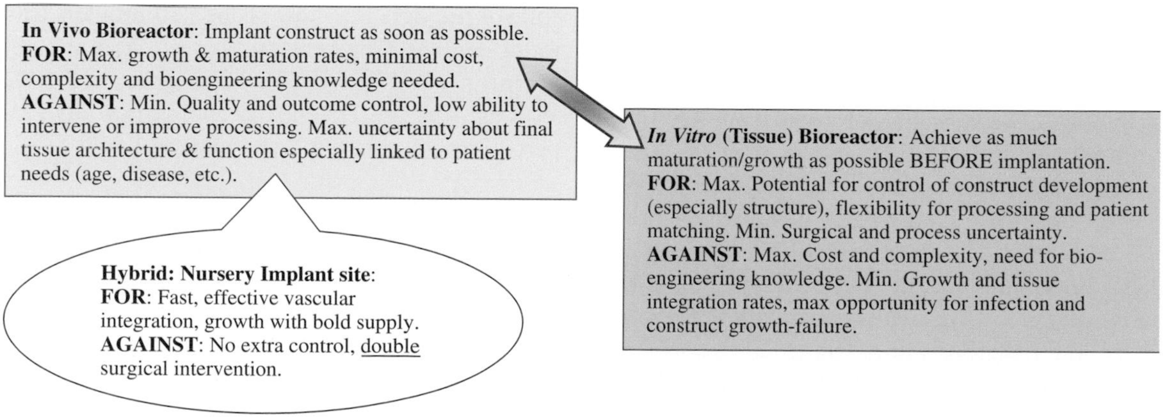

In Vivo Bioreactor: Implant construct as soon as possible. **FOR**: Max. growth & maturation rates, minimal cost, complexity and bioengineering knowledge needed. **AGAINST**: Min. Quality and outcome control, low ability to intervene or improve processing. Max. uncertainty about final tissue architecture & function especially linked to patient needs (age, disease, etc.).

In Vitro **(Tissue) Bioreactor**: Achieve as much maturation/growth as possible BEFORE implantation. **FOR**: Max. Potential for control of construct development (especially structure), flexibility for processing and patient matching. Min. Surgical and process uncertainty. **AGAINST**: Max. Cost and complexity, need for bio-engineering knowledge. Min. Growth and tissue integration rates, max opportunity for infection and construct growth-failure.

Hybrid: Nursery Implant site: **FOR**: Fast, effective vascular integration, growth with bold supply. **AGAINST**: No extra control, double surgical intervention.

Figure 8.16 *In vitro* versus *in vivo* bioreactors.

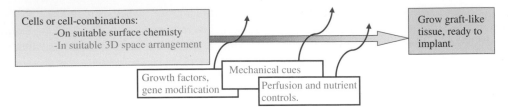

Figure 8.17 Overall summary diagram of bioreactor logic.

eliminate either form of bioreactor completely. In reality, we are discussing the timing where constructs are shifted from one state to the other (i.e. the length of the arrow in Figure 8.16).

This is where the fudge emerges. Each tissue and surgical procedure will require a different degree of pre-implantation 'control'. Where functional architecture can be generated quickly, constructs can be implanted shortly after construct assembly. Where greater levels of biomimetic complexity are needed, the process is forced towards a longer and longer *in vitro* bioreactor stage. In these cases, the technical challenge is substantial, as the cost and failure rates of existing systems will be impossible to carry into routine mass clinical use.

We need radical new concepts to reduce the time-dependency of current bioreactors.

How to maximize the cell-based contribution to tissue production

From the previous section, the question arises of: 'How could we develop and improve the innate ability of cultivated cells to produce the 3D structures we need?' In effect, this asks how far we can go *in practice* using cell-dependent biological cultivation technology. The question falls squarely in the tribal homeland of the cell biologist. It is at the heart of the most familiar of bioreactor philosophies but also the area least likely to give definitive quantifiable answers. It is based firmly on the assumption that, if we acquire just the right cell type or combination and give them cues they can follow, they will 'make' the tissue we want. The trouble is, as with the Hypocrites caveat, we shall never know if it is possible until we have done it.

Figure 8.17 illustrates one way this can be summarized. The current focus is on selection and pre-conditioning of the all-important cells in the construct. As we have considered earlier, this will involve the isolation and purification of cells, with promotion of differentiation towards defined cell types. Triggering cues here, as we have discussed already, may be nutrient, growth factor or mechanical in nature. In previous eras of tissue engineering, the focus has been on finding key growth factors or perfusion levels.

Finally, to return to our analogy of pumpkin cultivation, this strategy is like providing suitable nutrients at one stage for leaf production and then, later, shifting this to another to promote flower formation. After pollination (mechanical stimulation), fruit formation is triggered. The principle of this approach is consistently to use bio-technology stimuli to trigger each successive 'next *but natural*' stage within the bioreactor environment.

After each stage trigger, the natural growth process is left to take care of the complexities of completing the stage detail. We do not need to control the detail of flow shape or pumpkin fruit water content. The process triggers onset and the organism sorts out the detail (Figure 8.18). We shall use this process analogy to link into the next part – 'Process analyses' – in Chapter 9.

8.5 Overall summary – how can bioreactors help us in the future?

The problem and background concepts analyzed in this chapter are probably the trickiest and most intractable that we have to tackle. Not least, this is

Figure 8.18 Pumpkin bioreactor analogy: bio-artificial triggering of native sequence of cultivation stages. Cultivation conditions are maintained throughout with staged input of triggering cues, provided as the process controls. Between the process control/triggering stages, the biological system (i.e. the pumpkin plant) controls itself and its growth process, with the support of the basal bioreactor environment (culture conditions: perfusion, mixing, temperature etc.).

the focus point at which a number of theories of how tissues can be built meet with the cold water of what is presently practical. In the past, the big question has been 'How can bioreactors help us engineer tissues'? In fact, it is possible that we should flip this question[20] and ask: 'How can the field derive functional tissue bioreactor systems?' as this, in fact, is *the* core problem. In other words, we are not going to be able to engineer tissues until we can conceive logically of *how* to do it. This idea suggests that we may have even further to travel along the learning curve than we had imagined.

The strategy here has been to break the issues down into digestible lumps (questions of dynamics and monitoring are saved for the next chapter). In the first instance, we identified the existence of discomfort among the ranks of at least one of the tissue engineering tribes, with the assumptions of engineering tissues in 3D culture. This we could expand on to identify the presence of two competing but fundamentally different concepts (engineering and cultivation) of how to make 'things'. Both are applied to tissues, yet each has its own enormous logic flaw which would take a great deal of work to get past.

This became very clear when we tried to draw out parallels or analogies of making tissues based on how

either (i) engineering or (ii) cultivation technologies work at present. However, the building of analogies did lead us to a new concept, based on how most tissues really do grow. This is the idea that we should compare engineering tissues with how we achieve urban redevelopment, rather than growing or fabricating their components. This comes in a two part sequence; build and remodel.

At present, there are probably three threads of bioreactor thinking, which are freely intermixed:

(a) Minimal *in vitro* bioreactor use; implant cell-seeded support materials to mature/remodel *in vivo*.
(b) Maximize the culture and cell-based bioreactor role (cell-dominant, farming type).
(c) Maximize the engineering and fabrication contribution to the technology (i.e. minimize cell-dependence).

In effect, these three threads curently wrestle with the problem of 'when to tranfer', from (i) *in vitro* to (ii) *in vivo*. The idea is that if we knew where best to put that transition point we could design suitable 3D bioreactro culture systems to service it. **Early implantation** equals simple, inexpensive bioreactors; **late implantation** means we need increasingly complex sytems and processes. However, this point merges with our creeping suspicion that instead of either one or the other approach being better, we might really need *both*, within a *two-stage* process (ie. the 'build and remodel' urban re-development

---

[20]The origin of some of the simplest yet strongest of insights can lie in an inversion of a basic question, as in: 'Ask not what your country can do for you...' (J.F. Kennedy).

analogy). In fact, the early/late debate may be off-topic and not helping at all.

So, what if tissue bioreactors are best thought of as two-stage processes? Suddenly, there is a glimpse of a rational strategy. The template 'building' function is suited to engineering approaches, while growth and maturation (bio-functional remodelling) is best carried out by cultivated cells. Only then, as a third step, would the construct need to be implanted, for tissue integration. This now gives us two transition points, so – *a sequence* ! It implies the need for triggering changes leading up to each transition, which must be monitored and timed. In other words, we have derived an outline plan for a dynamic *process* to make tissues. This has to be progress over an all-or-nothing structureless tussle between imperfect 'alternatives'.

In the next chapter, we consider more fully the demands and opportunities of the fourth bioreactor dimension: time, sequence and process dynamics.

Further reading

1. Chaudhuri, J. B. & Al-Rubeai, M. (eds) (2005). *Bioreactors for Tissue Engineering. Principles, Design and Operation*. Springer, Dordrecht, Netherlands. [Comprehensive, modern textbook on tissue (3D) bioreactor theory and practice.]
2. Chisti, Y. (2001). Hydrodynamic damage to animal cells. *Critical Reviews In Biotechnology* **21**, 67–110. [The stirring and shear-damage question.]
3. Ko, H. C. H., Milthorpe, B. K. & McFarland, C. D. (2007). Engineering thick tissues: the vascularisation problem. *European Cells and Materials* **14**, 1–19. [The perfusion question.]
4. Cheema, U., Nazhat, S. N., Alp, B., Foroughi, F., Anandagoda, N., Mudera, V. & Brown, R. A. (2007). Fabricating tissues: analysis of farming versus engineering strategies. *Biotechnology and Bioprocess Engineering* **12**, 9–14. [The question of persuading cells to deposit enough collagen material to be useful.]
5. Freed, L. E. & Vunjak-Novakovic, G. (2002). Spaceflight bioreactor studies of cells and tissues. *Advances in Space Biology and Medicine* **8**, 177–195. [Coolest application of bioreactor technology: inspiring but questionable on economic grounds.]
6. Abousleiman, R. I. & Sikavitsas, V. I. (2006). Bioreactors for tissues of the musculoskeletal system. *Advances In Experimental Medicine and Biology* **585**, 243–259.
7. Chen, H. C. & Hu, Y. C. (2006). Bioreactors for tissue engineering. *Biotechnology Letters* **28**, 1415–1423. [References 6 & 7: Traditional review, including descriptions of a range of forms of (soft tissue) bioreactor.]
8. Naing, M. W. & Williams, D. J. (2011). Three-dimensional culture and bioreactors for cellular therapies (review). *Cytotherapy* **13**, 391–399. [Critical review, especially focused on uses for expanding cells for therapies.]
9. Carpentier, B., Layrolle, P. & Legallais, C. (2011). Bioreactors for bone tissue engineering (review). *International Journal of Artificial Organs* **34**, 259–270. [One of a selection of recent reviews of 'bone bioreactors', all from different perspectives and/or tribes. Clearly, bone is a popular target.]
10. El Haj, A. J. & Cartmell, S. H. (2010). Bioreactors for bone tissue engineering (review). *Proceedings of The Institution Of Mechanical Engineers. Part H, Journal Of Engineering In Medicine* **224**, 1523–1532.
11. Yeatts, A. B. & Fisher, J. P. (2011). Bone tissue engineering bioreactors: dynamic culture and the influence of shear stress (review). *Bone* **48** 171–181.
12. McCoy, R. J. & O'Brien, F. J. (2010). Influence of shear stress in perfusion bioreactor cultures for the development of three-dimensional bone tissue constructs: a review. *Tissue Engineering Part B, Reviews* **16**, 587–601.
[References 8–11: Selection of recent reviews of 'bone bioreactors' from different perspectives and/or tribes. Clearly bone is a popular target.]

The Bio-Boeing-with feathers and flapping (see Figure 9.14)

9 Towards 4D Fabrication: Time, Monitoring, Function and Process Dynamics

The last chapter ended on the somewhat problematical issue of the 'great bioreactor project'. In this chapter, we shall examine why there are, in fact, plenty of reasons to be cheerful – even optimistic – about the future for extreme tissue bioreactor technologies. However, it is clear that 'next generation concepts' will be essential for *controlled, stepwise fabrication* processes, including bioreactor culture stages. In effect, we would hope that multi-step controlled processing will replace the current, less effective, one-step (neo-agricultural) processes discussed in Chapter 8[21].

[21] At the risk of over-simplification, we are entering the era where we re-introduce the engineering into tissue engineering. This can take many forms, but the Trojan Horse approach here will use the theme of 'process dynamics and monitoring'.

Extreme Tissue Engineering: Concepts and Strategies for Tissue Fabrication, First Edition. Robert A. Brown.
© 2013 John Wiley & Sons, Ltd. Published 2013 by John Wiley & Sons, Ltd.

The first stop in this guest duet of 'dynamics and monitoring' will be process dynamics. Put simply, this refers to the changes that occur to our tissue construct as time progresses and (hopefully) as the process generates the structures we require. In fact, we are all pretty familiar with the *basics* of process dynamics. We normally think of these, for example, as increases *over time* in:

(a) resident cell numbers or differentiation state;
(b) changes in mechanical strength and stiffness;
(c) overall physical dimensions; or
(d) extracellular matrix material complexity and stability.

Respectively, we would see these expressed as changes in this-or-that unit of:

(a) live (viable) cell number;
(b) break-stress/stiffness modulus;
(c) μm of wall thickness/mm diameter; or
(d) protein concentration/cross-link density.

But in *all* cases, this should be expressed as 'per unit of *time*'.

Where our processing is rapid, ultra-rapid or just optimistic, this will be per hour, per minute or per second. Most likely, the processes measured over these short time scales would be non-biological fabrication and assembly. Processes which are monitored over periods of days, weeks or months are more likely to be biological, cell and culture-based. Over those time periods, it is increasingly unlikely that processes will be economically viable, except for specialist, high-value applications.

In other words, in this chapter, we are allowing *time* to become our master (as opposed to 3D structure or tissue bulk and composition). We have, then, reached the fourth dimension of extreme tissue engineering.

9.1 Controlling the dynamics of what we make: what *can* we control?

As we have seen, the dynamics of processes which are *predominantly biological*, such as cell culture, tissue bioreactor operation and cattle farming, are different from those in *engineering processing*. In particular, the level and type of controls in biology-based culture processes normally have a lighter touch. They tend to be lower resolution, operating with less detailed control, often using basic function outcomes, relatively wide tolerance ranges or qualitative measures (Text Box 9.1). The more remote

---

Text Box 9.1 How we make 'things' changes with the way we measure the making process (Figure 9.1)

Examples of how we measure the making process include:

- cell density would be monitored by actual counting of nuclei;
- break strength would be measured by physical clamping and stretching;
- meat quality (Figure 9.1) would be measured by weighing the meat : fat : gristle ratio.

In contrast, if the latter example were possible using an engineering process, we might expect to make more indirect, extrapolated measurements. For instance, cell density would be derived from real-time monitoring of CO_2 production, tensile properties from optical scanning of fibre content and meat quality deduced from an ultrasound scan of the cow's thigh. However, for these to 'work', we need to have such a high level of confidence in the process detail that it becomes possible to extrapolate *reliably* from remote (implicit rather than explicit) markers of what is happening.

By analogy, a school teacher taking the register *could* just count the empty desks or the coats in the cloakroom, or weigh the used milk cartons, if the children all behaved and dressed predictably. Needless to say, though, teaching children is a highly bio-social process, and wise teachers will go as far as positive facial recognition to exclude extremes due to truancy, kidnap or sibling-substitution. The result of this is that current monitoring tends to be occasional-slow-direct, rather than real-time and rapid using implicit or indirect markers. 'Indirect' monitoring is simply too unreliable for current, soft-control systems, where we do not know in sufficient detail just what is going on.

these measured parameters are from the final function, the more likely it is that they will correlate only weakly with that final function we want. As a result, there is a pressure to make the acceptable limits of these control tolerances ever wider – in other words we accept greater variability. In the case of free-range Limousin cattle, grazing on a mix of sunny and shady, flat and sloping pastures, we can see that measuring their tail diameter or leg length might be quick and easy to do, but it will give us only the roughest estimate of muscle : fat ratio or rump steak quality.

So, these *direct explicit*, end-function measures are characteristic of biological-based cultivation/ agricultural processes. In contrast, those controls which are realistically available tend to be *indirect, implicit* and almost 'passive' in nature. This seems reasonable and normal, as so many aspects of bio logical cultivation are complex and dependent on the cells/tissues or organism involved. The processes being measured, after all, are subject to biological variance and uncertainty – indeed, this can be their defining feature. For example, where and when bio-engineering and biotechnology processes develop 'engineering-like' forms of control, they cease to look like biological-cultivation and they start to resemble engineering-type fabrication and assembly processes, but using biological components.

For example, surgical implants can be made of (non-cellular) natural collagen protein sponges. These are made from natural components, but are purified, refined, structured and assembled using closely definable engineering processes – they are not grown. However, it is often not cost-effective to apply engineering-type levels of process control to biological cultivation processes. These light-touch control bio-cultivation systems commonly work quite well enough for what is needed.

To illustrate this, we can build analogies with the obvious dynamics that we find in the process of cattle rearing (Figure 9.1). Most of the basic, detailed controls and feedback processes for growing a Limousin cow to maturity (beef status) are inherent in the breed (genetics) and its habits/tolerances. This is a key part of the breed (gene content) that is built into the cattle. The

Figure 9.1 Limousin cattle grow themselves in a nutrient-filled bioreactor (i.e. a field), through a dynamic protein production process. The process engineer, or farmer in this case, monitors and adjusts the process in gross terms (food and shelter in the snow, antibiotics for infections and birth assistance) and generally balances the supply-removal of start material and end products.

farming process has developed to make the most of these and optimize how they operate *as a whole*. Farmers might monitor rates of growth by weight or physical dimension, muscle to bone ratios, rate of calving, bacterial load. In response, they can make adjustments (sometimes delicate) to the process to modify how these measures change. They might administer antibiotics, provide extra feed or supplements in winter, fend off parasites or kill off any toxic plants growing in the field.

However, it is important not to confuse peripheral, enabling controls with those which are primal and fundamental. Note that the cow grows on its own. This is hopefully optimal, converting grass into muscle proteins at a rate and in a shape (meat-cut) which is inherent to the breed. More particularly, it remains a Limousin cow in shape, composition and growth characteristics. These are the parameters that the cow controls through its genome. The farmer does not need, in fact cannot, do much to change these. Much as the beef farmer may wish it, the cow will only *ever* produce two hind legs and one rump. There is no room (and critically, no process-programme) for an extra set of prime rump muscles.

The point here is that for bio-production systems, from cell culture to farming, we generally do not even contemplate the detailed process controls which would be essential in a standard engineering processes. System complexity, and in many cases its inbuilt biological efficiency, make this impossible and/or unnecessary. The farmer never wakes up with the idea that he will change his process and produce Limousin cattle with 50 per cent more muscle in the rump region (though he may *dream* of five-legged cows). Cell culture labs might aspire to higher level of detailed process control but, unless they have the resources for decades of top level research into basic cell biology, their aims still tend to focus on optimizing what the biology offers (e.g. mineral content of the local grass species). They would rarely/never consider meddling with cell-division rates in the bone growth plates, average transcription rates or endoplasmic reticulum protein output in a major muscle block. Fiddling with the nuts-and-bolts of biological processes is simply not an option.

The contrast with the directors and managers of Toyota cars or Nokia telephones could not be greater. True, they are still limited by the laws of physics and economics, but these are basically very well understood (OK, except for the economics). Also, they still occupy much of their time 'optimizing' process dynamics to make sure the production lines do not run short of control pedals or batteries. But (and this is the *big* 'but'!) Toyota and Nokia have their names on the products; they design and fabricate all the parts in their products, from the most basic components to the box the product is delivered in (i.e. they work bottom-up).

If there is suddenly a need to make a cheaper, lighter mobile phone (Figure 9.2), a miniature computer-communicator or a pink electric car, it is in their power to produce it, so long as the bottom-up knowledge is in place.

Hence, the basic operating concepts of these two process types are poles apart. When we fabricate engineered devices, we *expect* to get right down to the minute details and to control every aspect of production – because we can. In contrast, when we grow biological products or devices, we *expect not* to have to control much of the nitty-gritty detail. The 'bio-' part (plus 500 million years of evolution) does it all for us.

As we have seen, this is a pefectly legitimate, reasonable approach when dealing with biological growth and *cultivation* systems, with their inbuilt bio-controls and feedbacks. However (and this is a big 'however' which some readers may have already spotted lumbering over the extreme tissue engineering horizon), this returns us to the key bioreactor fault line from the last chapter. – namely, the realization that we are not culturing/growing a complete *bio-system* in our bioreactors. The problem is that *isolated* cells are not playing with a full deck of cards. They cannot have the full system

(a)   (b)   (c)

Figure 9.2 Nokia have processes for fabricating mobile phones. The big difference, though, between this and cattle production, is that if Nokia wants to change that process to make a different phone, they can completely redesign the phone *and* the process, with intimate control of the detail. In this example, it shifts from one mobile format, a heavy, chunky structure (a) to a lightweight web phone (b), by altering (c) some basic parts.

of controls and information needed for building spatially defined 3D tissues.

And why ever *should* we expect a group of isolated cells to 'know' how, and in what shape, to make a knee joint? Real knees, after all, are made as an integral part of a single, coordinated, *time-based* process for making a whole leg, attached to a hip, attached to a spine, etc. as per the song. This means that strategies which leave process timing and dynamics to be decided later deserve a closer reality check. After all, leaving key steps till later, as a *refinement stage*, is only reasonable where we can be pretty certain that the process will get there, and will work at all.

A second danger sign lurks within the last chapter (under the heading 'Sterility (and scale-up)' – see Section 8.3.1). As a basic rule of tissue engineering, *it is never too soon to consider scale-up*. This rule is written into the obituaries of many biotech companies from the 1990s. It must argue very strongly that leaving process analysis as a downstream task (i.e. for someone else to do) is a luxury – an indulgence, even – that we should be very, very cautious of allowing ourselves to take.

Consequently, to conclude this section, it is still common in tissue enginering circles to tacitly consider that the dynamic and the time course of tissue fabrication are *matters for the future*. They can be left as 'someone else's problems' (the SEP principle[22]). After all, we are told, when it is difficult to get *anything* functional to grow, surely it is a reasonable strategy to first get *something* working, *then* to tackle the question of how long the process takes. Hopefully, the analysis here has highlighted the self-perpetuating danger built into this approach (Text Box 9.2).

Our analysis has gradually shifted from cattle rearing (indirect) to telephone manufacture (direct) – or from bioreactor cultures to tissue assembly processes. In each case, however, we have seen that the strategic landscape is *completely different* as we travel along the two tracks. We are forced to analyze how achievable our process controls will (ever) be and how we can objectively

---

[22] Source note: Arguably the first, but certainly the most famous description of the concept that 'people do not take any notice of a problem that can be assigned to someone else' came from the late Douglas Adams in his *Hitch Hiker's Guide to the Galaxy* series (Heinemann Press, London). *This proposes that someone else's problem (SEPs) are effectively invisible.*

Text Box 9.3 Direct and indirect tissue engineering

As we have seen previously, it is possible to distinguish between two general types of production process – *direct* and *indirect* – represented here by cultivation- and industrial-type fabrication processes. Clearly, the same distinction can be made between direct and indirect engineering of tissues. We can generate a working definition for direct (DTE) and indirect (ITE) tissue engineering (all good concepts need an acronym). This is based on the subjective judgment of which processes are largely under direct human control, against those where production is essentially under the control of non-human living systems (isolated cells, tissues or whole organisms). This is an imperfect division, as processes such as fermentation are sufficiently closely controlled to be thought of as direct, even though they are based on yeast synthetic processes.

In general, ITE systems work at developing better and better biological and cell-culture controls. DTE is where we devise fabrication and assembly processes for 3D tissue support materials which rely predominantly on human control processes, independent of cell activity.

However, in the special case of tissue engineering, there is a second characteristic which makes the processes doubly indirect. The use of temporary, normally synthetic 'scaffolds' to support the cells adds a further 'indirect' element, as the fabricated element is an extra pre-stage not found in nature. This is similar to the indirectness of fabricating a bronze statuette by a process of mould preparation (e.g. using a lost-wax technology), as opposed to directly cutting the shapes out of bronze. The corresponding ITE equivalent of this is to make an intermediate non-native cell support material (a 'scaffold' in the shape of the tissue), to be eventually replaced under cell action. In contrast, direct tissue engineering would aim to assemble the base parts of the tissue under human control with minimal cell involvement, at least in the first place.

This concept is helpful where we need to critically analyze our favourite process and bioreactor strategies.

---

measure (monitor) how fast we are moving towards our target.* The good news is that making this

> *If I want to estimate if it will be possible for me to cycle from Budapest to Amsterdam by next Tuesday, I can either (i) cycle all the way from Buda to the Dam and *know* the answer, or (ii) measure my average speed and *predict* the time it will take. Good planning uses option (ii).

analysis *and* monitoring its progress transforms the tissue engineering vista completely.

In effect, the target is no longer dominated by such a weak question (i.e. 'Can we get *anything* to grow?'), but rather it is elevated to the more challenging question, 'Does the rate of progress provide a reasonable tissue construct in a reasonable time period (with numerical estimates of what 'reasonable' might be)? With this, we move into the fourth dimension of tissue fabrication: the full use of *time* and *sequence* for engineering tissues.

9.2 Can we make tissue bioreactor processes work – another way forward?

So, can we use this analysis (see Chapter 8 and Text Box 9.3) to identify another more favourable model for engineering of tissues, directly: a new way forward? One available approach is to use the fourth dimension intelligently, to design dynamic/sequential processes which are hybrids of our two options. This is a pragmatic mix of direct and indirect tissue fabrication and processing – in other words, processing which would use the best of both *in sequence*.

For example, this might use fabrication (bottom-up engineering) in the early time stages, but shift in later stages of the sequence to biological cultivation of cells. This sequence could be used to generate levels of bio-complexity which are beyond our current technologies to generate directly.

Text Box 9.4 Process sequence: order, disorder and chaos

Sequences, as some people insist, 'are *everything!*' The picture on the left of Figure 9.3 shows a large building site in Chengdu, SW China (such sweeping-scale projects are more common in China than in Europe). An old, run-down housing area is being redeveloped (see Chapter 8, Section 8.25) to make way for a much needed sports and medical facilities. The former residents are getting new homes away from the busy city ring road which runs nearby. The sports and medical centre will go onto the red oval footprint (red line) and the planned new access road (black) is shown in the right hand picture.

From Chapter 7, we can recognize this as a good analogy of natural soft tissue growth remodelling cycles, where insertion of the new elements is linked with remodelling of existing structures to accommodate the structure changes. In natural processes, we barely notice the key element of sequence which is obvious here. In this building site, the previous occupants have already been moved to their new suburban homes and the site is being cleared of buildings. Traces of the new road are visible and being used by lorries to move the rubble away, with the foundations surveyed and marked into the levelled ground. The sequence of planning and process operations is clear to see. The quality of this essential sequence detail can be described, for both tissue building and urban development, as one of three levels: 'Ordered', 'Disordered' or 'Chaotic'.

The point here is that in making both cities and tissues, it is simple to distinguish (dis)order from chaos (one works; one does not). Identifying the difference between good order and less good order (disorder), however, is tricky. The difference is one of degrees; for example, as progress rate, number of hold-ups or final cost. In the end, the efficiency of our processes depends on where we find ourselves relative to this blurred line.

To illustrate, it is simple to see that arranging for the bulldozers, trucks and demolition cranes to arrive three weeks before the residents have moved out would cause *chaos*. Similarly, trying to lay the access road before the old flats had been moved away, or building on top of uncleared rubble, would also cause chaos. The process would stop dead. However, booking the roofers to arrive on the same date as the foundations are being dug would be just as inefficient as the roofers would be sitting around for three months, drinking tea in their cabins. Equally, late delivery of the drainpipes to the site will lead to sludge and delay every time it rains, with trucks and Town Council inspectors sinking into the bog that was once the access road.

To the casual or occasional observer, these sequences can look much the same (generally 'ordered'), but the whole project becomes a bit longer and there are more wasted materials. These are the differences between good sequence (order) and poor sequence (disorder). They translate to efficiency and inefficiency, and potentially success or failure for tissue production. We can either discover the difference (expensively) in hindsight, or employ or collaborate with process experts from the start.

The moral of the 'sequence' caveat, then, is that most non-experts, from any building speciality or tissue engineering tribe, can see chaos when it hits them, but improving disorder into order needs collaboration with specialists.

Figure 9.3

9.2.1 Blending the process systems: balancing the Yin and the Yang

Our new option, then, becomes one of balancing the 'Yin' of direct fabrication with the 'Yang' of cell cultivation. We have immediately avoided the 'either-or' extreme and are now focusing on making the best of each system by managing their *sequence* (Text Box 9.4). To develop such Yin-Yang process strategies, it is necessary to step back from the 'whole-process' and do an ILAS:

(i) IDENTIFY which are the important stages and sub-stages of production.
(ii) LIST the various possible timings and sequences.
(iii) ANALYZE options for how each stage could be integrated into a process sequence.
(iv) SELECT and prioritize the most promising process sequences for testing.

In other words, we are starting to divide into production *stages*, asking the question, 'Which can be done better by fabrication or cultivation?'

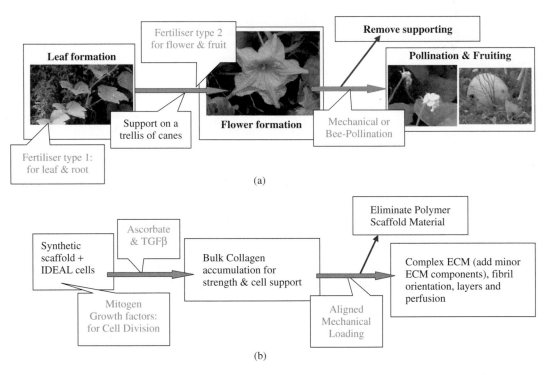

(a)

(b)

Figure 9.4 Translation of the pumpkin bioreactor analogy into a plausible scheme for bioreactor production of skin equivalent, using comparable 'trigger-and-growth' process. (a) Pumpkin bioreactor analogy: basic cultivation conditions are maintained throughout the process, with a sequence of staged inputs or triggering cues which work with and assist the growth process. For example, as the young plant is establishing, the fertilizer content promotes leaf and root development. As the stems get larger, some are mechanically supported for even illumination or wind protection. At a predetermined point (monitored trigger point), the fertilizer content might be changed to that which promotes flower and, later, fruit formation. In the meantime, flowers are pollinated and, as fruit forms, the supports are removed, allowing the pumpkins to lie on the ground and grow bigger. (b) A comparable sequence of triggers and supports can be envisaged for the cell production of skin-like tissue. Between each stage, the cells themselves are being required to perform their innate, programmed tasks of tissue synthesis and deposition. Key to both of these is the *sequence* in which the process supplies controls and enabling factors and then, in effect, sits back and waits for the natural cultivated 'growth stage' to complete. The control points can either initiate the next stage in the process, or speed up the rate at which it is completed. Importantly, in the case of biological production, these controls are unlikely to substantially alter the sequence of stages, as these are largely inbuilt to the growth process itself.

Is there then, an optimal sequence for these stages, and how and when can we move the construct from stage to stage? Let us first examine how this happens with familiar cultivation-dominated examples of tissue growth.

We can develop this (as seen in Figure 9.4a) as our familiar aspiration of pumpkin cultivation, drawn from the last chapter, with the process extended to highlight its parallels in a conventional tissue engineering plan for growing a skin equivalent (Figure 9.4b). As we have seen in the case of the pumpkin, the basic biological machinery of leaf and flower production is supported through both chemical and mechanical promoters (fertilizers to promote fruit formation and canes to support the leaves). New tissue formation is initiated, nurtured, accelerated and ripened in the case of the fruit, to the point of its harvest. This reflects surprisingly well the general plan for formation of a tissue such as skin using bio-culture (Figure 9.4b). It includes assembly of the chief biological and support components, amd provision of both chemical and mechanical factors which nurture and guide early tissue formation.

At the stage when accumulation of *useable tissue bulk* takes over, the support structures (scaffold) and growth additives are removed to promote maturation of the final tissue product. This final step sees the development of tissue/ECM (extracellular matrix) complexity. Both of these processes can be viewed as essentially a *sequence* of cell/tissue-based growth stages, each initiated or enabled by a series of external process triggers. The process supplies the triggering or enabling conditions. The plant (pumpkin) or cell-mass (skin) then performs its next programmed production task.

As we have established previously, such a strategy means that much of the fabrication process relies on the innate action of the cells under cultivation. We can now move our 'Yin-Yang' analysis along simply by asking which stages of the process our cells (i.e. currently the best available) can do *adequately*, and which stages are slow or problematic. Such weak points can be where the tissue output is poor (i.e. functional deficiencies), or where it is excessively slow to produce even minimal function. In other words, this tells us where the dependence on cell cultivation is leading to a log-jam point in the process.

Therefore in order to progress we still need:

(i) to set down criteria for minimal tissue properties; and
(ii) to identify effective ways of measuring them, *within* the process (i.e. without stopping the process or damaging the construct).

This is directly analogous to our example of a planned Budapest–Amsterdam cycle trip (see Footnote 3). While 'Just do it!' may be a good catchphrase for sports or for advertising a male fragrance, it is not a sound tissue engineering philosophy where we do not yet know how, or even if, the task will be possible.

In the example we have here, of skin engineering, it is clear that the generation of shape and complex 3D μ-structure is exceedingly difficult to get right outside the embryo. In biological terms, this is morphogenesis, a subject of serious scientific uncertainty (SSU). The subject of poor micro-architecture, characteristic of scars formed in most tissues during natural tissue repair, has been discussed extensively in previous chapters. It is clearly a limitation, given the *overwhelmingly* modest successes of engineering natural morphogenesis, even after 20 years of trying.

A second log-jam, identified previously, is the very limited ability of cell culture systems to generate and grow the *bulk material* part of connective tissues. Producing bulk ECM, needed for good functional mechanical properties, is a slow and energy-hungry process for cells in culture. The location of these two key log-jam points (μ-structured template and bulk native ECM), are shown in Figure 9.5. This flow diagram (derived from Figure 8.16 in Chapter 8) breaks down a typical tissue construct fabrication sequence into four stages, with the log-jam points marked in stages (*) and (**).

To summarize this section, a key (sometimes misunderstood) distinction is made in Figure 9.5 between having a cell-free scaffold which then must be cell-seeded, and having the cells pre-incorporated into the 3D support material as it aggregates around them. The latter is a 'passive' process involving no

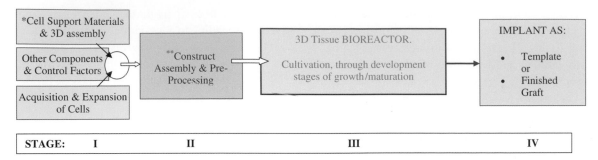

Figure 9.5 New scheme, evolved from Figure 9.4 (and see Chapter 8) for tissue bio-processing with a bioreactor stage in the process sequence. A strategy can now be drawn out which allocates the fabrication of parts to the tasks where it is best suited. In turn, the (cell-dependent) biological cultivation stages remain in the process where they cannot in practice be replaced. The first, *fabrication* stage can be described as production of components, including the 3D material that will support the cell content – in effect, the 'extracellular matrix' of the engineered tissue (*). This can be any biodegradable material but, as we have discussed previously, the closer it is to a native protein ECM, the closer the process is to direct tissue engineering. The second fabrication stage is seen in tissue template *assembly* (**), where the key basic tissue parts are put together to form a 3D equivalent of the eventual structure. It is assumed that this will be a simplified version, but how simple can we get away with? Sufficient 3D authenticity is needed for the eventual required tissue function. This is the point where the tissue is *constructed*, in this case without cell reliance, though cells must be introduced as the living component. The biological, cell-dependent processing comes after this assembly, in stage III. It involves all the bio-processes expected for growth and maturation to a functional tissue, performed through cell action in a cultivation bioreactor stage. The first fabrication and assembly stages (I & II) produce and fit together the 3D template, more familiar to embryologists as morphogenesis. The second, bioreactor stage is expansion and stabilisation, recognizable as tissue growth and maturation.

additional effort on the part of humans or cells. Cells are simply enmeshed in the structure of the biomaterial (in this case the ECM) as it aggregates, without any input from them at all – passively, to labour the point. Such constructs are clearly not cell-free, but neither would we consider that their fabrication is 'cell-dependent'. Therefore, such pre-seeded support materials clearly save us one complete stage. In effect, we have lost the drag of cell-seeding.

9.2.2 *Making the most of hybrid strategies: refining the timing and sequence*

This hybrid strategy, then, represents a new and potentially powerful approach which places the engineering (i.e. the direct fabrication technology) into those parts of the process scheme where they are most needed. These are the stages where cultivation technology is least effective. Previous chapters have examined mechanisms and materials which could be used for:

(i) direct fabrication of the initial cell support material;
(ii) direct assembly of the component parts into complex 3D tissue templates.

Current examples of these include:

(i) the use of directly assembled native fibrillar collagen supports (final ECM material of many tissues);
(ii) assembly of prefabricated matrix or cell layers into multi-layered 3D 'tissues' (i.e. layer engineering).

Examples of these direct fabrication processes are provided in previous chapters (Chapters 6 and 8), based on the example of collagen plastic compression and layer engineering. Other technologies to achieve these ends are also possible, and are being developed. At present, the detail of how direct tissue fabrication is achieved is less important than the fact that it *can* be achieved. It is the *availability* and *practicality* of such direct processes which allows us to design and analyze hybrid processing at all.

In fact, the process outline in Figure 9.5 is, by necessity, an over-simplification. In particular, Stage I is a catch-all for assembling the basic components for the process, cells, cell-support/3D scaffold and other controls such as growth factors. In practice, this would be likely to involve one or more

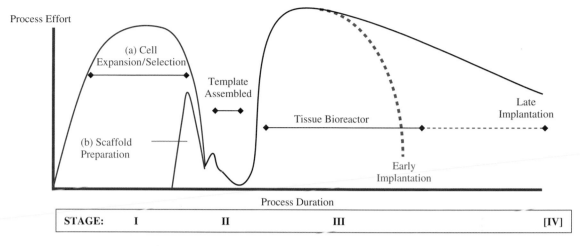

Process Effort

(a) Cell
Expansion/Selection

Template
Assembled

(b) Scaffold
Preparation

Tissue Bioreactor

Late
Implantation

Early
Implantation

Process Duration

STAGE:	I	II	III	[IV]

Figure 9.6 Process-Design graph (nominal axes) typically expected for indirect tissue engineering (ITE), based on a biodegradable polymer scaffold. This shows how the various stages might contribute to, and interconnect, during a conventional bio-cultivation process (derived from Figure 9.5). Typically, the first stage (red curves) would have two major components requiring process time and effort: (a) acquisition of the cells to be seeded; and (b) preparation of the polymer cell-support scaffold. Stage II is that part of the process where all of the components (cells, scaffold, adhesion and growth factors) are put together, and with a culture period to allow cell distribution, etc. Stage III takes the construct to some form of 3D tissue bioreactor with the purpose of getting the cells to deposit a biomimetic, collagen-rich ECM under the culture conditions provided. The aim here is to largely replace the biodegradable polymer scaffold. This may involve complex mechanical and/or growth factor cues, progressive perfusion and constant medium replenishment over prolonged periods. As indicated, this is likely to be the dominant stage in terms of time and, most likely, effort and cost. Early implantation of the construct (as a template tissue – blue dotted plot) would shorten this stage. However, where the requirement is for a functional graft-like tissue, this culture currently requires weeks or months for significant replacement of polymer scaffold function.

stages of cell acquisition, preparation, purification and validation. Traditionally, this could involve extended periods, particularly for time-dependent cell expansion in culture.

The refinement of our hybrid example can be illustrated in a different format, this time as a concept-graph of predicted 'process effort' versus 'time', in Figure 9.6. This is an evolution of the scheme shown in Figure 9.5. In this, our conventional process for producing a skin equivalent would probably use an *off-the-shelf* biodegradable polymer scaffold (therefore very short 'scaffold' preparation times). It would, though, need a significant and variable time for polymer surface treatment, cell seeding and (again) culture, in order to establish a useable cell density throughout the depth of the material.*

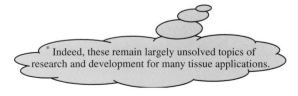

* Indeed, these remain largely unsolved topics of research and development for many tissue applications.

Only then would the construct really be ready for bioreactor culture, to grow a natural ECM with gradually developing mechanical properties.

If all went well, this cell-derived ECM would eventually replace the polymer scaffold: degradation rate of the polymer is key. The bioreactor stage could be very long (potentially months) for production of mature, mechanically strong, graft-like tissues. Shorter periods (days/weeks) are possible if the aim is to implant immature, limited-function templates or pre-tissues, which would mature *in vivo*. This form of design-plot provides a useful framework around which to:

- assess process sequence;
- predict (even to quantify) where problem stages are;
- identify opportunities to speed up the overall plan;
- design new sequences and envisage the linkages between stages.

In other words, even though such plots can be simple or qualitative, they are also invaluable tools for rational design, analysis and improvement of both content and sequence of our processes.

In the example analysis in Figure 9.6, we can predict that cell expansion and bioreactor culture are likely to be rate-limiting steps in the process. Cell expansion (part of Stage I) can be reduced by designing constructs to use:

- low cell seeding densities;
- immature epithelial sheets;
- rapidly dividing cell types/stages (immature progenitors, growth factor and gene activation); or
- allogeneic donor cells able to be pre-cultured en mass (so taking 'expansion' out of the process).

Currently, the bioreactor culture stage, particularly for connective tissues, which requires deposition of a dense, collagen-rich and mechanically functional ECM, will be long and costly. In this conventional process, there are few alternatives available to reduce the impact of Stage III, particularly where a stiff, synthetic polymer scaffold must be replaced. Indeed, the later periods are predicted to generate increasing problems (and hence complexities and delays), because cell synthetic activity will need to be high. This places corresponding demands on deep layer nutrient/oxygen delivery due to consumption and changing diffusion coefficient of the new ECM (see Chapter 8). If ECM synthesis and deposition are not to decline, then deeper zones must be perfused, perhaps by channelling.

If and/or when the process can be progressed to completion against rate-limiting factors, the result would be a connective tissue of relatively high, but as yet unknown, density. We cannot yet estimate, in practice, where or when our resident cells will simply stop depositing more or stronger matrix as a result of their inbuilt feedback mechanisms. We do know for sure that such negative feedbacks will operate at some point. This point seems, sadly, to be reached at disappointingly low levels of matrix density (often, many fold below tissue levels). While technical innovation in bioreactors will improve this situation, the gap is quantitatively very large, even using culture periods which we know are far too long. Even so, our plan still falls short of promising to generate local tissue-like μ-structure or compositional features (zones/layers) similar to those of native tissues. Producing these is again likely to increase the culture periods required.

Consider, then, how dramatically the process could be changed if we switch to *direct* engineering of the tissue, even if there were no change in the cell acquisition or expansion phases. In a direct engineering system, maximum effort would be invested into assembly of cell-material templates. As many as possible of the *most basic* structures and compositional features of the target tissue would be pre-fabricated at this stage. This is shifting the fabrication effort away from the cells within the construct and onto our engineering ingenuity. In other words, such a shift to direct engineering takes process-effort out of the (rate-limiting) bioreactor culture stage (III) and moves it into the construct assembly stage (II).

There are as many ways to achieve the detail of this new approach as there are possible versions of the 3D tissue templates we want to fabricate. However, in the present example, it is possible to identify that the most basic, important tissue mimetic

---

Text Box 9.5 Some tissues can, and some tissues can definitely not: early implantation

In practice, commercial bioreactors, including those for skin equivalents (such as Apligraf® or Dermograft® use relatively short bioreactor stages and follow a process-time plot resembling the dotted line in Figure 9.6. In other words, these processes are designed to lead to early implantation. As a consequence, at the time of implantation, constructs are either mechanically weak (with short survival times) or structurally dependent on the original polymer scaffold.

This is feasible for some tissues (e.g. skin) where function can develop slowly, but not for others, such as blood vessels or heart valves, where '*now*' is essential and non-negotiable.

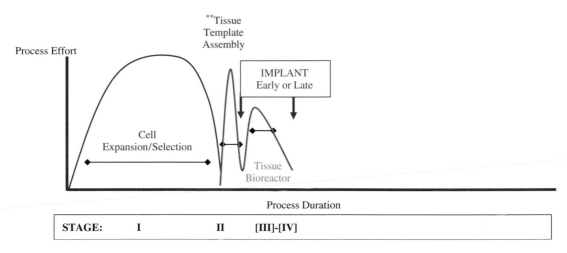

Figure 9.7 Process design graph (nominal axes) typically expected for direct tissue engineering (DTE), based on a native extra cellular matrix (natural polymer) material. Derived from Figure 9.6, in this process the cell acquisition and expansion (I) remains essentially unchanged, but these cells go directly into a tissue replica or template, prefabricated from native matrix proteins (e.g. by collagen, fibrin or, perhaps, polysaccharide material engineering). As a result, a simple but native 'tissue' has been assembled by the end of stage (II), ready to implant as a living graft. Two important differences are clear: (a) since (II) in this case is an assembly stage, rather than a cultivation stage, it is likely to be very rapid (minutes or hours at most); and (b) since the matrix is native, this represents a living tissue graft from the start, not a cell-seeded prosthesis, and so it can *immediately* act as a template to guide local natural tissue remodelling. The need for a distinct bioreactor stage (III) is now questionable, and it is only required to add cell-based complexity.

feature we introduce is the embedding of our cells (fibroblasts here) inside a dense, nano-fibrous network of native collagen fibrils. This example is the cells-embedded-in-collagen-gel system described previously. Ideally, constructs would be made as anisotropic/asymmetric layers, perforated by many μ-channels to improve perfusion (see Chapter 6). While the change from indirect to direct tissue fabrication would require new knowledge in some areas (e.g. in engineering native protein aggregates: collagen engineering), once achieved it would completely alter the process design graph, as shown in the shift from Figure 9.6 to Figure 9.7.

The new (direct engineering) process design, in Figure 9.7, still retains the cell handling and culture stage (I) for cell-seeding into the construct (Text Box 9.5), but cells are now produced as 'one of the starting components', as if they were a reagent rather than the primary producers. Indeed, it may be that this shift alone will generate new approaches in time. If we are going to ask these cells to perform other functions (now downstream remodelling rather

than building the initial tissue bulk), it may be that cell acquisition will be less difficult. Similarly, the stage of cell differentiation may be less critical and total numbers less demanding (Text Box 9.6).

However, the changes after cell acquisition show that the effort invested has shifted from cultivation to fabrication of the tissue bulk material, to completely alter the overall strategy. Recent examples of this new DTE approach – and particularly the use of a rapid cell-matrix assembly (stage II) – can make bioreactor culture completely unnecessary. In effect, the directly fabricated construct is a simple tissue in its own right, ready to be implanted as a living (if simple, immature) graft.

Such constructs, comprising living cells embedded in dense, native ECM, qualify as basic tissues, just as those which are produced by the end of conventional indirect engineering using 3D bioreactors, in Figure 9.6. As a result, the *direct engineering* process design can completely remove, or at least dramatically shorten, this major rate-limiting stage. At the same time, the implanted

---

Text Box 9.6 What do we need so many cells for, anyway?

There is an interesting question here, though it is outside the immediate topic of this chapter, namely, 'If we are shifting the process plan and the requirements of the bioreactor-culture, do we still need the same number or type of cells in our constructs?' Clearly, a lower cell-seeding number could translate into much shorter times for Phase I. Assumptions about the cell type, stage and the number of cells required to engineer any given tissue have tended to be aimed at the rate-limiting step of 'making tissue bulk'. The new, *direct* process plan shifts this emphasis and, with it, our expectation of how many cells will be needed to complete the plan.

For example, if we can *directly* fabricate dense collagen matrices for skin engineering, it seems likely that we will not need fibroblasts to lay down the high densities of collagen that make up the bulky dermal component. Rather, it would then be important to use smaller numbers of specialist calls to introduce cross-links, elastin or matrix-swelling components such as proteoglycans. Alternatively, non-fibroblasts might be able to introduce µ-channels or blood vessels to improves deep perfusion. In other words, the assumption that the cell-expansion stage (I) will remain the same, while stage (II) undergoes radical change, is probably not correct.

So – could 'less cell effort' also mean lower construct cell density, therefore less hypoxia/nutrient depletion, and so much faster cell expansion?

---

graft is immediately able to participate in local host cell-based remodelling. Hence, they can act as true tissue templates, in the same way as tissue grafts can be remodelled by the surrounding tissues. This is very different to the situation using synthetic polymer-based and prosthesis-like devices, where natural remodelling is, at best, delayed.

A second possibility is suggested by the two points at which implantation may take place, shown in Figure 9.6 – early and late. This suggests that where a bioreactor culture stage is retained (albeit a much reduced time period), it would have a very different purpose. The aim here is to increase the structural or compositional complexity of the construct. For example, resident cells might be encouraged to deposit elastin and minor collagen types for blood vessel walls, add collagen cross-links for strength in fascia implants or proteoglycans and calcium deposits for cartilage and bone, respectively. Such subtle bioreactor functions were largely not envisaged in Figure 9.6, where the first target is to replace the polymer scaffold with a bulk of ECM. In fact, in this 'new' bioreactor role, resident cells would be used to develop the construct complexity well beyond those envisaged in Figure 9.6. Consequently, we would not only expect dramatically increased process throughput, but also much more advanced tissue structure than first envisaged

9.2.3 A real example of making tissues directly

The example so far has been provided to demonstrate how it is possible to analyze a complete tissue engineering process. That analysis has allowed us to identify the weak and the rate-limiting stages, to highlight where the critical problems lie and to redesign the process sequence and timing accordingly. For completeness, it is important to explain that this was not a Utopian example which is implausible and so would never be of any practical value. In this example, the key change in Stage II (construct assembly) led to an increase in the process rate and allowed for removal of the problem Stage (III). This is not an imaginary ideal, but in fact has been demonstrated as feasible, indeed practical, as illustrated in Figure 9.8.

The tissue fabrication and assembly prototype described in this figure (produced as a proof of concept device) would effectively assemble and fabricate a simple collagen tissue in minutes. This device (and now others in commercial production) achieves the new Stage (II) of simple tissue template assembly by directly fabricating the bulk matrix, out of native collagen, around the required cells. *This collagen is aggregated and 'engineered' much as if it were a synthetic polymer* support, to give a living tissue-equivalent template – but in minutes.

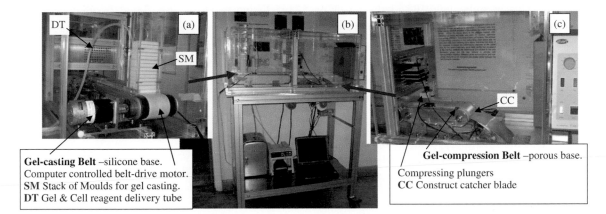

Figure 9.8 Photo of WG[x] machine, designed to carry semi-automated **living tissue assembly** shown in Figure 9.7. (a) shows a detail of the construct gelling belt with fresh gel moulds in a stack hopper, ready to drop and be dragged onto the silicone belt for filling (computer-controlled delivery of collagen, cells, particles, etc). Gels set (in about 30 min) as they warm on this belt and are transferred to a second, porous surface 'compression belt' (c). Once positioned, the plungers push down into the mould, compressing out controlled amounts of fluid (<5 mins) into the absorbent below. After compression, the now dense layers of living tissue construct are peeled off and stacked in the sequence to give multi-layer complex structures (the moulds are recycled). Timers and position-sensors, switches and motor drive feedback to the control computer (b) to regulate the content and structure of the layers. Importantly, many more collagen/cell/layering controls and components can be added into this base process (including perfusion channels), for fully customized end-tissues. The finished tissue emerges in minutes, as a predefined series of layers ready for use or culture. (*WG-device working nickname: *Wallace & Gromit machine*).

In the WG device design, the entire fabrication device (Figure 9.8b) is housed in a controlled chamber at 37°C, to promote collagen gelling. It comprises two processing conveyer belts. Belt number 1 (Figure 9.8a) is where gels are assembled in their moulds and set, in the sequence they will be layered together; on belt number 2 (Figure 9.8c), the gels are compressed, by plungers to give the required fluid removal. 100 μm thick layers are delivered from the second belt to form stacks of any predetermined sequence, to produce the required tissue construct, by repetition and stacking of layers in sequence.

As with most modern engineering fabrication, it is a *repetitive sequence* of many small sub-processes. In reality, this is repetitious to the point of mind-numbing tedium. And this is the key message which we can learn from the example. The processes and sub-processes are simplified to the point where they can be *fully defined* (and in this case made to work by non-specialists using Lego!). However, because we can *define*, precisely and reliably, *all* of the

timings, speeds, durations and volumes involved, we can then build the complexity back up through controlled repetition and sequence.

This is the reality of *directly* engineering tissues. The new speed of its operation means that some readjustment of expectation and planning is needed on our behalf. Two front-runner options open up:

(a) The '**bedside graft delivery**' concept now has a tangible tissue-fabrication-device as an exemplar. The effect of this might be imagined sitting next to the patient, delivering custom-made constructs for a surgeon to implant for minor reconstructions *as they are needed*. In addition to the 'as-and-when' attraction, there is the potential for huge product tissue versatility (in fact, reflecting the enormous variability in the detail of tissues needed by any given patient). In effect, it now becomes a feasible aspiration for the surgeon to dial up the detailed structure of each construct, customized to the patient's needs. This includes tailoring to the needs of the injury or disease (hand trauma, tumour resection, cartilage-bone degeneration) and to the patient's age, sex, ethnicity and cosmetic needs (e.g. between eyelid, cheek or foot skin).

(b) Alternatively, there is a '**mass production**' mirror-image logic which envisages a very large fabrication machine operating at a remote factory site, making, packing and shipping literally thousands of *identical constructs*. Where the cells do not need to come from the patient themselves (e.g. allogeneic cells), this would meet demand for off-the-shelf tissues (e.g. skin grafts for burns, major trauma or leg ulcers). This includes new 3D model tissue applications for animal replacement test kits for pharmaceuticals, cosmetics, chemical toxicity, hospital diagnostics, forensics and research (discussed in Chapters 4 and 5).

The gatekeeper step which opens these new (extreme) horizons is the minimizing/elimination of slow cultivation stages. This must reduce costs, with dumbing down (computer automation) of operations and applying a rocket to 'reproducibility'. So, examination of time and sequence can be the key to locating 'the box' we need to think outside.

To conclude, in those areas where strategies rely on biological controls (tissues in the *in vivo* bioreactor and cells in culture-rich bioreactors) we can take advantage of the faster, simpler and cheaper nature of bio-growth and cultivation, **where the products meet our needs**. However, because these are under rather limited human control, our options are limited where the results are not so good (e.g. slow, poor tissue quality, etc). The real step forward, then, comes where we replace cell-dependent production stages with equivalents which use cell-independent engineering approaches.

9.3 The 4th dimension applied to bioreactor design

9.3.1 Change, change, change!

So, picture the most successful bioreactor design you can imagine, carefully assembled to purr along, producing superbly functional tissue slices. With time, in culture, we would hope that these slices gradually increase in strength and complexity until the surgeon cannot tell where they came from. Importantly, though, where your bioreactor has achieved this happy end-point, there is one trick that, by definition, it *must* be managing to do. That essential element is *continuous change*.

As the construct changes from 'just assembled' to 'tissue-like', it must change its matrix properties (diffusion, mechanical, physical dimensions). It must change its cell properties (density, distribution, synthetic activity, perfusion level) and its fluid content (from high to low, protein poor to protein rich, oxygenated to hypoxic). So, in order to keep the bioreactor process functioning through the inescapable biological sequence in which new tissue develops, its running conditions must also change. These changes can either be in response to changes in the construct or, in some cases, they will *predict* and even *drive* construct development.

This is where we reach another crunch-point. How can our bioreactor conditions be designed to change together with, or ahead of, the needs of its cargo? Disappointingly, the simple answer is: **it depends**.

However, there is another more useful (if equally obvious) catch-all answer: **when the bioreactor sensors tell it to change!**

This sounds pretty trivial, but used correctly it is hugely helpful. At least a little humorous triviality here ensures that we *remember* to get the basics right. The simple secret is that there is at least as much philosophy as technology in good monitoring systems. In other words, the key decision of what to measure (and so how to measure it) needs cold logic and careful analysis*. This is not *necessarily*

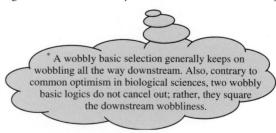

* A wobbly basic selection generally keeps on wobbling all the way downstream. Also, contrary to common optimism in biological sciences, two wobbly basic logics do not cancel out; rather, they square the downstream wobbliness.

the same as buying the most cutting-edge, laser-flashing devices, or expensive, emotionally satisfying marker systems – though it can be.

The problem is that the monitoring component has normally not been an integrated part of the philosophy of your treasured bioreactor, built in with steely forethought from the start of the process. More commonly, it is a near-panic-led afterthought to correct major process shortcomings, as and when they arrive. Good monitoring can make a mediocre process work and a good one excellent – but poor monitoring can allow the very best process to *fail*.

The aim here is *not* to list and critique all the possible monitoring systems available. Not only is this impractical and boring, but that information is already easily available. It is more important for you (as a future process designer) to understand the logic which can lead you to just the right technologics, literatures and equipment manuals. More ambitiously, it would lead you to understand *when* to install the monitoring system.

So, by way of a conclusion, two questions:

- When is change not an issue? (*A: When you switch off the bioreactor*).
- What sort of changes do we measure? (*A: changes of rate, magnitude, frequency and direction*).

9.3.2 For bioreactor monitoring, what are we really *talking about*?

The first thing is to demonstrate where this idea of measuring, sensing, monitoring comes in, and why.

This can best be done at the same time as dissecting out exactly what should make a good bioreactor work well. In this case, it can be useful to generalize, since so many biological processes operate in distinct sequences and stages. They have easily identifiable start-and-stop cellular and molecular events. Figure 9.9 illustrates how this can translate into a tissue bioreactor sequence, where each stage is 'triggered'. Once a new stage is initiated, the bioreactor continues to 'cultivate' under its new conditions until this stage is, in its turn, completed and the next stage-trigger is activated. The simplest version of this would be the increase of tissue dimensions, physical stability and/or complexity, over and above those of our initial assembled template. Such an 'expansion-stabilization' stage in bio-engineering corresponds to 'growth-maturation' in biology.

Where these stages overlap, or have indistinct beginnings and ends, then the process has to compromise in its identification of the 'trigger point. This involves locating the *least damaging or problematic* value of the monitored parameter for *both* the previous and the approaching stages (i.e. already triggered and about to be triggered shortly). For example, a cell density of between 1 and 5×10^6 per ml might be good for promoting an ideal cell differentiation during Stage 2 of the process. The next one, though (Stage 3: matrix deposition), turns out to work best at cell densities above 8×10^6 per ml. A good compromise 'Stage-2-to-3 trigger' value would then be in the region of 6×10^6 cells per ml.

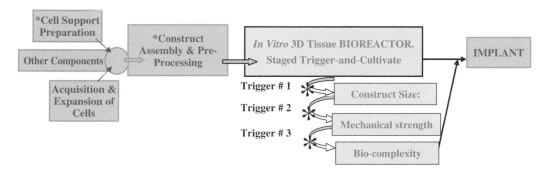

Figure 9.9 This diagram has evolved from Figure 9.5, focusing on and adding detail to the bioreactor stage, illustrating the 'trigger and cultivate' logic. In this example, changes in bioreactor conditions are triggered where: (1) the construct reaches a predetermined diameter or thickness; (2) its mechanical properties exceed a given value; (3) it gains a level of bio-complexity marked by the expression of one or more new proteins.

'Trigger-and-cultivate' processing is perhaps the most common we are likely to meet in cell systems. The construct goes through stages of growth or development, so the bioreactor needs to add factors or change parameters (e.g. mixing, temperature, pH, oxygen tension) as each stage is reached. This mirrors how natural systems develop, sometimes referred to as 'cascades' or 'cycles'. In effect, we must detect when that stage has been reached in order that the change in conditions can be triggered.

The example illustrated in Figure 9.9 has a relatively simple set of three trigger points using three measureable parameters in the construct, overall size, mechanics and marker protein expression. In this simple case, the system uses sharp step changes in bioreactor conditions (perhaps addition of a new growth factor or application of intermittent cyclic loading).

Where the control mechanisms are understood at a more subtle level, it might be better to grade one set of conditions into the next. For example, core O_2 perfusion might be gradually increased by graded fluid flows to meet the needs of greater cell consumption or longer diffusion path-lengths as the construct grows.

So, to summarize, monitoring at its core is essential to identify where a critical parameter has reached its trigger-point, and feedback is then needed to change the bioreactor conditions. This is the point where 3D tissue bioreactors show themselves to be either effective and well designed, or just likely to run out of steam. 'Running out of steam' is a term more colourful than precise; rather, what we see where the bioreactor needs better monitoring and feedback is a gradual decrease in the rate of change (development) of the tissue. In fact, the cells have their own feedback, so the construct rarely 'dies'; it just stops progressing! Better and more frequent tuning of conditions keeps change happening.

9.3.3 Monitoring and processes – chickens and eggs: which come first?

The big (or just easy to ask) questions here are how do we identify the key parameter(s) and how do we monitor them? In detail, of course, what should be measured depends on what is being engineered. Monitoring of the interior colour coordination in a dumper truck factory-production-line is going to be a lowish priority. Corrosion protection will be the winning factor for truck makers, though less so for manufacturers of city-shopper cars. We can, however, ask ourselves a few questions which will lead to a helpful design philosophy, whatever the nature of the tissue system.

1. What are the three or four main **sources of the variation** in the process (and do they interact)? This question tells us where we are on the spectrum of 'reasons why we are monitoring'. Processes with inherently variable or unpredictable outputs (e.g. bio-cultivation of human tissues) need wide tolerance ranges with reliable systems to *find and reject* the extremes.
2. What are the most important **functional** factors governing target tissue performance?
3. What would be the most **damaging features** if they developed (and what is the risk of this)? This 'global' question identifies how the final tissue construct really *must* and *must not* perform to be useful; and then, if the most important of these **functions** can be measured *directly*, or must be deduced from *indirect* measurements.
4. What are the most *prominent stages* in the development of the desired tissue?
5. Are there any *graded changes* and which could be candidates for measuring rates of change? Here we are considering the finer detail of 'bioreactor events', perhaps at the day-to-day level (and potentially minute-to-minute for some events). The simpler logic deals with these as parameter-switches, in effect go/no go, or keep/reject markers of good and poor constructs. Parameters measured here must be *clearly* either in or out of a pre-set range. The more sophisticated logic (when it is appropriate) leads towards measuring parameters which gradually change (e.g. fast or slow), so that the process can be tuned as it progresses.

The more analytical reader will notice that these questions represent a series with a rapidly narrowing focus, but along a single logic track. In more personal terms, it might run from 'Why am I here?' to 'What's for dinner?' via 'Who won the football?'

The first question determines where we sit on the most basic spectrum – namely, are we discussing monitoring a process which is *fundamentally variable* or one which is *defined and predictable*? A good parallel can be found in ceramics and pottery. In this case, we can be making machine-made tiles for the walls of hotels and banks, or individually hand-thrown mugs and bowls for Christmas gifts. The analogy here is that the ceramic tile producer and the hand potter both have to identify and understand just what their main user wants *most*.

Large areas of wall tiles build up into neat, symmetric patterns based on long (white grouted) straight lines between the tiles. This gives them (and so the customer's building) their grand beauty. Our eyes are exquisitely tuned to appreciate the perfect straightness of the lines, especially over very long distances. The concept here is that:

$$[\Omega(\text{Length of uninterrupted tiled wall}) \\ + \Phi(\text{Number of grout} - \text{lines})] \\ \times \gamma(\text{price per tile}) = W \text{ or } L$$

where W = wealth of the bank and L = luxury of the hotel.

The task of the tiler who constructs these patterns is relatively easy, provided the tiles themselves are all 'perfect'. Perfect in this instance means identical in dimensions, angles and thickness, *at all points and in all planes*. Even tiny inter-tile variations, fractions of a millimetre in length or degrees of corner angle, can make the job difficult or impossible. In turn, the customers/users are distinctly disappointed when the scale or precision of this pattern is degraded, reducing both W and L to all (the putative clients) who view the structure.

On the other extreme, customers of trendy gift shops and boutiques want, above all, a gift that is distinctive and unique. The happy recipient of such gifts feel especially honoured to own items as individual as they feel themselves to be. The contrast with 'corporate tiling' could not be clearer. But what does this teach us about process monitoring? After all,

the Mega-Ceramics tile fabrication machine makes perfect tiles and the craftsman potter makes quirky, individual cups and saucers.

The interesting, even surprising, point is that both industries need to be rigorous in their process monitoring. Surprise comes as the tilers might not understand why the potters bother, and the potters wonder what the tilers have to measure. The answer is both simple and illuminating (Figure 9.10). Even where the product is varied and 'random' (variation being a merit), some randomness can be damaging to the basic function. On the other hand, quaint shapes and paint patterns are good in a cup, but having the handle or bottom drop off is definitely not. There are functional limits even to quaint variation.

So, in the case of variable products, the processes must be monitored for *basic* functionality and how long that function lasts. In contrast, the tile manufacturer's machine allows very little measureable variation to develop, but the consequences of just a few rogue tile shapes or pattern colours are so damaging that careful monitoring of *machine* performance is essential. In the tiler's case, the real value of monitoring comes with the certificate of quality given to the customer and, more subtly, the very visible sale of cut-price reject tiles to 'less discerning' users. These two factors keep up the price that the tiles can demand for perfect consistency. In the potter's case, the value comes from fewer discussions with lawyers who specialize in scald injuries.

This analogy illustrates the important tissue engineering spectrum of 'inherent variability'. Many processes, such as those based in hand-crafts, biology and cultivation, tend to have relatively variable outcomes. This is especially true of processes dependent on human cells, which vary hugely. This hugeness then gets even bigger when the cell come from sick or injured people – in other words, patients. Our control over these processing variations is often minimal (which can explain the proportional and progressive disengagement, of our colleagues from the 'engineering tribes').

At the far end of this spectrum of process variability, monitoring is often set up to ensure that

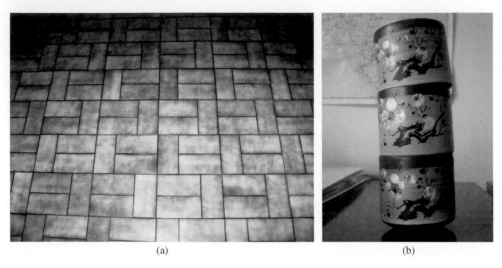

(a)   (b)

Figure 9.10 Identical floor tiles and variable cups. (a) The need for identical, non-varied tiles is clear from the perfect patterns and long straight lines of grout which make up the desirable wall and floor effects. (b) On the other hand, the distinctiveness of individually made cups and bowls is central to their value, with visible, quirky differences being deliberately introduced to each item.

key *functions* are fulfilled. They check that the construct performs within fairly wide tolerances and does not fail catastrophically. The monitoring, in this case, may be designed to measure a selected 'performance indicator' in the finished constructs. When the value of that performance parameter is above a predetermined cut-off, the pot (or tissue construct) is accepted and boxed ready for sale. However, immediately it falls below the triggering threshold, it is rejected and sent to be smashed (right hand side of Figure 9.10). In engineered tissues, this type of pass/fail monitoring might involve histological examination of the tissue structure, the number of living cells (as opposed to dead areas) or the ability to hold sutures during surgery. So we see the parallels with the accept/reject system which operates in making hand-crafted cups. In this case, variability is not a problem (in fact, it is a benefit), so long as the handles stay attached and tea does not dribble into the saucer.

As the detailed mechanisms of the process become better understood, there is a natural shift to the right, in Figure 9.11, in the types of monitoring used and the way in which results are applied to the process (i.e. process feedback). This is because greater process understanding opens the possibility of predicting events and intervening *before* the

product is complete. In other words, the aim is to tune the process as it is running, not after the event. This might, for example, involve speeding up or slowing down one of the process stages, making the constructs thicker or thinner or inserting more/less cells.

Clearly, there are considerable advantages where we go down this route. No longer does *every* product item, be it a tissue construct or a cup, have to be functionally tested (not a good situation – see below), but neither do all defective constructs *have* to go through the whole process before rejection. More importantly, this form of predictive monitoring allows us to introduce feedback changes to 'correct' or adjust conditions *during* the process. Process correction implies that we can also establish bands of acceptable construct performance, as opposed to the previous sharp 'fail/pass' line.

This evolution of how we aim to monitor the process is commonly based on increasing levels of understanding of the process itself. Interestingly, this it is more apparent in top-down processes such as in cultivation of cells or whole organisms, from farming to tissue engineering. Where it becomes possible to use low-variability components, and where the process becomes increasingly sophisticated, we can see that it is possible to

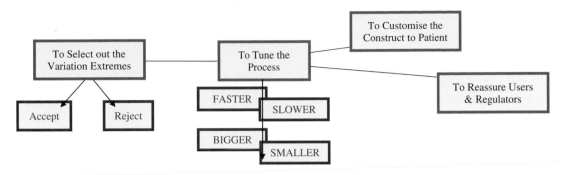

Figure 9.11 Diagram to illustrate the spectrum of targets for process monitoring. This demonstrates the tension which exists between 'simple' reject/accept of useful end products, as opposed to constant fine tuning of the process, depending on the level of 'inherent variability' in the process. The left of the scheme shows the approach of selecting on the basis of simple end-function competence. On the right, the logic is to feed the responses of our monitoring back into the process (e.g. bioreactor conditions), to fine tune it as it is running. Monitoring in the latter case will use parameters which are indirectly related to primary function and can be used to modify the process conditions. Though it is more complex to set up, process tuning leads to customization.

move further to the right of the spectrum in Figure 9.11.

It is possible to view this as moving up the process 'food-chain', in that the process itself can be monitored as the construct grows and matures. In tissue engineering terms, this can go hand in hand with the ability to:

(i) move away from dependance on autologous (patients own) cells, perhaps using pooled donor cells or re-programmed stem cell;.

(ii) use well-defined fabrication processing wherever possible, with less dependence on cultivation and biological production.

However, it also tends, in the case of tissue engineering, to take us further into the future.

The features of cell behaviour utilized here need to be understood to the level where rates of division/death can be predicted across a range of conditions and over time. Parameters such as percentage differentiation and response to reduced oxygen, for example, can be defined before cells enter the process, along with confidence limits. Where these parameters can be measured, they act as benchmarks for comparing constancy or variance of successive 'runs' or process cycles.

Process control tends to move further to the right in Figure 9.11, where we consider bottom-up engineering processes such as mobile phone or automobile production. Where this happens in engineered tissues, our monitoring control also takes on a special value and new uses. The upper branch in Figure 9.11 indicates the opportunity to replace the broad ranges of 'acceptable function' with *fine subdivisions* which lie within an acceptable range. This is important, as it moves us towards designing the final construct to *meet precisely* the patient needs, i.e. customized tissues. Where we reach this level, we have raised our target to well above the current aspiration, which presently aims to produce average, general or lowest common denominator tissues.

For example, we currently aspire to make skin grafts which are biomimetic yet *average*. This combination is at the same time both correct and disappointing, because the level of target biomimesis is so modest. 'Biomimetic' here means having two layers, one with fibroblasts in collagen, the other a surface covering of multilayered keratinocytes. Clearly, this very general definition of 'skin' can describe pretty well any part of the human body, so it is also 'average'. In contrast, the equivalent *customized* skin would be mimetic of a particular site, and definitely *not* average – resembling eyelid, forearm, back, palm or facial skin. It would be different for a child, a boxing champion or a pensioner! You have your own example of a skin to test out here. How many different 'types of skin' can you find over your body? We are looking for elasticity, orientation, stiffness, thickness, colour, hairiness (and other fluffy bits), pimples, dimples and creases.

Once we can monitor and fine tune the process, in ways that are common in conventional manufacturing, customization ceases to be a dream. By analogy, few of us would now expect to say to our bathroom fitter, just get me 'a toilet' (Figure 9.12). We think nothing of trawling through hundreds of types, sizes, colours, shapes and levels of embellishment, depending on our pocket-depth, mood and bathroom-location. It is truly surprising, then, that our aspiration has for so long stopped at 'just skin'. Perhaps an unexpected and happy side effect of sequence analysis in extreme tissue engineering will be new level of immodesty, as we see what is possible.

Finally, the last, far right consequence of process tuning, in Figure 9.11, is the extra value which comes with the tightest (and most expensive) monitoring (Text Box 9.7). In this case, the most sophisticated forms of monitoring and control data can be used *as an output in itself*. This is where the availability of such in process data can be used to reassure the product user (in our case the surgeons, patients or government regulators).

Mobile phones and precision lab equipment often come with data sheets declaring the accuracy and performance of that batch, or even that individual instrument. For the mobile phone, this might be the battery life or range of signal detection. For lab liquid handling pipettes, a data sheet would give tolerances for the volumes dispensed (accuracy) and reproducibility over many operations (though sadly we still await a coefficient of 'resistance to students'). Where the process is designed for pharmaceutical production, government regulators will keenly scrutinize the chemical purity of the drug, the content of the pill casings and the accuracy of the active ingredients in each tablet. At this level, the user reassurance that comes with such data means that process monitoring has, in effect, become a central part of the product (i.e. the tissue construct) itself. While much of tissue engineering is not yet at that stage, it would seem prudent to plan for its arrival.

9.4 What sort of monitoring: how do we do it?

Parameter selection, and the measurement of how parameters change *over time*, are just as important for successful bioreactor operation as they are for monitoring performance of an implant once it is in the patient. By now, the reader should be familiar with the habit in this book of *not* providing lists of what to do or not to do. True to form, we are not going to end with tables of how to measure this or that feature of your skin, bone or blood vessel construct. Rather, the theme of 'extreme tissue engineering' is to identify and analyze the concepts behind monitoring, so that the reader can tailor his/her *own* design rationally to the needs of their *specific* construct and its *particular* disease/injury application and anatomical site.

Listing these possibilities is too large a task for this volume, and it would be dangerously restrictive even if it were possible. As with the London Underground, it is only rational to work hard at understanding the platform and escalator signposts or route-planner – there are just too many track and train permutations to give people an instruction manual or SOP of how to use it.

9.4.1 Selecting parameters to be monitored

The previous discussion has concentrated on theoretical aspects; now we should look to the practicalities. What parameter(s) or characteristic(s) *should* we measure, and in what priority? Answer, of course is simple but again seems unhelpful; it depends on the function that the construct is designed to carry out. Skin must be water-proof and tough; nerve-repair guides must carry axon re-growth *fast* and in one direction. Conversely, it can be just as important to monitor for things that the construct definitely must *not* do:

- Thombus formation in small-bore blood vessels is a *major* 'no-no'.

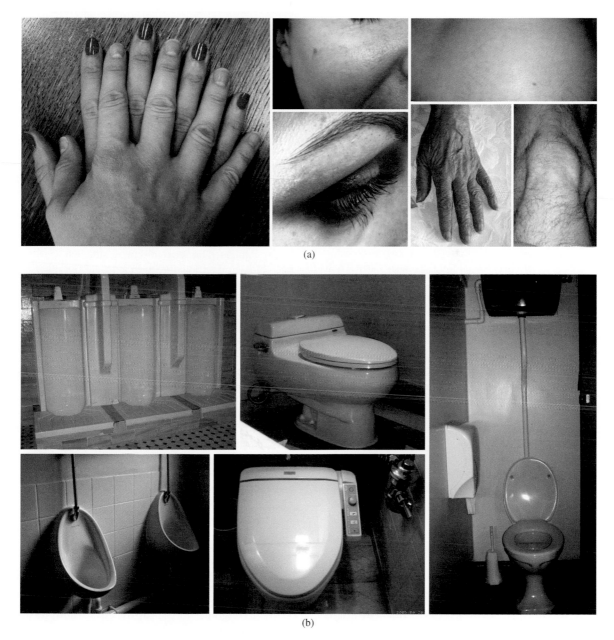

Figure 9.12 Of toilets and skin types: customization of variables and fine-tuning to function is already a default requirement in society. The question is, are we aiming too low with the notion of making a 'one-size-fits-all' or 'average' skin, any more than we would be in expecting an *average* toilet to suit all our homes and businesses? In the end, we need tissues which function *and* match their recipient. An octogenarian might dream of a *whole* new skin, but might not appreciate having a patch of 20-year-old tissue grafted into their 'old' hand. (a) shows a few skin types. From left, clockwise: young adult back of hand; cheek skin; mid-life back skin; hairy (non-footballer) knee skin; older person's hand skin; female eyelid. (b) shows a few of the forms of toilet in 'common' use. From bottom left, clockwise: typical modern, minimalist male urinal; traditional ornate (Spanish) urinals; modern US-style WC (China); classic (high cistern) early 20th century WC; high-tech, electrically heated combined WC and bidet (Japan).

Text Box 9.7 Evolving of process monitoring in top-down systems: milk

In the 19th century, a dairy farmer might reasonably have developed a milk production process in which the cow converted grass into milk in his smallholding or farm. He periodically extracted this cow-juice, bottled and sold it. Unfortunately, milk and cows being what they are, many people became ill from drinking this product. The situation was made worse because mothers fed cow's milk to their infants, thinking it better than their own (some things never change . . .). Clinical infections such as bovine tuberculosis, undulant fever or brucellosis were frequent and lethal. Poisonings such as 'milk sickness' (from cows eating poisonous weeds) or due to 'swill milk' (where cows were fed on distillery waste) were also seen. Early farmers may have fed batches of milk to the farm cat before selling to their best customers, to check for (and reject) the very worst batches. This is a gross, pass/fail test for acute toxic or infective defects, *assuming the cat is susceptible.*

As time went by and the infective nature was better understood, suspect batches of milk might be sent to the local dairy centre for culture on bacterial plates, identifying the infection type and titre (load or density). This made it possible to assess the milk quantitatively, based on tables of human tolerance to this or that level of each microorganism. Interestingly, at this point, it may have become possible to effect crude full-process tuning – or feedback control – by testing *the cows* for that infection and putting down those which tested positive (which is still the practice for foot-and-mouth disease).

With yet further understanding and technical investment, it is now conceivable to test the milk online for biochemical markers of the worst (i.e. key-marker)

infections or for known poisonous contaminants. At this stage, we might expect problem cows to be immunised or given antibiotics, rather than destroyed. This represents a progressive track back from direct, crude functional testing, through reductive bacterial culture testing and finally to indirect or implied, molecular testing. In addition to progressively improving the process, the milk-products and the herd, this is great news for the dairy cat.

Plausible as this scenario is, such fables often do not play out so perfectly in practice. The reality is that the improved understanding which develops along with better monitoring can provide simpler and cheaper alternatives to the *process* itself. In our milk example, the insertion of a Pasteurisation stage to the process largely eliminated bacterial infections, although microbiological testing developed in parallel for other reasons. Also, fencing off the cows from toxic weeds and using healthy foodstuffs prevented poisonous milk. Indeed, better process knowledge though development of monitoring systems commonly leads to major process changes which are, in fact, *less* complex than the original.

Interestingly, we *still* rely on the cow as the core self-monitoring, self-tuning and economic bulk grass-to-milk converter-machine. Where the cow can/will not meet the key-marker output measures (frequently because of infection), we still shoot it and start again. However, this economy-driven simplification of the monitor-feedback loop is only available to whole-organism culture processes. Delegation to the cow is not a luxury available to the tissue engineer. We must take on the mantle of the 'constant tailor', perpetually measuring and adjusting the process. So, as in other fields the constant tailor not only develops how and what he measures, but also what it means to the *process*.

- Urate crystal seeding into our urothelial constructs spells 'seriously uncool pain-in-the-bladder-region'.
- Immunogenic reactions to our favourite skin equivalent is rejection in any language.

This all sounds a bit glib, but notice it refers to 'what *the construct* does or does not *do*'. It is not based on what target tissue it is *supposed to be*, nor what it *looks* like. But the answer will be different for each implant (note: 'implant', not 'tissue') we may

choose to engineer in the future. As a result, this is probably the most specific answer we can hope to get at this stage.

The take-home-message is that we really must aim, at the earliest stage, to establish which parameter(s) are absolute 'must-haves' specifically for our new implant and its application. Clearly, once identified, these will help to define the one or two monitoring systems which need to be built into the process from the start.

9.4.2 What is so special about our particular 'glass slipper'?

If you are into making glass slippers, it is sensible to make *absolutely* certain they are a perfect fit for the feet of the wearer. So, let us plot out the logic we need (right from the start) to identify this special fingerprint analysis which will link our 'glass-slipper' engineered tissue to its unique implant target.

It is not easy to determine the real 'Number 1' where the range of possibilities is so very large. After all, this not only depends on the tissue itself but its eventual anatomical location, the state of the patient and his/her local tissue bed. A tendon can operate a pianist's finger by transferring feather-like loads with smooth precision; while in the foot, the same kind of tissue, in the form of the Achilles tendon, can drive a sprinter's full body weight forwards at speed. A nerve can serve facial muscles at one extreme, or sprout out from of the spinal roots at the other. A leg vein in a 60 year old patient and a pulmonary artery for a child may have many basic structures in common, but their primary demands are different. The age, injury and disease type – even the drug regime of the patient – can determine the primary must-have function. In the example above, reconstructive surgery of flexor tendons in the hand requires *relatively* modest mechanical loading, but tendons *must* glide freely or the hand cannot work at all. As a result, it is a primary requirement *not* to form fibrous adhesions. Achilles tendons must carry *very* large loads; but the problem of adhesions is minor.[23]

It is, therefore, not enough to go to classical anatomy or histological textbooks (Text Box 9.8) to identify the must-haves, because they:

(a) normally deal with mature (end point) tissues, while we need to measure the stages on the way to maturity;
(b) mostly describe static structures – but people move, so implants need to be dynamic;
(c) describe (in the main) healthy structures, while implants are for injuries, and patients take drugs.

Rather, the information we need comes from discussions with surgeons, pharmacologists, wound repair biologists and engineering collaborators – simple for any good tissue engineering team.

The detail of this *primary* must-have function still depends on a range of factors which are specific for each implant type, as shown in Table 9.1. Clearly, it is not always practical to select monitoring parameters and systems on a rigorous case-by-case basis, as implied by this table – there are just too many variables. However, it is not difficult to balance the general of the production and the specific of the implant needs. This involves compromise between practical process needs and efficient function of the construct. For example, it is simple to group the tissues and injury sites such that a platform process and monitoring system can be designed (e.g. for general nerve-guide implants). However, this can have the

---

[23]Note the examples have an extra layer. Contrast what the pianist needs as opposed to, say, a tyre-fitter. A face and a leg nerve reconstruction may be provided for David Beckham and Michael Caine – but which gets which matters to the result needed.

Text Box 9.8 Key tip: the Jumbo Jet principle

We are 'engineering' tissues here and it is not always important for the construct to 'look' exactly like its native bio-equivalent. After all, clinical needs can include performance of temporary or partial support of a body function. For example, an extracorporeal liver would support between transplants, and cardiac-assist muscles will supplement heart outputs. We sometimes just want to guide the natural repair process (nerve regeneration in the hand), and sometimes (e.g. in spinal injuries) we need to make tissues which never existed naturally. So, 'looking like the structure' of the native equivalent is often *not such a good monitoring target*.

After all, when we *functionally* fly like a bird to New York or Beijing, we are only too happy that the Jumbo Jet has *no* feathers and *minimal* wing-flap (see front piece p. 216).

Table 9.1 This illustrates the divergence of the 'absolutely-must-have' parameters for a few example engineered implants. These turn out to be a little like the needs of the distinctive hand-made pots.

Tissue type	Body location	Patient age	Example injury	Probable primary need
Tendon	Hand	45	Sharp trauma	No adhesion
Tendon	Achilles	25	Sports injury	Tensile strength
Nerve	Face (to mouth)	25	Surgical accident	Fast re-growth/cosmetic
Nerve	Spinal root	25	Road accident	*Any* nerve re-growth
Blood vessel	Leg vein	65	Disease	Non-thrombotic
Blood vessel	Pulmonary artery	12	Congenital	Mechanical strength
Skin	Cheek	18	Resected melanoma	Cosmetic match
Skin	Leg	65	Chronic wound	Strength and stability
Cornea	Corneal epithelium	28	Chemical burn	Maintain stem cells
Cornea	Full depth cornea	58	Cornea endothelial failure	Functional endothelium

potential to produce sub-groups of constructs for different patient groups and applications.

So, to summarize, precision in this early selection stage is a pivotal point in process design, where sound intellectual analysis leads to selection of the critical parameters. This early 'brain-work' will determine success or failure for years to come.

It is helpful here to understand the various categories of analytical approach which are available, and to practise balancing their advantages and drawbacks.

(i) Explicit/implicit, direct/indirect

Explicit/direct monitoring systems go straight for the functional focus, determining the *actual* key function. For example, this could be direct, explicit assessment of the lumen thrombogenicity of a blood vessel construct while it is being produced. For this, we might introduce whole blood (containing fibrinogen and platelets) and measure how much fibrin clot forms on the walls in unit time. This is as direct an assay as you could get, although it is difficult to implement repeatedly over time without damaging the construct.

An indirect/implicit test might monitor the production, over time, of endothelial cell markers such as anti-thrombogenic compounds or thrombolytic

enzymes (e.g. heparin-like molecules to block coagulation and plasmin to digest clots when they form). Such biochemical assays would be far simpler to design for real-time data collection, and much easier to repeat or re-analyze. More particularly, our excellent understanding of coagulation biochemistry, and the good correlation between such markers and function, would make interpretation of the output data pretty robust.

(ii) Destructive versus non-destructive testing

Where the aim is to engineer, for example, a tendon, it would seem reasonable to follow the development of an aligned fibrillar structure or acquisition of uniaxial mechanical strength. In this way, the effectiveness of the process, and its stage of development, could be judged and adjustments made to the cell culture conditions. However, the classical (gold-standard) method of determining tissue structure is through histology: fixation, embedding, thin slicing and staining of the tissue for microscopy. Similarly, a functional break-stress test involves clamping and loading the construct until it breaks in the middle. Obviously, while both of these would be excellent for giving direct measures of functional success, widely accepted in biology, the construct is destroyed, and so our knowledge is 'past tense'.

Destruction of the test material is a major problem for *in vitro* processing, as it is wasteful and not real-time, so not helpful for process tuning. Its results are indirect, and data must be extrapolated to constructs which are *not* destroyed. For clinical use, where constructs are grown within the body, or for assessment of construct progress post-implantation, it is normally out of the question. Try to imagine the reaction of a surgical patient if the implant team were to demand that they must remove the tendon six months after implantation, to make sure it was up to scratch!

While destructive testing is just about acceptable during the research phase, even here it is expensive, it is time-consuming, and it gives indirect data. As with other aspects of process design, it is good practice to plan for this early. In other words, the aim is for destructive testing to be minimized as an early requirement in the design (Text Box 9.9).

Adjusting to the technical need to find new, non-destructive monitoring approaches is, yet again, a matter of shifting our tissue engineering tribal thinking into 'extreme tissue engineering' mode. Where our aim continues to be to impress our biological tribe members and elders that we truly *have* made a tendon, skin, cartilage, blood vessel or muscle, we are likely to cling to the familiar (semi)-destructive methods. It is only when we try to sell this to team members from surgical or engineering tribes, and experience their gentle laughter and tough questions, that we wish we had thought ahead. The truth is, this is one of the *diagnostic points* distinguishing the naïve and the newcomer groups with limited inter-tribal mixing from experienced, habitual collaborators.

(iii) Invasive versus minimally invasive monitoring

This distinction parallels that of *destructive/non-destructive* monitoring. In the latter case, the sample may suffer damage and destruction, while in the former it is the patient who is not damaged. The less invasive the information-gathering step is to the patient (or the bioreactor), the more often and more easily data can be collected. However, the more invasive and the more destructive the test, the less ambiguous, more clear-cut is the meaning of the test.[24]

In both cases, then, the quality of the test method commonly has to be balanced against the damage done in getting that information. Procedures needing open surgery to collect large lumps (biopsies)

------

[24]The tension between these data collection modes mirrors the tension between surgeons and pathologists. The surgeon has the opportunity to do everything to save the patient, but often can't know what the problem really is. The pathologist knows the problem, *exactly* – but can't do a thing about it.

Text Box 9.9 Case study 1

We clearly *must* develop non-destructive monitoring techniques suitable to measure the primary functions at *some stage* in the tissue engineering process. Therefore, logically, this should be in the early process developmental stages, where they can be integrated and adapted most flexibly. In the tendon example, there are alternatives, such as minimally invasive optical fibre scattering measures. Quantitative fibril density and alignment data from such analyses can be used to follow structural development changes in real time.

This cannot deliver the mass of detailed information, especially around cell distribution, that is provided by histology. However, this point is exactly the message of our case study. We have already established the importance of *early* identification of the primary 'must have' parameter to be monitored. For tendon, this would be collagen density and alignment, *not* cell distribution, which might be a distant third or fourth in ranking.

Equally, our basic material-mechanics knowledge allows *excellent* extrapolation from such material-fibril parameters to stiffness and break strength without having to break the construct every time. In short, non-destructive testing of functional measures is frequently possible and is an early requirement.

of tissue are invasive. Minimally invasive collection would be performed down fine needles with a minimal scale of surgical intervention. As a general rule, the more an anaesthetic is needed (the bigger the volume of patient anaesthetized), the more invasive the test. Non-invasive techniques such as ultra-sound, optical or MRI imaging involve no physical entry at all into the patient's body.

Key, then, is that process monitoring is designed strictly to balance these tensions, rather than on grounds of tradition or familiarity for the host tribe.

Text Box 9.10 Case study 2

An example of real-time data collection would be the measurement of changing, real-time oxygen levels in the core of our cell rich skin construct. This can be monitored directly using a 300 μm diameter fibre optic probe (Figure 9.13), with minimal O_2 consumption. Together with data on the changing matrix density over time, and the sensitivity of your particular cells to hypoxia, it becomes possible to determine when and where cell stress or death *is about to occur* in the construct. In other words, core O_2 levels are converted by a simple computer model, into predictions of cell viability in space (3D location) and time (in the future).

This off-the-shelf technology would clearly transform our traditional 20th century tissue fermentor into a 21st century 3D bio-process system. In the former (20th century) approach, we establish when/where cells *had* died as the basis of accepting or rejecting the skin construct *at the end* of production (just as, in the niche potting industry, excessively wobbly mugs are smashed). In the 21st century approach, the O_2 sensor and computer model automatically feed data and its conclusions back to change the perfusion conditions in the culture chamber *before* cell damage or unwanted changes occur. In this case, functionally constant constructs emerge with their certificate of quality (just as precision tiles leave the ceramics factory).

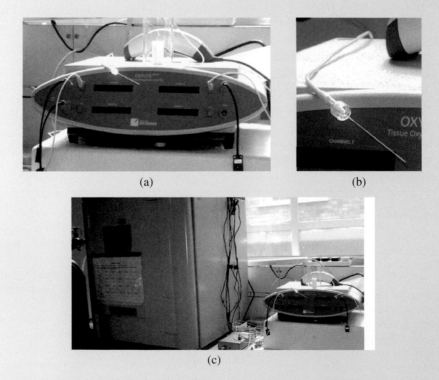

(a)   (b)

(c)

Figure 9.13 Real-time oxygen monitoring by optical fibre (Oxford Optronix, Oxylite 4000) (a). The fine fibre probe (b) is placed inside the construct to the monitor inside the incubator (c).

(iv) Real-time versus end-stage

Finally, many techniques, especially those used in research phases, are designed to sample constructs recovered at the end of the procedure. These often equate to destructive or invasive tests, and again translate poorly to processing or clinical implant monitoring. In this case, real-time monitoring must be our ideal as it provides many data measures over relatively short periods. It also identifies changes/rates of change *as they happen*. The value of this is enormous, feeding again into the all-important time dimension of the process (Text Box 9.10). First, in-process data allows for feedback and correction, or tuning of the process before it is complete (i.e. too late to change the result). Second, real-time (RT) data streams can be used to generate rates and trends which, in turn, can hugely improve the interpretation of indirect analyses. This can be the case where the trend or rate-of-change equations are pre-fed into computer models to predict the meaning of our indirect parameter measurements in terms of the function we want to know about (Text Box 9.11).

9.5 The take-home message

The last two chapters have taken us from Hippocrates to Limousin beef farming and jumbo jets.

But their take-home messages can be rolled together into a relatively simple whole. This suggests that the evolution of extreme tissue engineering bioreactors is starting to give us the confidence to wrestle control of our processes away from the cells and into our own hands.

Logic tells us that we *must* eventually do this, and new technologies tell us we *can*. This control helps us to re-think bioreactor design and, more radically, why and when we need them. The slavishly nurturing of our cells does not need to be a first priority. We can radically re-shape the time courses of our processes to design the fourth dimension and, with it, we can generate higher targets for our constructs and the way we monitor them.

Speed, reproducibility and the possibility of customization can revolutionize our ideas on how we employ even simple fabricated tissues. These new uses can range from mass-produced model tissues for replacing animal tests, to tailor-made bedside tissue implants. In fact, this 'new paradigm' was always embedded in the original tissue engineering idea. However, by thinking outside the cell-cultivation box (which we *must* do anyway), we can now explore **direct tissue fabrication** processes which can lift our targets.

In the same way, the driving concepts for human flight in the early 1900s moved away from images of Icarus and feathery-winged angels towards those

Figure 9.14 'Flying with feathers'. When a technology is new and our concepts are based mainly in the natural 'bio-world', it is inevitable that we are slow to appreciate how we might use non-natural, engineering equivalents. A good historic example of this is in heavier-than-air flight, where it is possible to plot the gradual disappearance of the shapes, flapping and feathers which characterize bird-flight. The real pioneer moments must have come where inventors and thinkers suddenly glimpsed how this or that aspect of a flying machine could be made to a *human design* or with *human-type materials*. This is where we are with extreme tissue engineering.

Text Box 9.11

Explain in one sentence, supported by a short analytical essay, why the evolution of process monitoring (from simple, direct and reactive to indirect and predictive) is more apparent in top-down bio-cultivation processes rather than bottom-up engineering and assembly processes. Use your own process examples and flow diagrams to illustrate the case.

---

which now allow us to construct shiny metal jumbo jets (Figure 9.14).

Further reading

1. Jones, S. (2005). On Darwin airlines. In: *The Single Helix: a turn around the world of science*, pp. 3–5. Little Brown, London.
   [Withering demolition of Creationist teaching in the UK using the tongue-in-cheek analogy of a naturally evolved airliner, flapping into its airport: in other words, *a feathery Boeing*.]
2. Vunjak-Novakovic, G. (2006). Tissue engineering: basic considerations. In: Vunjak-Novakovic, G., Freshney, R. I. (eds.) *Culture of Cells for Tissue Engineering*, pp 131–155. John Wiley, New Jersey.
   [Nice, basic entry into process thinking and tissue production sequences. Best used only as an illustration when it comes to detailed methodologies which use 'non-extreme', 1st generation technologies.]
3. Freed, L. E. & Vunjak-Novakovic, G. (2002). Spaceflight bioreactor studies of cells and tissues (review). *Advances in Space Biology and Medicine* **8**, 177–195.
   [Comprehensive analysis of the parameters which can affect bioreactors (at least at time zero), as needed to understand the effects of spaceflight.]
4. Freed, L. E., Guilak, F., Guo, X. E., Gray, M. L., Tranquillo, R., Holmes, J. W., Radisic, M., Sefton, M. V., Kaplan, D. & Vunjak-Novakovic, G. (2006). Advanced tools for tissue engineering: scaffolds, bioreactors, and signalling (review). *Tissue Engineering* **12**, 3285–3305.
   [Comprehensive if conventional analysis of bioreactor processing: leading to more recent . . .]
5. Haj, A. J., Hampson, K. & Gogniat, G. (2009). Bioreactors for connective tissue engineering: design and monitoring innovations (review). *Advances In Biochemical Engineering/Biotechnology* **112**, 81–93.
   [. . . advanced discussion, including the need for real-time and progressive analysis.]
6. Rice, M. A., Waters, K. R. & Anseth, K. S. (2009). Ultrasound monitoring of cartilaginous matrix evolution in degradable PEG hydrogels. *Acta Biomaterialia* **5**, 152–161.
   [Example of advanced non-destructive analysis: ultrasound-based.]
7. Mason, C., Markusen, J. F., Town, M. A., Dunnill, P. & Wang, R. K. (2004). Doppler optical coherence tomography for measuring flow in engineered tissue (review). *Biosensors and Bioelectronics* **20**, 414–423.
   [Example of advanced non-destructive analysis: optical-Doppler measurement of flow.]
8. Liu, J., Barradas, A., Fernandes, H., Janssen, F., Papenburg, B., Stamatialis, D., Martens, A., van Blitterswijk, C. & de Boer, J. (2010). *In vitro* and *in vivo* bioluminescent imaging of hypoxia in tissue-engineered grafts. *Tissue Engineering Part C, Methods* **16**, 479–485.
   [Example of indirect, image-based monitoring of cell oxygenation: contrasting with . . .
9. Cheema, U., Hadjipanayi, E., Tammi, N., Alp, B., Mudera, V. & Brown, R. A. (2009). Identification of key factors in deep O_2 cell perfusion for vascular tissue engineering. *International Journal of Artificial Organs* **32**, 318–328.
   [. . . example of direct, real-time, quantification of oxygen tension over time and position.]
10. Ziegelmueller, J. A., Zaenkert, E. K., Schams, R., Lackermair, S., Schmitz, C., Reichart, B. & Sodian, R. (2010). Optical monitoring during bioreactor conditioning of tissue-engineered heart valves. *ASAIO Journal* **56**, 228–231.
    [Example of imaged-based monitoring of cardio-vascular tissue growth, in contrast to . . .]
11. Syedain, Z. H., Meier, L. A., Bjork, J. W., Lee, A. & Tranquillo, R. T. (2011). Implantable arterial grafts from human fibroblasts and fibrin using a multi-graft pulsed flow-stretch bioreactor with noninvasive strength monitoring. *Biomaterials* **32**, 714–722.
    [. . . example of real-time monitoring of the development of vascular construct mechanics: dynamics over about two months' growth in culture.]

10 Epilogue: Where Can Extreme Tissue Engineering Go Next?

Half of my students advised me, 'Under no circumstances do a crystal-ball-gazing, what-the-future-holds section!' Tradition has it that when this comes from wrinkly established scientists, it is at best embarrassing and at worst off-putting.

This has merit. There is clearly a tendency on one hand to predict wildly optimistic or over-ambitious progressions for your view and ideas (after all, you will be retired before anyone proves you wrong!). But on the other hand, crusty guys who are joined at the hip to the core concepts *can* actually have a privileged view of the log-jams and off track opportunities before they become obvious.

I decided to take-the-advice-but-not (a Manchurian compromise). Here is the section on 'futures' – but notice how short it is. This is because this particular question needs little analysis to find its answer.

10.1 So where *can* extreme tissue engineering go next?

The simple, first-level answer is 'Nowhere – it will end'. Nothing in science stays *extreme* for very long. The 'extreme' part implies that it is at the frontier – rough, partly understood, partly surprising, presently untamed. This is good for science and bad for technology and translation, whether industrial or clinical. From history, we know that either such areas are tamed and become useful, or they remain ambiguous, with limited capacity to translate to our target aspirations. Ambiguity, surprise and randomness are the stuff of sports and the arts. Olympic tissue engineering is not going to happen soon, although tissue engineering in the Olympics is coming and may already be with us.

No, extreme tissue engineering (ETE) is a useful concept now but, by its very nature, it will need to be reassessed in years to come, as it will either have become *not particularly extreme* (i.e. useful and tamed – and so a rubbish title for the 2nd edition), or '*chronically extreme*' (and thus not a useful, translatable scientific field). However, given its present trajectory, I cannot see the latter (becoming not useful) as remotely likely to happen.

Now, we know perfectly well what happens to successful fields in science. They become channelled, socialized and subdivided. So we are likely to see sub-divisions, perhaps into 3D tissue model-, graft- and drug release depot-engineering; cell-rich and matrix-rich tissue engineering; human, veterinary and plant tissue engineering! Given time and enough productive success, some of these sub-divisions will establish their own international societies, journals and perhaps even the odd specialist journalist.

Extreme Tissue Engineering: Concepts and Strategies for Tissue Fabrication, First Edition. Robert A. Brown.
© 2013 John Wiley & Sons, Ltd. Published 2013 by John Wiley & Sons, Ltd.

By then, ETE will be gone, because its concepts will have become useful and familiar – so not extreme. At that point, it will be one of you, the readers of *this* extreme, who will be able to tackle the new 'extreme' that you will inevitably find.*

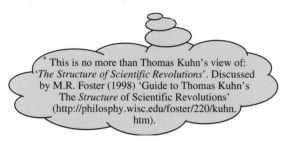

* This is no more than Thomas Kuhn's view of: '*The Structure of Scientific Revolutions*'. Discussed by M.R. Foster (1998) 'Guide to Thomas Kuhn's The *Structure* of Scientific Revolutions' (http://philosphy.wisc.edu/foster/220/kuhn. htm).

To briefly summarize, there may be signs emerging of a strategic track which could take us towards that horizon. This comes from an increasing awareness that we should *actually take control*. That means *making* the 3D tissue structures ourselves, rather than relying ever more on our cells. This is the idea that we should have confidence in the abilities of our tribes, *in collaboration*, to fabricate things directly: ourselves.

However, 'direct fabrication' involves learning:

- how to layer and assemble *native* tissue layers, zones and domains, not as one big, porous lump;
- how to build up structures as a continuum across different scale-hierarchies, not as a series of distinct levels;
- how to analyze and recapitulate our target tissues in terms of their (an)isotropy *in all three planes* rather than the two planes we find in textbook histological slices.

It means discovering:

- how bio-simple we can get away with, rather than how bio-complex we can make things;
- how to support the cells with a tissue-fabric or matrix of an appropriate native protein, so that they *are* tissues from t_{zero}, rather than temporary templates;
- how to make the tissue with its cells *in place* from t_{zero}, rather than toiling with seeding stages;
- and so how to fabricate, *but* only using non-lethal conditions.

In effect, these can be summarized as setting our goals so that **our cells get a job with vacations, rather than a sentence of hard labour.**

These are undeniably high targets, but at least where we fall little short, the tissues that we *can* make will still be impressive. We shall have avoided the trap of aiming low and still missing.

This, then, was the Manchurian compromise, between 'do' and 'don't' offer a glimpse into the future. The prediction of this wrinkly scientist is that *you* or *your colleagues* will define the next stage of extreme tissue engineering, providing that *this* stage stimulates enough of you to dream outside of you speciality.

Hopefully, both groups of my students will be happy with this compromise epilogue.

Index

Extreme Tissue Engineering: Concepts and Strategies for Tissue Fabrication, First Edition. Robert A. Brown.
© 2013 John Wiley & Sons, Ltd. Published 2013 by John Wiley & Sons, Ltd.